Oceanography

Charles L. Drake
Dartmouth College

John Imbrie
Brown University

John A. Knauss
University of Rhode Island,
Kingston

Karl K. Turekian
Yale University

Holt, Rinehart and Winston

New York   Chicago   San Francisco   Atlanta
Dallas   Montreal   Toronto   London   Sydney

Library of Congress Cataloging in Publication Data
Main entry under title:

Oceanography.

Includes index.
1. Oceanography. I. Drake, Charles L. II. Title.
GC16.026   551.4'6   77-13571

ISBN 0-03-085644-2

Printed in the United States of America
8 9 0 1   0 7 4   9 8 7 6 5 4 3 2 1

Preface

Few people stand at the edge of the sea without feeling awed by its size and power, and without wondering what lies hidden beneath its surface. This book explores and explains this world of the oceans. It is aimed at the undergraduate who has an interest in the sea: how it works and how it affects our lives.

Study of the ocean must start with a description of its different aspects: its currents, waves, tides, depth, temperature, chemistry, and biology. Much of this information is a legacy from the explorers, inventors, natural philosophers, adventurers, and scientists of the past. But oceanography has undergone a spectacular development since World War II, and this book is our attempt to provide an overview of present knowledge.

A compilation of facts does not by itself constitute a science. To understand the ocean we must view facts as clues to help us find out what is going on. The main emphasis of this book lies here. In writing it, we have gone beyond mere description and tried to identify the most important processes at work. Some are going on now, before our eyes, at rates fast enough to be observed directly. Included here are studies of tides, waves, and other movements of the surface waters. But other processes, such as those which involve the deep circulation of the ocean, go on too slowly to be analyzed directly, and must be inferred by indirect methods. The slowest processes of all are those which change the climate of the Earth, the position of the continents, and the shape of the ocean basins themselves.

The organization of the book follows the development of these ideas. The first three chapters set the historical, geographical, and cosmic framework for the study of the oceans. Chapters 4 through 8 deal with the movement of the water filling the ocean basins—how it interacts with its basin as well as the atmosphere above, and how it responds to astronomical influences. Chapter 9 analyzes other processes going on in the ocean by studying the distribution of chemical elements. Chapters 10 and 11 examine the animals and plants which live in the sea, the processes which maintain them, and the impact they have on the ocean. Chapter 12 asks how large a role the ocean could play in feeding an expanding human population. Chapters 13 through 15 are concerned with deposits accumulating on the ocean floor: their sources, their complex routings, and the types of information they contain about oceans of the past. Geophysical studies leading to the enunciation of the plate-tectonic model for the evolution of the ocean basins and Earth's outer spheres are discussed in Chapters 16 and 17. Chapter 18 then discusses the chemical cycles between continent and ocean in light of this global model. Chapter 19 focuses on the climatic history of the Earth as revealed through the study of marine deposits. Chapter 20 brings our focus back to the continental margins where human involvement with the sea is most obvious, and Chapters 21 and 22 discuss the economic resources of the oceans and the problems associated with the multiple demands on these resources by different kinds of users as well as with the expression of diverse international interests.

The authors wish to thank Ms. Peggy Middendorf of Holt, Rinehart and Winston for her help in bringing this book to completion.

CONTENTS

OCEANOGRAPHY

H.M.S. Challenger (left) and track of the *Challenger* expedition, 1872–1876 (right).

Ocean Science

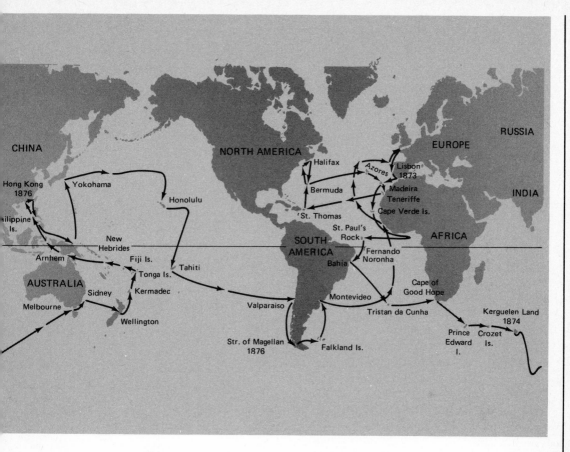

The map shows the route of the HMS Challenger expedition with the following labeled locations:

CHINA, Hong Kong 1876, Yokohama, Honolulu, Philippine Is., NORTH AMERICA, Halifax, Azores, Lisbon 1873, Bermuda, Madeira, Teneriffe, St. Thomas, Cape Verde Is., EUROPE, RUSSIA, INDIA, AFRICA, New Hebrides, Arnhem, Fiji Is., Tonga Is., Tahiti, Kermadec, St. Paul's Rock, SOUTH AMERICA, Fernando Noronha, Bahia, Cape of Good Hope, AUSTRALIA, Sidney, Melbourne, Wellington, Valparaiso, Montevideo, Tristan da Cunha, Kerguelen Land 1874, Prince Edward I., Crozet Is., Str. of Magellan 1876, Falkland Is.

1

At 10:00 A.M. on December 21, 1872, *HMS Challenger,* a steam corvette of 2306 tons, sailed from Spithead, England, on a voyage of discovery that would last three and one-half years. This was no ordinary voyage of discovery; its aim was not to find new lands but to explore the vast expanses of the oceans and the *terra incognita* that lay beneath. Starting with George Nares as captain and six scientists under the direction of C. Wyville Thomson, the expedition traveled

CHAPTER

68,890 nautical miles took, 492 soundings and 263 sets of temperature measurements, dredged the bottom of the ocean at 133 stations, and made 151 biological trawls. By 1895 an international team of 76 scientists under the supervision of John Murray had analyzed these observations and prepared a 50-volume report which included chemical analyses of the water, records of currents, information on the depth and nature of the sea floor, and descriptions of over 3000 new species.

The *Challenger* expedition was the beginning of ocean science, because it was the first time that marine data were gathered systematically over wide areas using special equipment for observation and collection. The expedition set a pattern for all subsequent oceanography, and therefore laid the groundwork for our modern knowledge of the sea. In recognition of this accomplishment, a modern research vessel—equipped to pioneer the drilling of mile-long holes into the floor of the deep ocean—was named *Glomar Challenger*.

Among the fundamental questions for which we seek answers in this book are some once regarded as impenetrable mysteries. Where does the sea water come from? Why do currents flow? Why do waves form and break? What makes the tides rise and fall? Why are the ocean basins shaped the way they are? In addition, we will use our scientific knowledge to approach some of the practical questions that are so much in evidence today: How much food can man expect to get from the sea? What untapped resources of minerals and energy does the sea hold?

OCEANOGRAPHY AS
A SCIENCE

The goal of oceanography is to understand the ocean. The first step toward this goal is to assemble systematic and accurate observations of the sea and to make them generally available. In oceanography, these data are published as charts, graphs, tables, verbal descriptions, and photographs.

But no matter how numerous or accurate these observations are, they only describe what the ocean is like. They do not by themselves explain how it came to be that way. As in other sciences, the process of understanding the subject involves a set of creative and purely intellectual acts. How effective these actions are depends primarily on the investigator's skill in identifying and symbolizing a small number of essential *simplifying concepts*. These concepts—often called hypotheses, theories, or natural laws—are better thought of as conceptual *models*. The models are of many types. Examples include such concepts as force, process, and energy. If we think of science as a game, the object of the game is to explain as much as we can with a minimum number of models—to simplify without distorting.

Thus the student of oceanography should not only examine the facts presented on charts and diagrams, but also learn and use the conceptual models which have over the years proved useful in understanding the causal structure that lies behind the mass of raw observation. These simplifying concepts are derived in many ways and take many forms. Many come from the more general (and therefore more basic) sciences of physics and chemistry, and include gravity, energy, the atom, and the molecule.

There are four major subdivisions of oceanography, based mainly on the nature of the simplifying concepts used. *Physical oceanography* applies the basic principles of physics, chemistry, and its close cousin, meteorology, to understand how and why seawater moves. Phenomena studied include waves, tides, and currents. A specialized branch of physical oceanography is concerned with shore processes, the interaction of forces of sea and land. *Chemical oceanography* applies the basic principles of chemistry to the description of the composition of seawater. Vertical and horizontal variations in chemical properties are important in identifying and investigating the operation of many oceanographic processes. *Biological oceanography* is the study of life in the ocean, including the interactions between organisms and their physical environment. *Geological oceanography* applies the principles of geology and geophysics to define the depth and shape of the ocean basins, to study the material on and below the sea floor, and to decipher the history of the ocean.

MARINE RESOURCES AND TECHNOLOGY

Today, as in the past, mankind depends in many ways upon the sea. For every nation located on a coast, the ocean provides a playground, an inexpensive means of communication and transport, a source of food, and the opportunity for developing its military arm at sea. Where fish are abundant, or deposits of petroleum or valuable minerals occur in its coastal zone, that fortunate nation can also develop these resources.

As the search for minerals and fossil fuels becomes more intensive, seabed resources that lie outside these territorial zones are coveted by many nations, and international disputes arise. To date, these arguments have not been resolved (see Chapter 22).

The general principles of any branch of ocean science can be used to develop a technology to deal with practical problems. Such applications include ocean engineering, petroleum exploration, seabed mining, environmental oceanography, fisheries and food technology, and military technology.

OCEAN CAREERS

Students considering oceanography or marine technology as a career may wonder how these fields are financed, where job opportunities are, and what kinds of training to undertake. In the United States most of the effort in marine science and technology is funded by the federal government. During 1977, for example, the total expenditures by various agencies and departments amounted to about $950,000,000. Of this amount about $776,000,000 was spent by the Departments of Defense, Commerce, Transportation, and Interior. The total expenditure for oceanographic research amounted to about $145,000,000. Approximately half of this ($76,000,000) was spent by the National Science Foundation. Oceanographic research is carried out at many universities with funds provided primarily by the National Science Foundation.

As a guide to obtaining proper academic preparation, a student should first determine in what general field his or her interests lie. Once the decision is made, the best academic preparation is normally to major in an appropriate science on the undergraduate level: physics, chemistry, biology, geology, or engineering. Specialized courses in oceanography are normally offered only in graduate school.

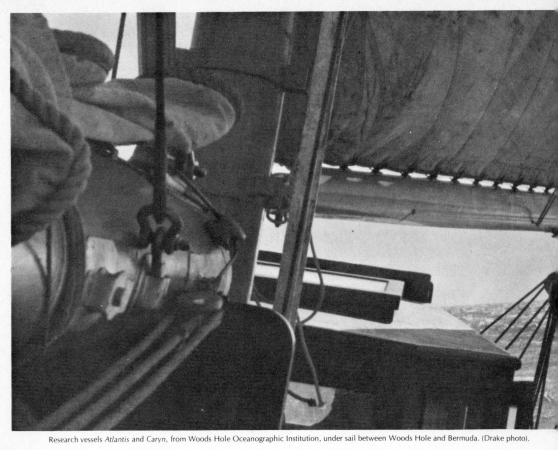

Research vessels *Atlantis* and *Caryn*, from Woods Hole Oceanographic Institution, under sail between Woods Hole and Bermuda. (Drake photo).

The History of Ocean Exploration and the Development of Oceanography

2

CHAPTER

In many ways the most striking part of the manned space program was that a few persons could view the Earth from afar, in its total perspective. These few have shared this experience with the rest of us through photography and brief, but poetic descriptions and have shown how the Earth's surface is dominated by water, with the continents as giant islands making up only about a quarter of the total surface (Figure 2.1).

In earlier and simpler times such a pic-

ture was not available. Man was huddled together on one Earth island and surrounded by a limitless sea or one whose limits spelled disaster to the seafarer. With a limitless sea, or even a very large one, the attitudes and habits of people must be quite different from those of people who recognize its finite size. Freedom of the seas, for example—a tradition dating from the times when the oceans were considered virtually limitless—came to mean freedom to use (or abuse) the seas and their resources without restriction, a concept no longer tenable as the numbers who use the sea and as their abilities to do so increase.

Until quite recently, the sea was traditionally used for transportation, hunting, fighting, and disposal of wastes—uses that generally involve only the surface or near-

FIGURE 2.1   View of the western hemisphere in August, 1974, taken from NASA's Synchronous Meteorological Satellite (SMS-1), which is operated by NOAA's National Environmental Satellite Service. (NOAA photo.)

surface waters. The sea also provided security. In earlier times island nations, such as England, could more easily defend themselves against invasion, and the United States could cite the Atlantic as a reason for staying out of European wars and as a reason for Europe not to interfere with nations in the Americas.

The recorded history of ocean exploration and ocean science can be divided into several periods. The first is the saga of the Phoenicians and their heirs; the second is the efforts of the Europeans of the Middle Ages and the Renaissance which led to the period of intense worldwide colonialism. During these periods a description of the ocean and its processes came primarily from explorers. Finally, the age of geographic exploration ended; scientific interest in the oceans emerged and culminated in the spectrum of oceanographic sciences we have seen develop over the past one hundred years.

EARLY EXPLORATION

The first historic description of far-ranging ocean expeditions is recorded by Herodotus. He noted that prior to 600 B.C., Phoenician sailors sailed down through the Red Sea and along the east coast of Africa. It is said that three years after the expedition started, it returned to the Mediterranean by way of the Strait of Gibraltar; thus this was the first circumnavigation of a continent. The feat was not repeated until the fifteenth century by Portuguese sailors, so most of Africa remained a mystery to the ancient world.

The Phoenicians also sailed through the Strait of Gibraltar into the Atlantic—reaching Britain, perhaps in search of the tin that had been mined in Cornwall since about 2000 B.C. and was needed for the extensive manufacture of bronze in the ancient world. Sailing through the Arabian Sea and across the Indian Ocean, they went east through the Malacca Straits, perhaps to Malaya for tin as well, and on into the Pacific. Because the Phoenicians were traders, their motive in making these explorations was to enlarge their commercial enterprises, but in doing so they also enlarged on the knowledge of the geography of the oceans. Such wide-ranging ocean trading continued through the Roman era and has continued to this day. The small Arabian and East African *dhows* (a type of sailboat) carry people and goods all over the Arabian and Red Seas.

For reasons unrelated to the problems of eastern trade, Irish monks, driven by piety, a spirit of adventure, and the desire for solitude, sailed far to the west in the North Atlantic. They reached Iceland before the arrival of the Norse, and some believe that St. Brendan actually reached the east coast of North America.

Among the Norse who settled Iceland was Eric the Red, born in Norway. His career in exploration was determined at least partly by his behavior in Iceland. He murdered a neighbor in an argument and was exiled to a small island off the west coast of Iceland. From this island he sailed west and explored Greenland, establishing the first settlement there in Igaliko Fjord. His son, Leif Ericson, is credited with finding Baffin Bay, Labrador, and Newfoundland, and Leif's brother, Thorvald, had the distinction of being the first European to be killed by Indians—in Labrador.

The rise of Moslem power in the Near East and its spread across north Africa and into Spain caused concern about the safety of traditional land trade routes. The established passage to the East became a virtually impassable battleground for almost 500 years and resulted in a search for different and better routes to the Indies or the spice lands. From the middle of the fifteenth to the middle of the sixteenth centuries, this search, principally by Spain and Portugal, the major maritime nations of that time, spurred one of the greatest periods of geographical discovery.

In the 30-year period from 1490 to 1520, navigators from southern Europe doubled the known surface of the world and a whole hemisphere was linked inexorably into European history. In 1488 Bartholomew Diaz reached the southern tip of Africa and, having been there, named it Stormy Cape. His sponsor, Prince Henry of Portugal, who stayed home, renamed it the Cape of Good Hope. In search of a route to India, Vasco da Gama rounded the cape in 1497 and sailed up the east coast of Africa, repeating the Phoenician voyage some 2000 years earlier. During this same period Columbus sailed to the Bahamas and the Caribbean. He was using Ptolemy's scale of the earth, with a circumference of 18,000 miles (33,500 km), which put the Indies some 6000 miles (11,100 km) closer to Europe than it actually is. Around 250 B.C., Eratosthenes had made a surprisingly good measurement of the circumference, about 24,000 miles (44,500 km); one wonders whether Columbus knew the correct size and figured he could not promote a voyage of this distance or whether he really believed Ptolemy was right.

The first soundings, or measurements of depth, appeared on a chart produced by Juan de la Cosa in 1504, and in 1585 they were included in maps by Gerard Mercator, father of the mercator projection. By 1600 the major land areas of the world had been defined and reasonable charts had been drawn. The major unknowns were the polar regions. Most maps of the time showed a great continent in the south and extended the northern lands to the pole—but with straits leading both east and west from the Atlantic to the Pacific. These were the basis for explorations such as those of Henry Hudson, who sought passages to the Orient through the river and the bay that bear his name; and those of the great English explorer Captain Cook, who from 1768 to 1780 mapped much of the South Pacific, including Australia and New Zealand. He also rediscovered the Sandwich (Hawaiian) Islands. Although Cook failed to find the Antarctic Continent (first seen by the Russian explorer von Bellingshausen in 1820), he did infer the presence of land on the basis of the glacier-produced icebergs in the area.

MAPS AND CHARTS—SIZE AND SHAPE OF THE OCEANS

A perception of the dimensions of the ocean basins is a function of the ability to draw limits to their boundaries and to determine their depths. We accept this ability today without much thought, but to the ancients, who lacked this knowledge, the oceans were perceived as being geographically limitless and bottomless. It is only in the last half millennium that the geographical extent of the oceans has become understood and only in this century have their depths been systematically plumbed.

Maps of coasts and nearby waters have a very ancient history. Mapmaking is one of the oldest of the graphic arts and is common to all primitive peoples. Eskimos have constructed maps of their coasts that compare well with modern charts. The Micronesians of the South Pacific, who lived in the trade-wind zones where winds and swells are remarkably constant, navigated by stick charts which represented the directions of waves produced by these winds or reflected by islands in the area. They knew the intersecting patterns of these waves with sufficient accuracy to allow them to make long journeys between the islands with small boats (see Chapter 8).

The Babylonians are credited with being the first to divide the celestial sphere into 12 signs and later 360 degrees. Strabo, in the first century A.D., attributed the art of land surveying to Egypt, where annual flooding of the Nile removed boundaries and landmarks and created the necessity for new surveys. The first formal geography, produced by Hecateus of Miletus about 500 B.C., was in the form of a seaman's guide, giving the number of days for sailing from one place to another. The earliest surviving pilot book, the *Periplous of Scylax,* appeared about 350 B.C. and described routes, headlands, currents, landmarks, anchorages, and other information important to the navigator.

It is improbable that any navigation charts as such existed until the advent of the magnetic compass. Navigators piloted by use of the stars and by such signs as the color of the water, the nature of the bottom, the fish, the birds, and other natural phenomena. Distance was judged by the speed of the vessel, first measured by the time it took a chip of wood thrown into the water from the prow of a moving vessel to reach the stern. Since that time, devices for measuring the speed of a ship have been called *logs*. These were improved by attaching a triangular piece of wood to the end of a line and measuring the amount of line that ran out over an interval of time. The measurements were made by counting the number of equally spaced knots on the line that passed by during a given time as measured with a sandglass. From this came the term knot (1 nautical mile per hour; 1.15 statute miles/hr or 1.85 km/hr) as a measure of speed, used first by ships and now by aircraft as well.

LATITUDE AND LONGITUDE

The early Greeks conceived of the world as being in the middle of a flat shield surrounded by a body of water. Although their knowledge of world geography was limited, they were content to leave a blank at the point on the map for which no knowledge was available. Ptolemy, in the second century A.D., made enormous contributions to world geography, but where he had no knowledge he filled in the blanks with theoretical conceptions. Had he been a lesser man, these additions might not have mattered, but his access to the library at Alexandria, Egypt—the greatest of his time—his association with ship masters and travelers, and his ability to synthesize this information gave him great stature. So great was his influence that some of his speculations were not questioned for 1500 years. His work was far in advance of previous efforts in the field. He brought an order into the geographical data of his time and introduced both the concepts and the names of latitude and longitude.

In this system, the Earth is divided north-south by a system of parallels of latitude running east-west and designated 90° at the two poles. Although the length of a degree of latitude varies a small amount with distance from the equator (Figure 2.2), it is very nearly equal to 60 nautical miles, so a minute of latitude is about one nautical mile ($0°01'$ = 1 nautical mile = 6080 feet = 1853 meters). The Earth

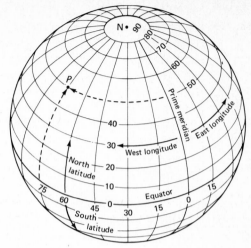

FIGURE 2.2   The geographical grid of latitude and longitude. Latitude is measured in degrees north and south of the equator; longitude is measured in degrees east and west of the prime meridian. Thus point *P* is designated 50° N, 75° W.

is divided east-west by meridians of longitude that range from 0° at the prime meridian through Greenwich, England, and east and west to 180° at the international date line in the Pacific Ocean. The meridians converge at the poles; hence the length of a degree of longitude varies from zero at the poles to about 60 nautical miles at the equator.

MAP PROJECTIONS

The earliest charts used for navigation were of the Mediterranean Sea. These charts were unrelated to latitude and longitude, but carried a scale of distance and a compass rose defining direction. A navigator would select the appropriate direction and, using stars as reference points, would follow a course from one point to another. The charts ignored the fact that the Earth is spherical. This did not matter greatly when navigating the Mediterranean, but when voyagers ventured farther to sea, it became plain that charts were needed that related to latitude and longitude, not merely to bearing and distance. This created the problem of expressing on a flat surface a representation of the Earth's surface that was useful to the navigator.

The problem was resolved by Gerard Mercator, who in 1569 adopted the projection that bears his name (Figure 2.3). On this projection, meridians and parallels are straight lines forming a rectangular grid. Meridians are equally spaced, but parallels increase in spacing from the equator toward the pole. On the Mercator projection the scale increases greatly at high latitudes, thus, for example, making Greenland appear to be larger than South America, whereas in fact it is only one-eighth as large. This projection has the enormous advantage that any straight line drawn on the map is a line of constant compass direction, or a rhumb line. Thus a navigator could draw a line on his chart between two points and steer a course along the indicated direction with confidence that he would arrive at the proper destination. Charts and plotting sheets on a Mercator projection are the standard type used for navigation today.

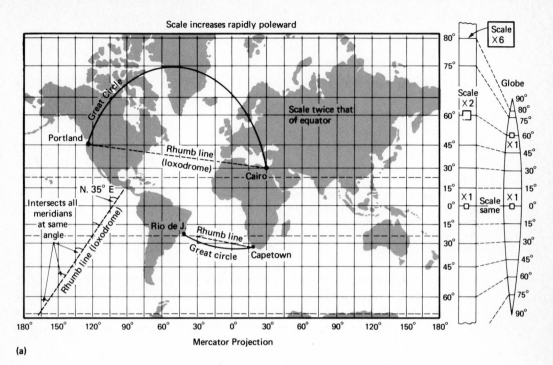

(a)

(b)

FIGURE 2.3 A rhumb line on a great circle chart will be a straight line only if it happens to correspond to a great circle such as a meridian or the equator. (A rhumb line, or a line of constant direction, is a straight line on a Mercator projection (a), but it does not represent the shortest distance between two points of the Earth's surface. The shortest distance is along a great circle.) If a navigator wants to lay out in advance the optimum track from one point to another across the ocean, he will first draw a straight line on a great circle chart (b); this will give the shortest distance between two points. He will then determine the latitude and longitude of points at intervals along this line and transfer these points to a Mercator projection chart. On this chart, the line segments connecting these points approximate the curved path that represents the shortest distance, but each of these line segments is oriented along a true compass direction. Thus by steering the vessel along these gradually varying courses, he comes close to the shortest track from the one place to the other.

Mercator projections have a disadvantage in that straight lines from point to point, although true compass directions, are not necessarily the shortest distance between the points. Since the Earth is a sphere, the shortest distance between two points on its surface is along a great circle. A great circle is the tracing on the Earth's surface of a plane that passes through its center. This plane cuts the sphere exactly in half. Meridians of longitude are a particular set of great circles in which the circles intersect at the poles, and if one navigates directly along a meridian on a Mercator projection, the track followed will be the shortest distance on the Earth's surface between the two end points. If one navigates along a parallel of latitude, other than the equator, the distance traveled will not be the shortest distance (Figure 2.3a).

To determine the shortest track between two points, a different type of chart is required—the great circle chart, constructed on a gnomonic projection (Figure 2.3b). A characteristic of this projection is that any straight line drawn on a map represents a great circle on the Earth's surface and hence the shortest track between the two end points.

NAVIGATION—FINDING ONE'S POSITION AT SEA

Regardless of map projection, a map is no better than the information displayed, and the early charts of the world suffered because of man's inability to navigate. Generally, the most sucessful of the early explorers were those who could push the crude techniques of the time to the very limit. Even until very recently, determining the exact location of a bottom-mounted instrument or a small seamount required great care and skill and sometimes a little luck.

Early navigation was done by pilotage—steering of the ship from one landmark to another. In the open ocean the ship's position was determined by dead reckoning—estimating the speed and knowing the time elapsed and the distance traveled along the course or courses steered. The uncertainty in position increased steadily with time, because of currents, drift, and errors in steering or measuring of distance. Thus it was necessary to be able to determine the true position at intervals in order to correct the dead-reckoning position.

The Greeks and Phoenicians learned to use the sun and the polestar and sailed far from shore and at night. Latitude was relatively easy to ascertain, since it could be measured by determining the altitude of the polestar or of the sun at midday or of any star when it crossed the meridian. The greatest single aid to navigation was the magnetic compass. With it the navigator could proceed by dead reckoning for long periods, making necessary adjustments to his position by determining latitude from the sun and the stars.

Devising a method to determine longitude (one's east-west position) was much more difficult than measuring latitude. The principle of determining longitude by measuring the time of local apparent noon, when the sun is at its maximum elevation above the horizon, was known in the sixteenth century. The difficulty arose because there was no clock or chronometer accurate enough to measure this time. In 1714, as a result of many maritime disasters due to bad navigation, the British Parliament offered an award of £20,000 to anyone who could establish a method of determining longitude within an error of 30 miles after a six-week voyage. John Harrison, a self-taught Yorkshire carpenter, won this award, and his chronometer marked the beginning of time observations of longitude at sea.

Celestial navigation methods improved through the years with the development of the sextant (an instrument for measuring the elevations of celestial bodies above the horizon) the provision of radio time signals for correcting chronometers, and the publication of nautical almanacs that give the positions of the sun, moon, and the planets or the stars; until nearly the middle of the twentieth century it was the only reasonably accurate method of navigating in the open ocean.

ELECTRONIC NAVIGATION AIDS

One of the difficulties with celestial navigation is that a person must be able to see the sun or the stars to determine a position. In areas with a high percentage of cloud cover, such as the polar regions, navigation by celestial methods is very difficult. Thus there developed a search for electronic navigation methods to resolve this difficulty as well as to improve accuracy.

The use of electronic navigation aids did not start auspiciously. The first such devices were radio-compass stations that were the forerunners of the radio beacon stations of today. These stations had a radio receiver connected to a loop antenna that could be rotated 360° and was affixed to a dummy compass. A ship desiring a bearing would contact the station on the proper frequency. The loop antenna would then be rotated to the proper direction and the bearing transmitted back to the ship. Unfortunately, in those days there was an inherent 180° ambiguity in the bearing—that is, for example, the station could not tell whether the signal was coming from the northeast or the southwest. This difficulty was to lead to one of the greatest disasters in the peacetime history of the United States Navy (Figure 2.4).

The development of radar was a great improvement in near-shore navigation aids, and on the open sea, served to prevent collisions between vessels or with ice. Radar, an acronym for **ra**dio **d**irection **a**nd **r**anging, provided the navigator not only with a bearing to an object or navigational point, but also provided a distance or range. Since radar employs an ultra high frequency signal, its range is limited to line of sight. It is invaluable for inshore navigation or for location of navigational hazards within its limited range of a few tens of miles.

All modern electronic navigation systems can be used in fair weather or foul, but such systems, except a few, have the disadvantage of limited range (Figure 2.5). Even with the exceptions, a large number of shore installations would be required in order to approach worldwide navigational coverage of the oceans. This problem has been resolved with the introduction of satellite navigation. The navigator can

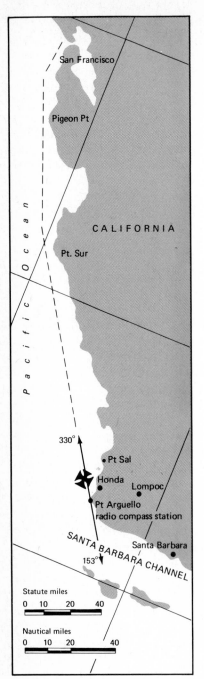

FIGURE 2.5   Most modern electronic navigation systems are of two types: those which employ a hyperbolic triad and those which use a ranging pair of shore stations. Some systems are composites of these. (a) In the hyperbolic triad system a master is coupled with two slave stations. Each of these transmit radio signals and the lines of equal time delay between the master and each slave can be represented by a hyperbola on the Earth's surface. Two intersecting sets of hyperbolae will be generated and the crossing points of the two hyperbolae determined by a receiver on the vessel can be translated into a position using suitable charts or tables. With a ranging pair, (b) distance is measured from two shore stations, and the intersection of appropriate distance circles marks the position.

FIGURE 2.4   On September 8, 1923, nineteen of the fifty-five destroyers in active service with the United States Navy were steaming in formation at high speed from San Francisco to San Diego, California. Weather was bad and visibility was limited. Navigation was the responsibility of the flagship, and the others were to follow the leader. A bearing was requested from the radio-compass station at Point Arguello, near the mouth of the Santa Barbara Channel, that showed the vessels to be close to the dead-reckoning position, and subsequent bearings indicating that the squadron was considerably to the north and east were discounted. At about 9:00 P.M., the flagship ordered a 55° course change to port to take the squadron into the Santa Barbara Channel, and within minutes the flagship drove onto the rocks at a speed of 20 knots. She was followed by six others, but the remaining twelve, three of which struck the rocks, managed to stay clear.

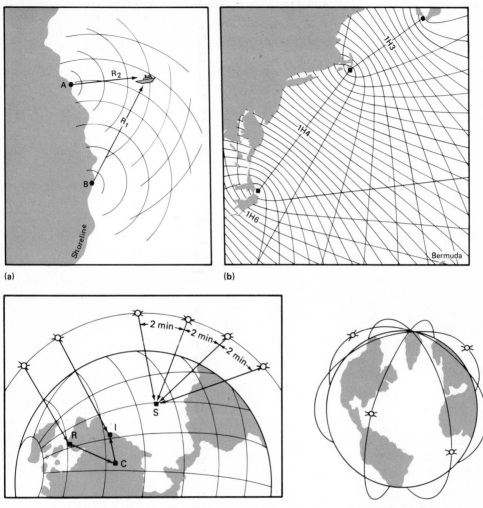

(a)

(b)

Transit Navigational Satellite System

Four Satellites in Polar Orbit

FIGURE 2.6 The system consists of up to four satellites, four tracking stations (R), two data injection stations (I), a computing center (C), and shipboard navigational equipment (S). The satellites are in orbits at altitudes of 500–700 miles and orbit the earth about once every 100 minutes. Every two minutes the satellite transmits synchronization and identification signals and information describing its orbit and the orbits of the other three satellites. Data on the orbit are determined by the tracking stations during several passes, and corrections are made as the satellite passes over the injection station. Each transmission from the satellite is timed so that the end of the data transmission coincides with the even two minutes of Universal Time, so it also serves as an accurate time signal.

   The position obtained from the satellite is based on the shift in frequency (Doppler shift) that occurs whenever the relative distance between satellite and receiver changes. Such change can be measured by a receiver on a ship any time a satellite passes at an elevation angle between 10° and 70° and will be due to movement of the satellite in orbit, movement of the ship on the earth, and rotation of the earth. From the estimated (dead-reckoning) position of the ship, the ship's course, speed and antenna height and the satellite information, accurate positions can be obtained in any weather conditions.

determine his position from a single pass of an earth-orbiting satellite in all weather conditions and with much greater accuracy than is possible with the use of celestial navigation. A position can be determined at least every two hours with an accuracy of the order of 0.2 nautical miles (Figure 2.6). Because of its accuracy, the frequency of fixes anywhere in the world ocean, and its all-weather capability, the satellite navigator is becoming a standard on oceanographic vessels.

EARLY OCEAN SCIENCE

Most of the earliest knowledge of the oceans came from ship captains and explorers, whose interest in the ocean was primarily governed by its relevance to the navigation and safety of the ship. Thus information on the tides, currents, and shallow-water depths has been accumulating for centuries, but it took the development of the transocean telegraph cable in the mid-nineteenth century to provide enough information on the depth of the ocean to generate the first chart of the ocean depths that was not mostly fantasy. At least part of the reason for the slow development of ocean science was the great difficulty in generating information, as can be appreciated in the following report of an attempt to measure the temperature in the deep ocean in 1818 by lowering instruments from a ship. "From the great friction we found much difficulty in getting it [8600 meters of line] onboard again. It took a hundred men just one hour and twenty minutes to do."

On the other hand, tidal information and speculation as to the cause of tides date back to the ancient Greeks. The tidal range is low in the Mediterranean, but as soon as sailors ventured through the Pillars of Hercules (Strait of Gibraltar) the tides were to be reckoned with. The relation of tides with phases of the moon was known to Pliny, and the variation in tidal range in different parts of the world was remarked upon by every careful observer. But Cook's observations of the almost complete absence of tides in Tahiti and other South Pacific Islands came too late for Daniel Defoe. The opportunity for Robinson Crusoe to rescue material from his shipwreck was made possible by a magnificent tidal range such as that found off Defoe's own English coast rather than the almost imperceptible rise and fall of the tide in most Pacific Islands (see Chapter 8).

Knowledge of ocean currents also engaged the interests of the early explorers. The strong current that flows into the Mediterranean through the Strait of Gibraltar was reported by Pliny, and the reversing currents of the Arabian Sea (Figure 7.4) have, since the ninth century, apparently been known to Arabs, who timed their trading voyages to Africa to take advantage of currents. Although crude surface current charts were developed in the eighteenth century, among them Benjamin Franklin's chart of the Gulf Stream (Figure 2.7), the development of generally useful current charts awaited the solution of the longitude problem. As soon as navigators had a way of determining longitude as well as latitude, they could correct their dead-reckoned positions daily and determine how far the currents and winds had drifted them off course. Shortly after the chronometer came into general use, help-

Reproduced from Transactions American Philosophical Society Vol 2,1786

FIGURE 2.7 Ben Franklin's chart of the Gulf Stream. (From *Transactions of the American Philosophical Society,* 1786.)

ful surface current charts began appearing in the sailing directions published in France, England, and the United States.

In 1850 Matthew Fontaine Maury, a naval lieutenant, made a major contribution by organizing information about current charts on an international basis. Maury was not only an important synthesizer, he also had a fine sense for what today is called public relations. His book *The Physical Geography of the Sea* went through many editions and reached a large audience of nonscientists. This was the period of the clipper ship, and Maury made much of the fact that with the new sailing charts, which showed the winds and the currents, the sailing time from Boston to San Francisco was reduced from 183 to 135 days and round trip from England to Australia from about 250 to 160 days. The contribution of the naval architects, who were responsible for these fine-lined ships with their magnificent sails, was not noted by Maury, but perhaps the architects also ignored the contributions of Maury's new sailing charts when they wrote about these reductions in travel times.

The thousands of ship-drift observations that poured into the various hydrographic offices of the world soon produced current charts of great fidelity, and the surface current charts available at the end of the nineteenth century have been little improved since then.

Except for tides, currents, and shallow-water topography, knowledge of the ocean proceeded slowly. As early as the seventeenth century it was generally known that deep-ocean water could be colder than surface water. For example, in 1665 Robert Boyle (later to be known primarily for his laws about gas behavior) reported on the coldness of the water below the surface. Some of this evidence came from salvage divers, and some apparently from travelers in the tropics who reported that a bottle of wine could be cooled by lowering it on a line beneath the ship. The difficulty of making these observations, however, combined with the technical problems of accurately measuring temperature at depth, postponed any significant study of deep-ocean temperatures until about the time of the *Challenger* expedition.

Similarly, there was relatively little detailed information on salinity. Again Boyle summarized what was known at this time; namely, that the ocean is uniformly salty and that the degree of saltiness does not apparently change appreciably from place to place. Knowledge of the ocean chemistry, however, depended upon the development of chemistry itself. As the techniques and knowledge of analytical chemistry grew, so did the knowledge of ocean chemistry.

Before the days of the steam engine, measuring the depth of the ocean by sounding line was tedious and back-breaking work. Good charts were developed in the shallow waters where grounding was a constant danger; but once a ship was in the deep ocean, there was little incentive for soundings other than man's curiosity. The development of the steam winch removed the manual labor of pulling in the mile of sounding line, and the development of sounding wire (piano wire) in the latter part of the 19th century also helped; and as noted previously the development of the submarine cable changed the issue of knowing the ocean's depth from one of curiosity to one of some practical importance. The bathymetric map of the North Atlantic (Figure 2.8) produced by Maury in 1857 was based upon some 800 soundings. The main features of the North Atlantic are there, the continental margin, the deep basins, and a median ridge, but detail is lacking.

Knowledge of the biology of the ocean developed slowly in the pre-*Challenger* era. In the early nineteenth century many believed that there was no life in the cold, dark, deep ocean, and the absence of light seemed to be ample theoretical reason for no further investigation. In the early 1800s the British naturalist Edward Forbes made dredge hauls in the Mediterranean and suggested that the limit of life was 300 fathoms (about 600 m); that limit was not to go unchallenged long. The Norwegian biologist Michael Sars had samples from a depth of 450 fathoms (about 1000 m) and a telegraph engineer, G. C. Wallich, found evidence of life on a sounding line retrieved from a depth of 1200 fathoms (about 2500 m). However,

FIGURE 2.8   Matthew Maury's bathymetric map of the North Atlantic Ocean, based on approximately 800 soundings. (From Findlay, 1873.)

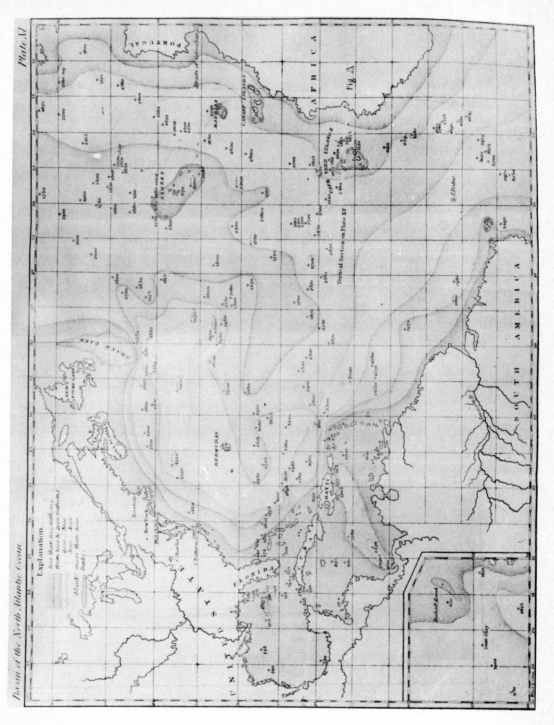

it was the work of William Carpenter and Wyville Thomson that finally caught the imagination of the scientific world. In two cruises off Scotland in 1868 and 1869 they brought up living material from dredge hauls as deep as 2400 fathoms (about 5000 m). Perhaps more important, they found markedly different faunal populations in dredge hauls a few miles apart. We now know they had dredged on opposite sides of what is called the Wyville Thomson Ridge, which extends between Scotland and the Faeroe Islands and separates the cold Norwegian waters from the warmer North Atlantic deep waters.

DEVELOPMENT OF MODERN OCEANOGRAPHY

The time was ripe for a major effort, and the British, through the efforts of Thomson and the Royal Society and with the cooperation of the Royal Navy, took advantage of it. It is generally agreed that modern oceanography dates from the *Challenger* expedition of 1872–1876. The *Challenger* was a spar-decked corvette of the Royal Navy and was fitted out by the Royal Society so that Thomson could examine the ocean, its contents, and its bottom for reasons of pure curiosity. This was the beginning of systematic examination of the ocean both geographically and with depth.

The *Challenger* was not a large vessel—2306-ton displacement, 226-feet overall, 36 foot beam. She was rigged as a sailing vessel but had a 1200 horsepower auxiliary engine. Amidships was a steam winch of 18 horsepower, which was used to pay out and haul in the hemp lines used for bottom sampling. Four thousand fathoms (7300 m) of this line, 2-3 inches (5-8 cm) in diameter, were flaked out on deck, taking up most of the open space. The sheaves, or pulleys, over which this line was paid out were supported by a large number of big rubber bands that helped compensate for the movement of the vessel because of wave action or temporary catching of dredges or trawls on irregularities of the bottom (Figure 2.9). It took 10 to 12 hours of hard work to make one trawl or dredge in 3000 fathoms (5500 m) of water.

She covered almost 69,000 miles (130,000 km) in the Atlantic, Pacific, Indian, and Southern Oceans—as far south as the edge of the Antarctic continent. She made 362 stations in which soundings, dredgings, and trawling were undertaken, and the properties of seawater measured. Enormous collections of biological specimens were obtained, including over 700 new *genera* and many thousands of species of animals. For example, before the expedition 600 species of a single-celled organism called *radiolaria* were known and the *Challenger* found 3500 *new* species. The cruise was even lucky enough to discover the deepest place in the oceans, now known as the Challenger deep, in the trench off the Marianas. The results of this voyage were published over a period of 20 years by some of the most distinguished scientists of the day.

The *Challenger* established the pattern for the next period of exploration—major expeditions of some duration, sometimes years, during which large quantities

FIGURE 2.9   The dredging and sounding arrangements on board the *Challenger* during her globe-encircling cruise, 1872–1876. (From Thompson, 1878.)

of data and samples were collected. These data would be returned and distributed widely through the scientific community to be worked up and published over a period of many years.

Although many deep-sea expeditions followed the *Challenger,* two deserve special mention. The first of these was the German *Meteor* expedition between 1925–1927 (Figure 2.10). The rationale for this expedition came from Fritz Haber, the 1918 Nobel laureate in chemistry, who conceived the idea of extracting enough gold from seawater to pay the German war debt. This idea was abandoned

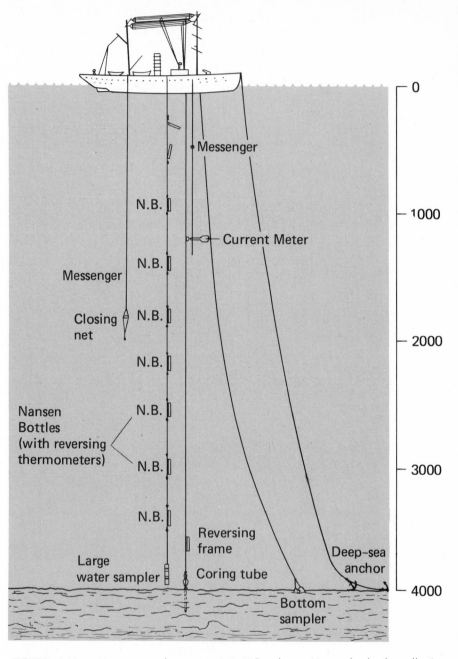

FIGURE 2.10   *Meteor* on a deep-sea station. N.B. refers to Nansen bottles for collecting samples of water at different depths. These bottles are tripped and closed by messengers attached to each bottle, which then slides to the next bottle down and trips it in turn. (After *Meteor* Reports, 1932.)

in 1927 when it was found that the amount of gold in seawater was far too small to be extracted profitably (see Chapter 17), but by then the expedition had been carried out. The *Meteor* made 13 crossings of the South Atlantic between 20°N and 60°S and, using the best available techniques, measured with great accuracy the properties and movements of the water masses, the topography of the ocean floor, and the nature and distribution of sediments on the sea floor (Figure 2.10). The *Meteor* expedition showed that hydrographic measurements must be carried out systematically and with standardized instruments if they are to be of use in determining the circulation pattern of the oceans.

The second expedition deserving mention consists of the expeditions carried out by the *Discovery* series of vessels from Britain. The British, because they were at that time deeply involved in open-ocean whaling in Antarctic waters, sent *Discovery I* (Figure 2.11), and later *Discovery II,* south to determine the oceanography,

FIGURE 2.11   The Royal Research Ship *Discovery*, a 198 foot, 1600 ton barque. She worked in the Antarctic Ocean from 1925–1927 and was replaced by *Discovery II* in 1930.

topography, and biological productivity of the Southern Ocean, where the whales were. These vessels operated on a basis different from earlier oceanographic cruises in that they were permanently assigned to oceanographic duty and made a continuing series of cruises, not a single long cruise. This mode of operation is more like the modern pattern in which vessels are constructed for or permanently assigned to oceanographic work and make cruises on a continuing basis, year after year. *Discovery III,* built after the decline of the British whaling industry, operates in this mode in all oceans, not just around Antarctica.

OCEANOGRAPHIC DEVELOPMENT IN THE UNITED STATES

As in other fields of science, oceanography in the United States generally lagged behind development in Europe, although not as much as one might expect for a young country. Benjamin Franklin found that Yankee traders consistently made better time crossing the ocean than did British mail packets. He learned that the Yankees were riding the Gulf Stream going to Europe and sailing far enough south to avoid it on their return. Franklin's is the first known chart of the Gulf Stream (Figure 2.7). Franklin also gave instructions on how one might determine, by measuring the water temperature, whether or not one's ship was in the Gulf Stream. He became curious about the role of oil in calming waves and for some time carried a small quantity of oil in the head of a cane so he could test its effectiveness on ponds and lakes as the occasion arose. At the time of the Revolutionary War Franklin wrote to the commanders of all United States warships asking them not to interfere with Captain Cook and his ship *Resolution,* which "may expand the progress of all sciences useful to the human race." Franklin's gesture did not find favor with Congress and his request was overruled; but, for whatever reason, Cook's ship did not incur hostility during the war.

The role of Matthew Fontaine Maury has already been noted. Maury was a naval lieutenant in charge of the Navy's Depot of Charts and Instruments, which has now become the Navy Oceanographic Office. Maury joined the Southern cause at the outbreak of the Civil War, and the most important oceanographic work after his departure was done not by the Navy but by the United States Coast and Geodetic Survey (now the National Ocean Survey of the National Ocean and Atmospheric Agency). In the late nineteenth century the United States Coast and Geodetic Survey developed methods for measuring the currents of the Gulf Stream that were not equaled for a half century. It began charting not only the shoreline and harbors, but also the continental shelf. It also provided Alexander Agassiz with the survey ship *Blake* (Figure 2.12) for three dredging cruises in 1878–1880 in the Caribbean and along the U.S. east coast. Agassiz and Charles D. Sigsbee introduced steel cable to oceanographic work, redesigned the winches, and brought dredging up to modern standards.

The Department of Terrestrial Magnetism of the Carnegie Institution of Washington was concerned with the magnetic field of the earth and wished to measure it at sea. As a result, the vessel *Carnegie* was built, in 1909, of nonmagnetic ma-

FIGURE 2.12   Systematic studies of the Gulf Stream and nearshore waters were begun by the U. S. Coast Survey before 1850. One of the most famous of these programs was the dredging expedition of the 148 foot *Blake* (1887–1880) under the direction of Alexander Agassiz.

terials—wood, copper, and bronze (Figure 2.13). It was a sailing vessel, with an auxiliary gas engine made of bronze. The engine blew up in Samoa in 1929 and *Carnegie* was lost, but not before she had made many important contributions to oceanography in fields far beyond her original purpose.

In 1927, the National Academy of Sciences, recognizing the German efforts of the *Meteor* expedition, established a Committee on Oceanography to examine the United States effort. The committee came up with a series of recommendations that led to the establishment of the Woods Hole Oceanographic Institution and the Oceanographic Laboratory of the University of Washington, and provided additional facilities at Scripps Institution of Oceanography, which had begun as a biological field station. Support for implementing these actions came from the Rockefeller Foundation and the Carnegie Corporation. In 1930, Woods Hole constructed in Copenhagen a 142-foot steel ketch, *Atlantis* (Figure 2.14). Scripps had a schooner, *E. W. Scripps,* which was originally a yacht (Figure 2.15). These institutions had limited staffs and depended upon summer visiting scientists for much of their efforts. Oceanographic cruises were of short duration and were limited to a radius of a few hundred miles. By World War II, there were a number of small institutions in the United States and Europe; the state of oceanography could be judged by the fact that virtually everything known about the subject in 1941 could be pub-

FIGURE 2.13 Designed by H.J. Gielow and built by the Tebo Yacht Basin Company of New York in 1909, the *Carnegie* was the first specifically constructed nonmagnetic vessel in which every effort was made to avoid the use of steel. Her primary mission was mapping out the magnetic forces as they prevail over the oceans. She was active in this role until her tragic loss in 1929 at Apia, Samoa from a gasoline explosion. (Courtesy Carnegie Institution of Washington.)

FIGURE 2.14 The *Atlantis*, a 142-foot steel ketch built in Copenhagen in 1931 for Woods Hole Oceanographic Institution. In 1964 *Atlantis* was sold to the Consejo National de Investigaciones Cientificas y Technicas, Buenos Aires, Republica de Argentina, where she was renamed *El Austral*. (Courtesy Woods Hole Oceanographic Institution.)

FIGURE 2.15 The *E.W. Scripps*, originally the 104-foot luxury yacht *Serena* of Hollywood actor Lewis Stone. In 1937 she was purchased by Mr. Robert P. Scripps and presented to Scripps Institution of Oceanography for use as a research ship. Disposed of in 1935, she was used for the movie ''Around the World in Eighty Days'' and later became a South Seas island schooner. The *E.W. Scripps* now rests on the bottom of the Papeete Harbor in Tahiti. (Courtesy Scripps Institution of Oceanography.)

lished in a single book, *The Oceans,* by H. V. Sverdrup, Martin W. Johnson, and Richard H. Fleming.

World War II changed the picture significantly. The Pacific War was a naval war conducted among islands and reefs that were uncharted or poorly charted. In wartime, events are usually considered more important than economics, with the result that much is learned and many devices are perfected in far shorter time than would be the case in the normal course of human events. This was the case in World War II, and many of the instruments and techniques developed for wartime use were applied to the expanded oceanographic research efforts that followed the war; oceanography prospered and has grown in many directions. In the United States much of the work was done by universities and today is supported by the Office of Naval Research and the National Science Foundation.

THE ERA OF INTERNATIONAL PROGRAMS

By 1957 it had become clear to everyone that the vastness of the ocean and the developing interest by nations in the sea transcended national concerns, and collaboration in basic scientific exploration has been gradually increasing on the international level.

The first such organized effort occured in 1957–1958 and was called the International Geophysical Year (IGY). IGY developed from two earlier International Polar Years, 1882–1883 and 1932–1933, which were organized for the purpose of making synoptic measurements in high latitudes primarily of phenomena in the atmosphere and especially the upper atmosphere. The IGY expanded the scope of activities to include oceanographic research, among others. It became possible to plan long cruises, coverage of the oceans greatly increased, and many of the blank spots were filled in. Perhaps more important, international cooperation was encouraged and machinery was set up for the international exchange of oceanographic data.

The Deep Sea Drilling Program (DSDP, now the International Program of Ocean Drilling, IPOD) began in 1964 with a successful program of drilling to depths of up to 1000 m in waters off the southeastern United States. In 1968, *Glomar Challenger* (named after the British *Challenger*), a dynamically positioned deep-sea drilling vessel, was constructed and commissioned. With the primary support of the National Science Foundation over 400 holes have been drilled in water depths reaching 6000 m and more than 1000 m into the bottom sediments and 600 m into the underlying hard rock. Scientists from 21 countries have participated in these cruises, and many others have participated in program planning and selection of drilling sites. Deep-sea drilling has greatly extended our knowledge of the sediments and rocks of the sea floor and, through studies of the microfossils in the sediments, of the evolution of ocean life and the circulation of the seas in ancient times.

The International Decade of Ocean Exploration (IDOE) is concerned with problem-oriented research in four areas: (1) environmental quality and the factors that affect it, such as the effects of adding toxic substances to the seas; (2) environmental prediction, which is concerned with geochemical or oceanographic baselines from which prediction can be made; (3) living resources, the factors that control their productivity, and the manner in which they may be intelligently harvested; and (4) seabed assessment, or the determination of the nature and location of the mineral resources beneath the sea.

A number of large-scale projects, involving many institutions and investigators and often participation by many countries, have been initiated by IDOE. The Geochemical Ocean Section Study (GEOSECS) is a program to study the distribution of the chemical properties of seawater for use in understanding large-scale oceanographic processes. Major expeditions in the Atlantic and in the Pacific were conducted by United States vessels (Figure 2.16), and German and Japanese traverses were made to complement the coverage. The results have sharpened our understanding of the modes and rates of ocean circulation and the chemical properties of the ocean.

The Mid-Ocean Dynamics Experiment (MODE) was initiated to study the large-scale eddies and mixing that goes on in mid-ocean. The latest experiment, POLYMODE, covers an area of the western Atlantic from Newfoundland to the Antilles and involves United States, British, Canadian, and Soviet investigators.

The Seabed Assessment Program is designed to study the processes by which minerals and hydrocarbons are formed and concentrated in the marine environ-

FIGURE 2.16   One of the newest of the U. S. oceanographic research vessels is *Endeavor*, operated by the University of Rhode Island. She is 177 feet long, displaces 922 tons, carries a crew of 12, and 16 in the scientific party. U. S. universities operate about 20 such ocean-going research vessels, from between 150 to 250 feet long.

ment. Activities in this area are worldwide and range from studies of formation of metals at mid-ocean ridges to manganese nodules on the sea floor and the structure and geologic properties of continental margins.

The Coastal Upwelling Project is designed to gain knowledge of the upwelling process, because it is by this means that nutrients are brought up from the depths to replenish those consumed by organisms. Although the upwelling areas are limited in extent, their extraordinary productivity makes them exceedingly important.

The International Geodynamics Project was initiated by the International Council of Scientific Unions and is a 50-nation effort to understand better the implications of the plate tectonics model with regard to formation of the geological features found at the Earth's surface and to the ability to locate resources. Because the forces that deform the Earth's surface are of global scale, it will require an international effort to truly understand them. An example of the types of activities included in this program is the French-American Mid-Ocean Undersea Survey (FA-MOUS) project in which a detailed study was made of the central valley of the mid-Atlantic ridge using the submersibles *Archimede, Alvin,* and *Cyana* (Figure 2.17).

Much can be expected from these major international programs, but it is well to remember that most of the major breakthroughs in science have been made by individuals, working alone or with a few colleagues. Without dimming our enthusiasm for large-scale and international programs, we must at the same time try to protect the individual investigator whose intuition may lead us along the most fruitful paths of investigation in the future.

FIGURE 2.17   The three submersibles *Archimede, Alvin,* and *Cyana* drawn to scale. All three can accommodate three passengers. The difference in size between *Archimede* and the other two is because *Archimede* was designed to go to greater depths (11,000 meters). The heavy sphere requires flotation material that is provided by gasoline in the main part of the hull.

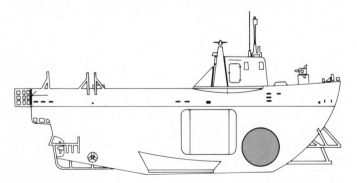
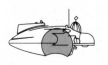
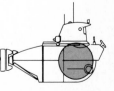

SUMMARY

1. Geographical exploration of the oceans began in a search for reliable trade routes and culminated with the search for the southern continent and the Northwest passage.

2. Geographical position on the Earth's surface is determined in terms of latitude north or south of the equator and longitude east and west of Greenwich, England.

3. Navigation began with pilotage, using landmarks on shore for reference. In the open sea, navigation was conducted with reference to celestial bodies, a method now supplemented by electronic aids and satellites.

4. Early ocean science was concerned primarily with currents, waves and tides, near-surface phenomena, and the sea floor in shallow waters.

5. Modern deep-sea oceanography began with the *Challenger* expedition, 1872 –1876.

6. Major development of oceanography in the United States occurred after World War II, with the aid of instruments and techniques developed during the war.

7. Oceanography is an international science; in recent years it has been abetted by a number of major international programs.

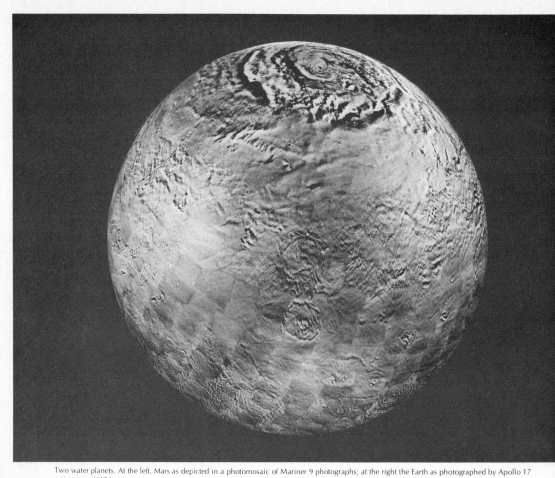

Two water planets. At the left, Mars as depicted in a photomosaic of Mariner 9 photographs; at the right the Earth as photographed by Apollo 17 astronauts. (NASA.)

Earth:
The Planet
With an Ocean

3

Of all the planets in our solar system only the Earth has the combined properties of warmth and volume of water to permit the existence of vast oceans. The small amount of water on the surface of Venus exists as a gas, because the temperature on the planet is much greater than the boiling point of water; the small quantity of water on Mars ex-

CHAPTER

ists in the ice caps and as permafrost, because of the low surface temperature of that planet. In comparison, 71 percent of the Earth's surface is covered by liquid water, and this we call the ocean. To any extra-terrestrial visitor, the ocean would be the dominant feature of our planet; thus it is altogether fitting to think of the Earth as the "ocean planet." We will now examine the general features of the "ocean planet" in the context of the solar system. How is the Earth different from the other planets? What do the planets have in common? How did the Earth get its ocean? How does the terrain hidden by the oceans differ from the dry land? These are some of the questions for which we seek answers.

PLACE OF THE EARTH IN THE SOLAR SYSTEM

STRUCTURE OF THE SOLAR SYSTEM

Our sun is a medium-sized star, one of 100 billion in our galaxy. We know that our star has a planetary system, because we live on one of its planets. We do not know how many other stars have planetary systems associated with them, because planets by nature are not self-luminous nor do they make up a large enough fraction of the mass of a star-planet system to be detected visually. There is no reason, however, to believe that there is a paucity of planets in the universe and, given the size of the universe, it is not unlikely that other ocean planets exist somewhere.

The solar system contains the sun and nine planets whose names are well known: Mercury, Venus, Earth, Mars, Jupiter, Saturn, Uranus, Neptune, and Pluto. In addition to these, there are other members of the solar system: the comets, the asteroids, moons around some of the planets, interplanetary dust, and possibly additional planetary bodies beyond the orbit of Pluto.

The plane of the orbit described by the Earth's revolution around the sun is called the plane of the *ecliptic*. The planets (and the asteroids, whose orbits lie between the orbits of Mars and Jupiter) describe orbits whose planes do not show

TABLE 3.1   Properties of the Planets

	Sun	Mercury	Venus	Earth	Mars	Jupiter	Saturn	Uranus	Neptune	Pluto
Mass (Earth = 1)	329,000	0.054	0.81	1[a]	0.11	314	94	14.4	17.0	0.05?
Radius (kilometers)	695,000	2,439	6,050	6,370	3,400	71,000	57,000	25,800	22,300	2,900
Density (grams per cubic centimeter)	1.41	5.42	5.25	5.52	3.96	1.33	0.68	1.60	1.65	3?
Albedo[b]	—	0.06	0.73	0.39	0.26	0.51	0.50	0.66	0.62	—
Effective temperature ($°K$)[c]	6,000	616	235	240	220	105	75	50	40	40
Surface temperature ($°K$)	—	616	600	300	230	130	—	—	—	—
Observed and expected gases in atmosphere[d]	H, He, O, Fe, N, Mg, C, Si, and others	—	CO_2, N_2, Ar, H_2O, HCl, HF, O_2(?)	N_2, O_2, Ar, CO_2, H_2O, HCl, HF, and others	CO_2, H_2O, N_2, Ar	H_2, CH_4, NH_3	H_2, CH_4, NH_3	H_2, CH_4, NH_3	H_2, CH_4, NH_3	CH_4
Distance from the sun (A.U.)[e]	—	0.39	0.72	1	1.52	5.2	9.5	19.2	30.1	—

[a] The mass of the earth is 5.976×10^{27} g.
[b] The *albedo* is the fraction of the incoming sunlight that is reflected.
[c] Temperatures are given on the absolute or Kelvin scale. 0° on the Kelvin scale is $-273°$ on the Celsius (or centigrade) scale.
[d] The symbols for the molecules listed are:

CO_2	=	carbon dioxide	HCl	=	hydrogen chloride
N_2	=	molecular nitrogen	HF	=	hydrogen fluoride
O_2	=	molecular oxygen	CH_4	=	methane
Ar	=	argon	H_2	=	molecular hydrogen
H_2O	=	water	NH_3	=	ammonia

[e] A.U. = "Astronomic unit," the mean distance of the Earth from the sun (149.6×10^6 km).

great departures from the plane of the ecliptic, whereas the comets describe orbits which apparently obey no orientation rules.

The major properties of the planets are shown in Table 3.1. On the basis of density and total mass, the planets are divided into two groups, the "inner" or "terrestrial" high-density, low-mass planets (Mercury, Venus, Earth, and Mars) and the "outer" or "Jovian" or Jupiter-like low-density, high-mass planets (Jupiter, Saturn, Uranus, and Neptune). Pluto, the most remote visually verified planet, cannot be unequivocally categorized, but it appears to resemble the terrestrial planets.

The sun is the major source of heat received by the surfaces of the planets, and the farther away a planet is from the source the less heat is received by it. The surface temperature of a planet is approximately related to the intensity of the sun's heat impinging on the planet. The thickness and composition of the atmosphere will actually determine the average surface temperature.

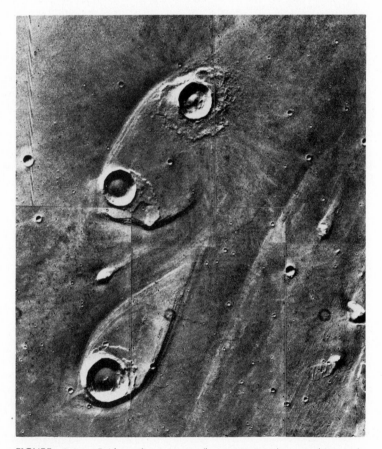

FIGURE 3.1   Evidence for past water flow on Mars is shown in this mosaic of photographs taken by the Viking 1 orbiter from a height of 1600 km on June 23, 1976, prior to the descent of the Viking 1 lander. The features showing teardrop-shaped islands of erosion-resistant material formed by the impact craters indicate water flow from the bottom to the top of the photograph. One shoreline of the channel is inferred to be at the lower right. (NASA photograph.)

Of all the terrestrial planets, aside from the Earth, only Mars shows evidence of water in condensed form on the surface. Mars has ice caps made up primarily of water and there are features on the planet's surface which indicate stream or subterranean water flow activity during the planet's history (Figure 3.1). It is this possibility of available fluid water, however small and transitory, that makes the search for life on the planet more than an idle exercise. The recent results of the Viking landers moved one step closer to answering this question. On the basis of the experiments performed on the two Viking landers it appears that life as we know it does not exist on Mars.

The Jovian planets are composed of hydrogen and helium in condensed form. The composition of the surface gases of these planets is controlled by the presence of large amounts of hydrogen; thus methane and ammonia are the expected gases and have indeed been observed on the planets.

ORIGIN OF THE OCEANS

The question of how the Earth got its ocean is linked directly to how the solar system was formed. Despite our growing knowledge about the way stars work, based on laboratory and theoretical studies, we still do not have an unambiguous model for the formation of stars and solar systems.

Three fundamental observations about the nature of the cosmos provide a starting point for our speculations. (1) The overwhelming composition of the cosmos is 75 percent hydrogen and 25 percent helium, with the rest of the chemical elements of minor importance. (2) Gas and dust are distributed throughout the cosmos. (3) Rotation and turbulence are everywhere observed in the universe (for example, galaxies form pinwheel arrays, stars rotate, planets revolve).

In addition to these general observations we also know that our solar system is about 4.55×10^9 years old. With these constants we can construct a scenario for the formation of our little corner of the galaxy.

Star formation is going on continuously in the galaxy. At any time, the gas and dust in a part of the galaxy can form gravitational nuclei by turbulent action that accrete to form protostars. As the accumulating protostar forms, it invariably inherits a rotation, because the parts from which it is made are rotating. With continuing growth it gets hot enough to begin hydrogen fusion and becomes a full-fledged star. At the same time, a disk of matter perpendicular to the axis of the rotating star develops. This rotating disk from which planets are formed becomes hot enough to vaporize virtually all the elements, but not hot enough to undergo fusion. As the disk, primarily composed of hydrogen, cools by radiation, crystals of compounds stable at the high temperature condense. Virtually no volatiles such as water and carbon-rich compounds will condense until very low temperatures are reached. Low temperatures were attained rapidly at the distances of the Jovian planets; thus they swept up even hydrogen gas. But the terrestrial planets were hot too long and remained too small to accumulate gases directly from the solar nebula.

How then were the volatiles added to Venus, Earth, and Mars on which we can observe water, carbon dioxide, and other gases? The answer appears to lie in the fact that as the planetary nebula cools, the volatiles and water condense or react with the exposed high-temperature grains available, to form a complexion of oxidized iron grains and silicate minerals which contain water, such as the common clay minerals found in soils. Moreover, organic compounds were synthesized in the nebula at low temperatures, and these also were absorbed on grains. This assemblage has indeed been found in a class of meteorites called *Type I carbonaceous chondrites*. These are characterized by the fact that they contain 5 percent organic carbon and 20 percent water (tied up in minerals). This special class of meteorites has all the properties necessary to add an ocean and an atmosphere to a planet. Some people feel that carbonaceous chondrites are related to comets, since comets are also known to be enriched in water and organic compounds.

Thus it appears that this carbonaceous chondrite material, in falling on the growing planet in its late stages, could provide the necessary volatiles for the terrestrial planets demonstrated to have atmospheres. The time scale of the planet formation process is in dispute. Some people feel that the formation of the sun, the condensation of the particles from high temperatures to low temperatures, and the formation of the planets all occurred within about 100,000 years or less. Others feel that about 100 million years elapsed between the first two stages and the final formation of the large terrestrial planetary bodies.

Whatever the time scale, the Earth received its water and gases associated with solid phases. When did these substances escape the solid carriers and, as gases, reach the Earth's surface to form the oceans and the atmosphere? Was the major outgassing of the Earth virtually instantaneous, or was it a slow process continuing to this day? Phrasing it in another way: Was the ocean the same volume at the formation of the Earth as it is now, or has there been an increase in the volume throughout geologic time?

Although arguments have been made for a continuous degassing of water from the Earth, the evidence seems to point to an early episodic degassing. We know that as mud or rocks are heated to temperatures above 500°C, they will begin to lose their water and other volatiles. If venting to the Earth's surface occurs, this water will be released and thus the tendency to store water at depth where the temperature is high will be slight. Indeed the deepest crustal rocks that we have observed in the roots of ancient mountains are "dry" rocks with very little water as part of the rock-forming minerals.

There is, however, evidence that even today gases are being released from the interior of the Earth, mostly through volcanoes. Most of these are "recycled gases" once on the Earth's surface associated with sediments sucked down to the interior by large-scale crustal processes (see Chapters 16 and 17). However, as a result of some elegant studies of rare gases (helium, neon, argon, krypton, and xenon) in seawater and volcanic rocks of the deep ocean, we are reasonably certain that some of this material from the Earth's interior is "new." The rate of supply or the flux of this "new" material is probably small compared to that needed to supply the present-day atmosphere and oceans, even if the process were to operate at its present rate for the 4.5 billion years of the Earth's existence.

MAJOR FEATURES
OF THE EARTH

If we consider the superficial features of the Earth, we note that there are certain distinct units. There is an *atmosphere,* composed primarily of nitrogen and oxygen gas. Water is strongly represented not only in the oceans but as rivers, lakes, ground water, and glaciers. We shall call this the *hydrosphere.* Clearly both the mountains and plains and the ocean bottoms are composed of rocks or their derivatives, and this general grouping we call the *lithosphere.* It should be noted that the term ''lithosphere'' has come to have two meanings. In the sense used here, it refers to the solid Earth, but it is also used to describe the rigid outer shell of the Earth above a more plastic interior (see Chapter 16). The *biosphere* is the domain of life, and life is known to occupy a variety of niches in land, air, and water.

This classification provides a framework in which to discuss the major features of the Earth. The masses of the ''spheres'' are shown in Table 3.2.

The average composition of the atmosphere is shown in Table 3.3. We all know that the water content of the atmosphere ranges from virtually zero over the driest of deserts and the coldest parts of the Antarctic ice cap, to higher concentration levels dictated by the local temperature and pressure of the air mass.

The hydrosphere is dominated by the oceans (Table 3.4). The second most important reservoir for water is the ice caps of Antarctica and Greenland, with ground water following close behind in volume as a water reservoir. In comparison rivers, lakes, and atmospheric water are very small in volume, although our experience with these bodies may make us more aware of these parts of the hydrosphere than the volumetrically more important ones.

TABLE 3.2   Major Subdivisions of the Earth

Subdivision	Mass (grams)	Composition
Lithosphere	5.98×10^{27}	Rocky mantle and crust and iron-nickel core
Hydrosphere	1.4×10^{24}	Water and dissolved materials
Atmosphere	5.12×10^{20}	Nitrogen, oxygen, argon, and others
Biosphere	$\sim 7 \times 10^{17}$	Organic compounds in living organisms

The biosphere is represented in virtually every part of the Earth's surface. Some organisms are carried high in the atmosphere by wind and others live in the deepest recesses of the sea. And even in parts of remotest Antarctica, where the closest approach to an abiotic environment exists, at least one organism has effectively intruded—man. Although the total mass of life (biomass) is small compared to the other three major parts of the Earth, life is an important component in determining the composition of the atmosphere and hydrosphere and the rate of destruction and redistribution of the components of the crustal lithosphere at least.

TABLE 3.3   Composition of Water-Free Air at Sea Level

Component	Content (percent by volume)
Nitrogen	78.084
Oxygen	20.9476
Argon	0.934
Carbon dioxide	0.0314
Neon	0.001818
Helium	0.000524
Krypton	0.000114
Xenon	0.0000087
Hydrogen	0.00005
Methane	0.0002
Nitrous oxide	0.00005
Ozone	
Summer	0 to 0.000007
Winter	0 to 0.000002
Sulfur dioxide	0 to 0.0001
Nitrogen dioxide	0 to 0.000002
Ammonia	0 to trace
Carbon monoxide	0 to trace
Iodine	0 to 0.000001

TABLE 3.4   Relative Distribution of Water in the Earth's Surface Other Than in the Oceans and Excluding Rocks

Reservoir	Percent	Volume (cubic kilometers)
Polar ice and glaciers	75	29×10^6
Ground water at depths less than 750 m	11	4.2×10^6
Ground water at depths greater than 750 m, but less than 4000 m	13.6	5.3×10^6
Lakes	0.3	120×10^3
Rivers	0.03	12×10^3
Soil moisture	0.06	24×10^3
Atmosphere	0.035	13×10^3
Total nonoceanic water	100	39×10^6
Volume of oceans		1350×10^6

    The lithosphere is the major mass of the Earth. The interior structure of the Earth (Figure 3.2) has been determined by the geophysical methods described in Chapter 15 and shows a strong zonation. The outermost zone is called the *crust*. The crust is made up of the complex of rocks that can be obtained easily by geologists from rock outcrops or drilling operations. These are classified as *igneous*, derived from the molten state; *sedimentary,* derived by the weathering and erosion of other rocks and accumulating as deposits; and *metamorphic,* the result of secondary heating, pressure, and chemical reconstitution on preexisting rocks. There are rocks, as one can expect from the innate complexity of nature, that fall between these categories. The ocean basins beneath the sediment cover, as we shall see, are composed of igneous rocks called *basalts.* This rock type and related ones make up the volcanic islands of the oceans, including such well-known ones as the Hawaiian Islands and the Aleutians in Alaska. They are also found on the continents. The dominant rock type on the continents beneath the sedimentary rocks lies somewhere in composition between the igneous rock called *granite,* used commonly as a building stone, and the range of metamorphic rocks produced from sediments and igneous rocks. As a convenience of nomenclature the continents are sometimes considered to be "granitic," and the ocean basins "basaltic," in composition. Granitic rocks are generally rich in potassium and poor in magnesium and iron. Basaltic rocks are poor in potassium and generally rich in magnesium and iron.

Below the crust of the Earth is the *mantle,* believed to be composed primarily of iron and magnesium silicate materials. Below the mantle is the *core,* believed to be composed of iron-nickel by analogy with certain meteoritic types. At least part of the core appears to be molten.

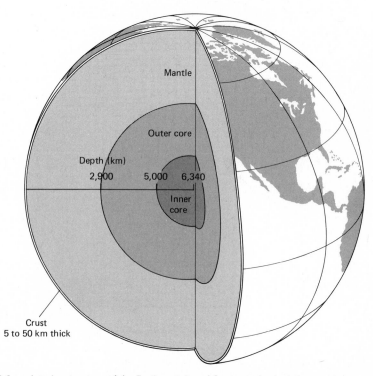

FIGURE   3.2   Interior structure of the Earth as inferred from geophysical data.

OCEAN DEPTHS

About three-quarters of the Earth's surface lies under the ocean surface and only one-quarter is exposed on the continents and on islands. Some of the latter is also covered by water, as lakes, streams, or ice. If we take into account that the shallow continental shelves are more closely related to the continents and include their area with the exposed part, then about 65 percent of the Earth's surface is oceanic and 35 percent is continental.

The continents and oceans are not evenly distributed over the surface of the Earth. If we moved the North Pole so that it were at London, England, the "northern" hemisphere would be almost half continent and half ocean, whereas the opposite, or "southern," hemisphere would be 90 percent ocean.

We can observe the varying topography of the Earth from vantage points on land, in the air, and from satellites. The structure of mountains and plains, of river canyons and rolling hills, of swamps and deserts—all can be mapped in the greatest detail and virtually all the features can be confirmed by geologists with hammer and compass. The bottom of the ocean cannot be directly observed except in the clearest of relatively shallow water. Light is absorbed in water with a great enough efficiency that at a depth of several hundred meters the oceans are for all practical purposes pitch black. It is only with the help of remote sensors of a variety of sorts that we can know the details of the ocean floor. How have these techniques helped us to understand the ocean bottom?

ECHO SOUNDING

A fundamental tool of oceanography is the echo sounder. It provides a method of acquiring knowledge of the depth or the topography of the sea floor that is important in all investigations of the oceans. The basic principle of the echo sounder is shown in Figure 14.1. A sound pulse is transmitted by a ship at the surface; it travels to the sea floor, is reflected by it, and returns to the surface. Because the speed of the vessel is slow compared with the speed of sound in seawater, this is essentially a vertical reflection.

If the time the sound pulse travels to the bottom and returns is accurately measured and the velocity of sound in seawater is known, then the depth is given by

$$\text{Depth} = \frac{\text{reflection time}}{2} \times \text{velocity of sound in seawater.} \quad (3.1)$$

The velocity of sound in seawater varies with temperature, salinity, and pressure (see Chapter 4) from about 1410 m/sec (4630 ft/sec) for Arctic surface water to the highest values near the sea floor of about 1530 m/sec (5020 ft/sec).

BATHYMETRIC MAPS

The measurement of ocean depths is called *bathymetry,* and maps made by contouring the bathymetric data are called *bathymetric maps,* which are analogous to topographic maps on land. Early bathymetric maps showed only the main features of the sea floor, because data points were few and not always accurate. With the development of the Precision Depth Recorder and with an increase in the abundance of high-quality soundings, it became possible to construct both bathymetric and the more interpretive physiographic maps of the sea floor (Figure 3.3). Reasonably accurate physiographic maps of the ocean basins of the world now exist (Figure 3.4, pages 46–7). It is unlikely that any major new topographic features will be found on the sea floor in the future, but it is very likely that many smaller features, such as volcanic mountains or "seamounts," remain to be discovered.

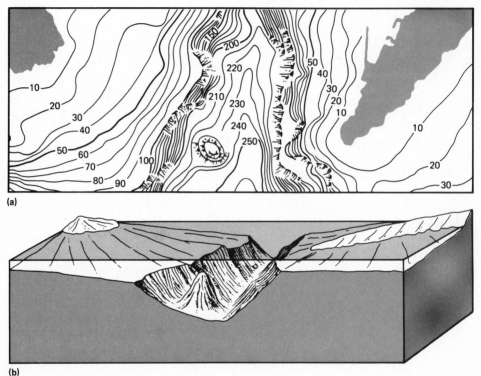

(a)

(b)

FIGURE 3.3   (a) Bathymetric chart and (b) corresponding physiographic block diagram. The depth contours are in arbitrary units. (From Bowditch, 1966.)

DISTRIBUTION OF ELEVATION AND DEPTH ON LAND AND SEA

If we construct a plot of the elevation above or below sea level against the percentage of the Earth's surface that is within a given range of elevations (Figure 3.5), we find that about 30 percent of the surface is within 1 km of sea level and 54 percent is in water depths between 3 and 6 km. This reflects the first-order division of the

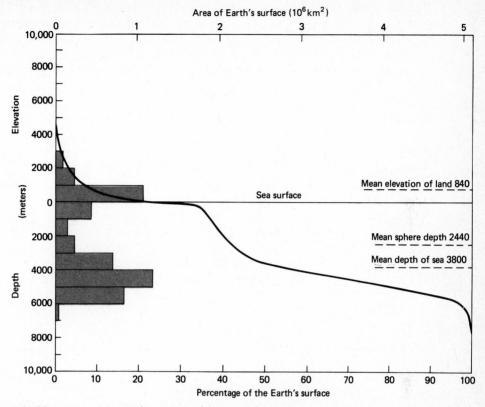

FIGURE 3.5   Hypsographic curve and frequency histogram of elevations above and depths below sea level at 1000 m intervals. (From Sverdrup, Johnson, and Fleming, 1942.)

Earth's topography into continents, with an average elevation of 800 meters, and the deep ocean basins, in which the dominant depth is about 5 km. Of the remainder, about 8 percent of the surface is between 1 and 3 km below sea level, and less than 1 percent is deeper than 6 km.

NATURE OF THE OCEAN BASINS

We can further divide the oceanic topography into second-order features, *continental margins,* the *deep ocean floor,* and the *major oceanic ridge systems* (Figure 3.6). The continental margins include the continental shelves and shallow epicontinental seas that are now flooding continental areas, as well as the area between the shelves and the deep ocean basins. On some margins there is a gradual slope from the edge of the shelf to the basin; in others the transition is more abrupt and is marked by steep escarpments; and in still others the margin includes *deep sea trenches.* In places the deep ocean floor has flat topographic features called *abyssal*

FIGURE 3.4 Physiographic diagram of the world. (From B.C. Heezen and M. Tharp, published by the American Geographical Society.)

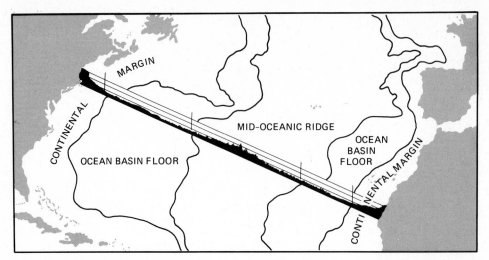

FIGURE 3.6   Generalized cross section across the Atlantic Ocean showing the major topographic features of the oceans. (After Heezen, Tharp, and Ewing, 1959.)

plains, but commonly it appears as gently undulating *abyssal hills.* Within the deep ocean floor province are numerous seamounts, islands, and island groups and rises.

The major ocean ridge systems together compose the longest continuous geologic feature on the surface of the Earth, stretching some 60,000 km (see Figure 3.4). The ridge system is cut by a large number of faults or fracture zones. Although the axis of the ridge appears to have a very sinuous shape, most of the sinuosity is caused by offsets along these fracture zones.

FACE OF THE DEEP SEA FLOOR

Many thousands of photographs have been taken of the floor of the world ocean. These photographs have revealed the nature of the life and the rock materials on the deep sea floor and have changed ideas about the nature of the sea bed. Although *Challenger,* and a few ships before her, had dredged animals from the deep sea floor, their abundance, nature, and habits were not well known. Sediments and rocks had been recovered from the bottom, but their *in situ* relationships were a mystery.

One of the most striking findings from underwater photography was the abundance of evidence of deep sea life in the form of tracks, burrows, or the animals themselves (Figure 3.7a). Most photographs showed some sign of life on the sea floor. Another surprise was the evidence for bottom currents sufficiently strong to erode and transport sediments. The physical oceanographers had predicted that such currents must exist near the sea floor, but the photographs (Figure 3.7b) revealed ripple marks, signs of scouring, and bending over of stemmed organisms, which indicate that these currents are sufficiently strong that they can significantly

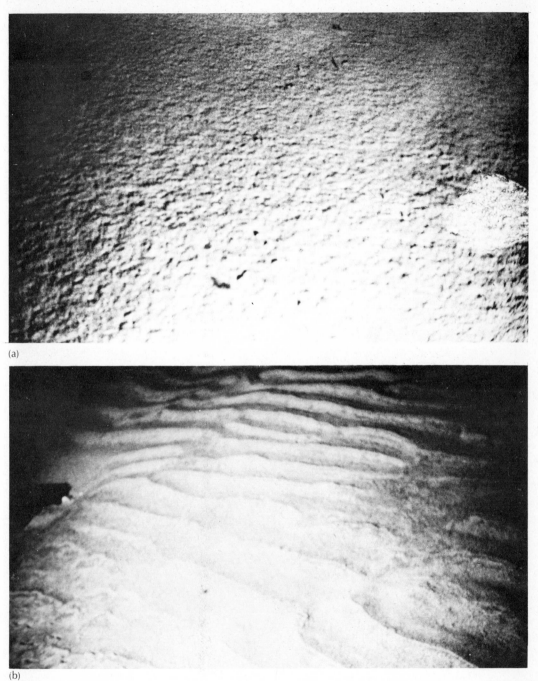

(a)

(b)

FIGURE 3.7   (a) Burrows in the sediments at a depth of 5000 m in the Kalamata Trench off Greece. Attempts to recover the animals that dug these burrows by the submersible *Archimede* were not successful. (Drake photo.) (b) Current ripple marks on the sea floor off San Diego, California. Such ripple marks have been found at depths exceeding 5000 m. (Photo by R. Dill.)

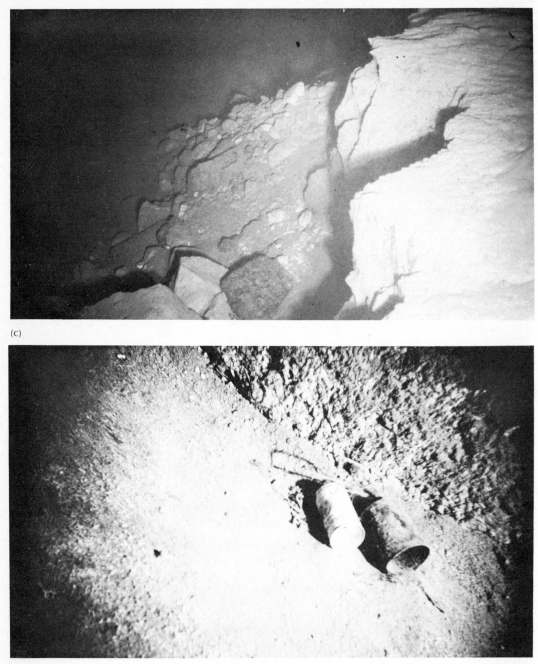

(c)

(d)

FIGURE 3.7 (c) Steep scarp on the inner (landward) wall of the Kalamata Trench off southwestern Greece. Photo taken at a depth of about 2700 m. (Drake photo.) (d) Tin cans on the sea floor in the Gulf off Baja, California. By examining the type of can, its depth of burial, and the date of introduction of the type of can to the area, geologists are able to estimate rates of sedimentation. (Photo by R. Dill.)

shape the sea floor. Another sign of this sediment transport is the muddy water frequently found in underwater photography. When the submarine *USS Thresher* sank in 2.5 km of water on the continental rise off the coast of Maine, attempts were made to locate the wreck by photography. These efforts, and later dives by the bathyscaphe *Trieste,* were hampered by the large amounts of fine sediments suspended in the water and carried along the margin of the continent by strong bottom currents.

Most of the sea floor, as viewed by underwater photographs, is covered by a blanket of fine sediment, altered to varying degrees by the action of animals and currents. Rock outcrops and rubble are found near the axis of the major ocean ridge system on the flanks of seamounts and islands, on steep escarpments, and, frequently, on the walls of deep sea trenches (Figure 3.7c). In high latitudes, where icebergs are or have been numerous, ice-rafted material is common and rocks, boulders, and finer materials can be carried great distances and deposited on the deep sea floor. It is also regrettably true that the ocean floor as the final receptacle for debris carried by large floating objects has felt the fine hand of man, and all too often photographs reveal the less attractive aspects of civilization (Figure 3.7d).

SUMMARY

1. The Earth is the only planet in the solar system with an ocean.

2. The ocean was probably derived from the heating of water containing minerals accumulating near the end of the Earth's accumulation history. The closest examples of this conjectured material are the meteorite class, Type-I carbonaceous chondrites, and comets.

3. The Earth can be divided into the *lithosphere,* the realm of rocks; the *hydrosphere,* the realm of water; the *atmosphere,* the realm of gases; and the *biosphere,* the realm of life.

4. Most of the water of the Earth is found in the oceans. Reservoirs of secondary importance are the ice caps and ground water (and water in silicate minerals).

5. The *continents* are generally *granitic* in composition, and the *ocean basins* are underlain by *basalt.*

6. The topography of the ocean basins is determined with high-precision echo sounding.

7. The ocean basins are divided into the continental margins, the deep ocean floor, and the major oceanic ridge systems.

Water evaporating from a warm ocean into the cold air condenses and forms "sea smoke." (Official U.S. Navy photo.)

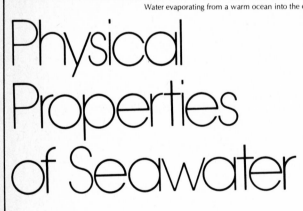

Physical
Properties
of Seawater

4

CHAPTER

Water is an extraordinary fluid. It is the "universal solvent"; all elements will dissolve in water at least a little. Water has the highest heat capacity of any common fluid. That is, more heat is required to raise a given mass of water 1 degree than that required for almost any other fluid; similarly, more heat is required to evaporate a given mass of water than that required for any other common fluid. Life is strongly influenced by these and other properties of water. For example, as we will see in Chapter 5, the high heat capacity of water means that the seasonal variation in our air temperature is less than it would be otherwise. This chapter describes some of the most important properties of seawater.

TEMPERATURE

The oceans are cold. The water is warm only at the very surface; it becomes colder in the lower sections—the farther down, the colder. Surface temperature charts such as Figure 4.1 are a measure of but a thin puddle of warmer water found near the surface. Only 8 percent of ocean water is warmer than 10°C, and more than one-half is colder than 2.3°C. The histogram of Figure 4.2 is a

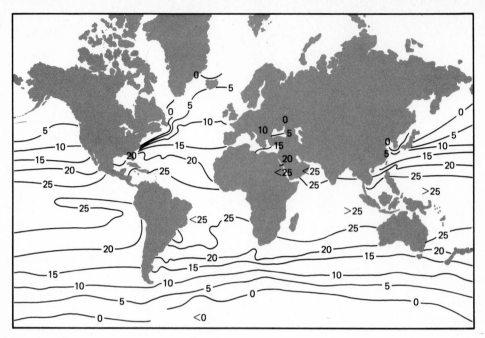

FIGURE 4.1 Annual mean sea surface temperature over the world (°C).

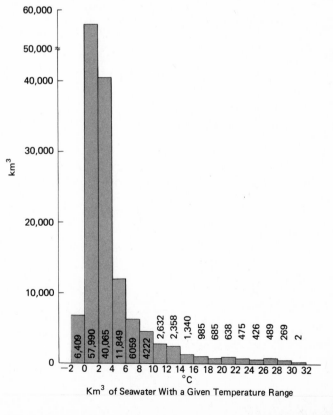

FIGURE 4.2 Distribution of temperature in the world's oceans. The histogram represents the number of cubic kilometers of seawater in each 2° temperature range. Approximately 75 percent of the ocean water has a temperature between 0° C and 4°C. (After Montgomery, 1958.)

good indication of ocean temperature. To a very simple approximation, the temperature distribution in the ocean can be expressed as a cold body of water with a thin layer of warmer water on the surface. The surface water is warmer in the tropics than it is in the higher latitudes, as shown in Figure 4.3.

The change in temperature with depth is not uniform. A thermometer lowered through the warm surface water will generally record a *rapid* decrease in temperature with depth rather than a *gradual* decrease. Much of the ocean is characterized by a warm surface layer of rather uniform temperature, followed by a region of rapidly decreasing temperature, which separates the surface layer from the cold abyss. The surface layer is often referred to as the *mixed layer;* the region of rapidly changing temperature is called the *thermocline* (Figure 4.4). Because the temperature of the surface layer is warmer in the tropics than it is at higher latitudes, whereas the deep ocean is uniformly cold, the characteristics of the thermocline change with latitude. The largest thermoclines are found in the tropics. The processes leading to the development of the thermocline are discussed in Chapter 5. In some parts of the deep ocean, particularly in the ocean trenches, the temperature slowly increases with depth (Figure 4.5). The increase is partly related to the warming of the bottom water by the heat escaping from the interior of the Earth.

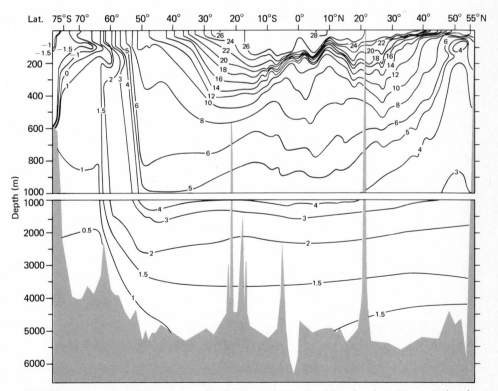

FIGURE 4.3 Average temperature distribution in the Pacific. Note the change in vertical scale at 1,000 m. To a first approximation the ocean is uniformly cold except for a thin layer of warm water— warmer in the low latitudes. (After Reid, 1965.)

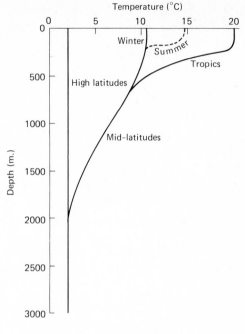

FIGURE 4.4 Typical temperature depth curves for the tropics, mid- and high latitude. Seasonal changes can be expected in the top 100 m.

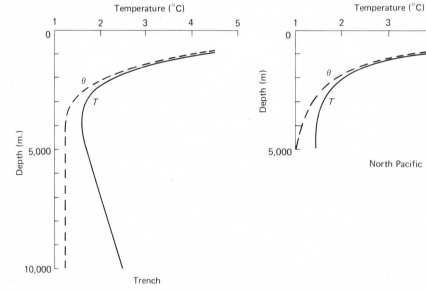

FIGURE 4.5 In the deep trenches and in certain other areas, the *in situ* temperature *(T)* increases near the bottom because of the warming of the bottom water from the heat escaping from the Earth's crust. Also shown are the *potential* temperature profiles (θ). *In situ* temperature is simply the actual temperature of the water at the point it was measured. *Potential temperature* is the water temperature if its temperature were measured at the surface of the ocean. For water at a depth of a few thousand meters the difference is several tenths of a degree, and since oceanographers need to measure temperature to hundredths of a degree in order to learn about the processes occurring at depth, the difference between *in situ* and potential temperature is critical at these depths. The difference is related to the compressibility

SALINITY

The most characteristic feature of seawater is its saltiness. The dominant dissolved substance in seawater is sodium chloride—or common salt! The gradient between the salt water of the ocean and the brackish water of estuaries and salt ponds that ring the seas is characterized by sharp faunal and floral boundaries. Roughly speaking, the total amount of dissolved material in seawater is its *salinity*. More precisely, salinity is defined as the "total amount of solid materials in grams in one kilogram of seawater when all the carbonate has been converted to oxide, the bromine and iodine replaced by chloride and all organic matter completely oxidized." The median salinity of the ocean is 34.69 g/kg of seawater, or 34.69‰—in such expressions "‰" stands for parts per thousand. The composition of seawater with a salinity of 35‰ is shown in Table 4.1.

TABLE 4.1   Composition of Seawater*

Component	Concentration (grams per kilogram)
Chloride	19.353
Sodium	10.76
Sulfate	2.712
Magnesium	1.294
Calcium	0.413
Potassium	0.387
Bicarbonate	0.142
Bromide	0.067
Strontium	0.008
Boron	0.004
Fluoride	0.001

* For a salinity of 35‰; defined as the mass in grams of the dissolved inorganic matter in 1000 g of seawater after all bromide and iodide have been replaced with chloride, and all bicarbonate and carbonate converted to oxide.

No one routinely measures salinity according to the definition. Until about 1955 nearly all salinity measurements were made by determining the amount of

of seawater. All fluids compress a bit when squeezed. For example, the volume of a cubic meter of seawater would be reduced by slightly more than 2 percent if it were lowered from the surface to a depth of 5,000 m, where the total pressure is more than 500 times that at the surface. In the process of being compressed, the temperature of the water increases by nearly one-half a degree, because the compression occurs without an exchange of the heat with the surroundings. The process is called *adiabatic*. One of the most striking examples of the difference between potential and *in situ* temperatures can be found in deep sea trenches. The potential temperature in the trench is isothermal, whereas the actual measured temperature increases with depth.

chloride ions in a unit mass of seawater and by relating this "chlorinity" to salinity by an empirically derived formula. The presently accepted relationship between the two is

$$\text{Salinity} = 1.80655 \times \text{chlorinity} \qquad (4.1)$$

This formula carries the explicit assumption that the ratio of dissolved salts in seawater is a constant. Except for small variations in the calcium concentration, this has been shown to be true. Thus the relationship holds to about ±.02‰ in total salinity, which is also the standard of precision usually assigned to this method of chemical titration.

Because the ocean salinity is relatively uniform, measurements of salinity must be made with great precision. For example, as can be seen in Figure 4.6, the total salinity range of 75 percent of all the water in the ocean is between 34.50‰ and 35.00‰. In the Pacific Ocean the salinity is even more uniform; nearly half the wa-

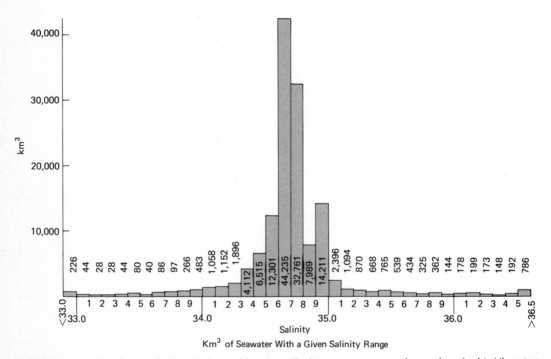

FIGURE 4.6   Distribution of salinity in the world's ocean. The histogram represents the number of cubic kilometers of water within each 0.1‰ range. Since 1960 most salinity observations have been made by measuring electrical conductivity. The latter is a function of temperature as well as salinity; thus a conductivity meter must also measure temperature with great accuracy if it is to provide a measure of salinity. An uncertainty of ±.01°C in temperature translates approximately to an uncertainty of ±.01‰ in total salinity. Present instruments can provide an accuracy of ±.003‰ on samples collected and measured in a laboratory. Continuously recording *in situ* measuring devices are almost as accurate. The theoretical limit for using electrical conductivity to determine the salt content is a function of the constancy of the ionic ratios of the dissolved salts. It is about ±.0001‰. (After Montgomery, 1958.)

ter is between 34.6‰ and 34.7‰. For many purposes we can simply assume that the ocean is of constant salinity. However, our understanding of the details of ocean circulation is often predicated on small salinity differences. For example, the salinity in the deep Pacific changes from about 34.70‰ in the South Pacific to 34.68‰ at 40°N. Oceanographers have agreed that this small change can best be explained by assuming that this water is moving northward and being diluted by less saline water from the overlying column.

DENSITY

The density of seawater is determined by its pressure, temperature, and salinity. Fresh water at 20°C has a density of 1.0 g/cm³. Seawater is heavier; for example, water of the same temperature but with a salinity of 35‰ has a density of about 1.025 g/cm³. As the water becomes colder it becomes even heavier; for example, seawater of a salinity of 35‰ and a temperature of 2°C has a density of about 1.028 gm/cm³. Both fresh water and seawater become heavier as they are compressed; for example, the pressure at a depth of 5000 m will further compress the seawater so that its density is about 1.050 g/cm³.

In oceanography, density (ρ) is seldom written out completely. The following convention is used:

$$\sigma = (\rho - 1)\ 10^3 \qquad\qquad (4.2)$$

Thus, for $\rho = 1.02750$, $\sigma = 27.50$. Oceanographers generally calculate density from temperature, salinity, and pressure measurements rather than measure it directly, and they are often interested in the density of seawater as a function of temperature and salinity alone. When they make density calculations which ignore pressure, they refer to the density as σ_t or σ_θ values depending upon whether they use the *in situ* or potential temperature to calculate density (see Figure 4.5). For many purposes the two are interchangeable, because the differences are very small.

Generally σ_θ (and σ_t) increases with depth, and when this occurs the water is said to be *stably stratified*. In a stratified ocean it is difficult to mix water *across* lines of constant density; it is much easier to mix water *along* lines of constant density. In physical terms it requires work, an increase in potential energy, to move material *across* lines of constant density. The only work required to move material *along* lines of constant density is the energy to overcome the friction of the water, and seawater is very "slippery." Two examples will suffice. In the early fifties the United States exploded an atomic bomb underwater in the eastern Pacific. Part of the program called for tracking the radioactive water resulting from the explosion. The radioactivity from the blast spread out as a very thin layer, or lens. After 40 days it was found to cover an area of 40,000 sq km, but was only 60 m thick. An example based on natural phenomena is the overflow water from the Mediterranean which moves out through the Strait of Gibraltar and can be traced over much of the North Atlantic. The vertical exaggeration in the scale of Figure 4.7 can be misleading. In effect the lens of Mediterranean overflow water is about 1000 m thick and some 2000 km long.

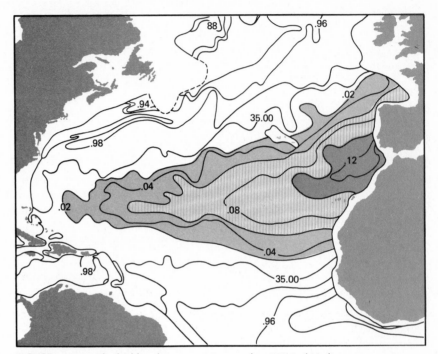

FIGURE 4.7   The highly saline tongue (greater than 35‰) of Mediterranean waters can be seen spreading out across the Atlantic after it comes through the Strait of Gibraltar. The water spreads as a thin layer a few hundred meters thick at a depth of approximately 1000 m. (After Worthington and Wright, 1970.)

SEA ICE

At any given time, about 3 or 4 percent of the ocean is covered with sea ice; sea ice differs from freshwater ice in a number of ways—for example, process of formation, freezing temperature, and physical properties. Fresh water freezes at 0°C. The freezing temperature of salt water decreases as the salinity increases. For the salinity range of 30‰ to 35‰ the freezing point varies from −1.6°C to −1.9°C (Figure 4.8).

At some risk of oversimplification, the formation of sea ice can be considered simply as freezing of the fresh water and leaving the salt behind in brine pockets. As the temperature reaches the freezing point, ice crystals of pure water are formed, which "surround" the unfrozen water. This unfrozen water is enriched in salt left behind by the frozen crystals, and results in a further lowering of the freezing point of the water in these brine pockets. If the frozen crystals do not com-

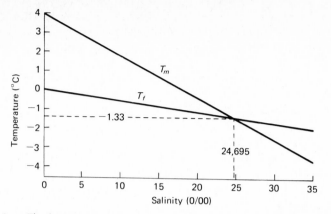

FIGURE 4.8   The freezing point of water (T_f) decreases from zero for fresh water to about $-3.5°C$ for a salinity of 35‰. The temperature of maximum density (T_m) decreases with increasing salinity from a maximum of nearly 4°C for fresh water. Once the salinity is 24.7‰ or greater, seawater acts as a "normal" liquid; that is, its density continues to increase as it cools until it freezes.

pletely surround the salt-enriched unfrozen water, this water will sink and mix with the seawater below. If the freezing is slow enough, nearly all the enriched brine will escape and the salinity of the sea ice will be approximately zero. Rapid freezing entraps most of the brine resulting in a total sea-ice salinity close to that of the surrounding water. Most sea-ice salinities range from 2‰ to 20‰, with the older ice averaging lower salinities because with old ice the leaching out of the brine is enhanced by alternate melting and freezing as the air temperature changes. If the temperature is low enough, the salt itself will begin to crystalize. Sodium sulfate begins to crystalize at $-8.2°C$ and the important sodium chloride at $-23°C$.

Because sea ice has brine pockets and complex salt crystalization patterns, it is less easy to characterize its strength and similar physical attributes than is the case for freshwater ice. Generally sea ice has about one-third the strength of freshwater ice of the same thickness. However, the strength of sea ice may be comparable with that of freshwater ice for old sea ice (with very low salinities) and for sea ice at a temperature well below the crystalization of sodium chloride.

LIGHT IN THE OCEANS

The oceans are dark as well as cold. At depths greater than 100 m, the strongest light seen at midday by an underwater explorer is the occasional bioluminscent flash of passing fish, or zooplankton. In the silt-filled estuaries such as Chesapeake Bay and Long Island Sound it is often difficult to see any distance in even 10 m of water. Unlike the atmosphere, which is relatively transparent to all scales of electromagnetic radiation, the ocean is opaque. Neither long-wave radio waves nor the short-wave ultraviolet radiation can penetrate very deep in the ocean.

In any fluid medium, including seawater, the loss of the sun's radiation follows very closely a relationship called Beer's law, which says that the amount of energy absorbed in a given distance is proportional to the amount present. As a consequence we can characterize seawater in terms of its *transmittance*. For example, if the transmittance of a given body of water was 90 percent per meter we can easily calculate how much energy reaches a given depth. If 100 units of radiation reach the surface, 90 percent (or 90 units) will reach the first meter. Eighty-one units will reach the second meter, which is 90 percent of the remaining radiation (.9 × 90 = 81), and 72.9 units will reach a depth of 3 m (.9 × 81 = 72.9).

The transmittance of water varies with wavelength, and even though water is relatively opaque to sunlight, the visible part of the spectrum is much more transparent than either shorter or longer wavelengths (Figure 4.9). The difference be-

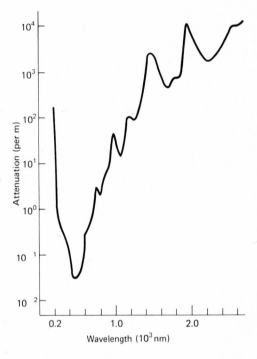

FIGURE 4.9   Attenuation coefficient of ultra violet, visible, and infrared waves as a function of wavelength. The wavelengths are in nanometers (1 nm = 10^{-9} m). The visible spectrum is from 400 to 700 nm. For an attenuation coefficient of 0.1, the transmittance is 90 percent. For an attenuation coefficient of 1.0, the transmittance is 37 percent; for an attenuation coefficient of 10, the transmittance is .0045 percent; and for an attenuation coefficient of 100 the transmittance is 37 × 10^{-42} percent. (After Morel, 1974.)

tween fresh water and seawater is of little consequence; the clearest ocean water has transmission characteristics similar to those of distilled water. Of course, transmittance in the visible spectrum decreases as the water becomes more turbid. As discussed in Chapter 5, less than 1 percent of that part of the sun's energy which reaches the sea surface penetrates as deep as 100 m into the ocean. The transmittance for three ocean and nine coastal water types are shown in Figure 4.10.

As might be expected, the eyes of creatures living in the ocean have adapted to that part of the spectrum with the greatest transmittance. Since land animals evolved from the sea, it is not chance that man's "visible spectrum" is that part of the radiation spectrum with the maximum transmittance in the ocean. It has been reported that there can even be adaptation within ocean regions. *Euphausia pacifica* (a small shrimplike creature) found in the open ocean has maximum light sensitivity to 465 nm ("nm" stands for "nanometer," one billionth, or 10^{-9}, meters), the wavelength of maximum transmittance for clear ocean water. A different variety of the same species found in Puget Sound is most sensitive to 495 nm, which is close to the maximum transmittance for coastal water.

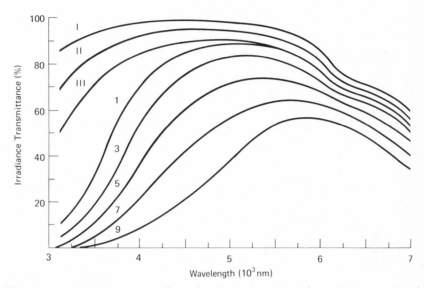

FIGURE 4.10 The transmission at all wavelengths for three types of oceanic water (I, II, and III) is greater than for all types of coastal water (1, 3, 5, 7, 9). As the water becomes more turbid, however, the wavelength of maximum transmission shifts toward longer wavelengths. In the clearest ocean water (I) the maximum transmission is at about 465 nm (in the blue green). In the most turbid waters (9) the maximum transmission occurs at about 575 nm. The primary reason for the difference is that dissolved organic matter in the ocean selectively absorbs the shorter wavelength. As a result, the wavelength of maximum transmittance shifts toward longer wavelengths in coastal regions (After Jerlov, 1968.)

SOUND IN THE SEA

Because of the opaqueness of the ocean to the transmission of electromagnetic energy, we are denied the use of radio transmission and radar as a tool for working in the ocean. Submerged submarines receive radio messages by either floating an antenna on the surface or having the radio transmitters use radio frequencies with such long wavelengths that the Beer's law relationship of the preceding section does not apply. The ocean, on the other hand, is much more transparent to sound transmission than is the atmosphere, and because of the peculiar sound-velocity pattern in the ocean, sound can be transmitted extraordinarily long distances through the ocean. A few pounds of TNT detonated off Hawaii can be heard by an underwater microphone (a hydrophone) off San Francisco. Even more dramatic: A depth charge exploded near Australia was monitored by hydrophones near its antipode off Bermuda.

The velocity of sound in the ocean varies with pressure, temperature, and salinity. A reasonable average value is 1500 m/sec or 5000 ft/sec, which is four to five times faster than the speed of sound in the atmosphere. Sound velocity in-

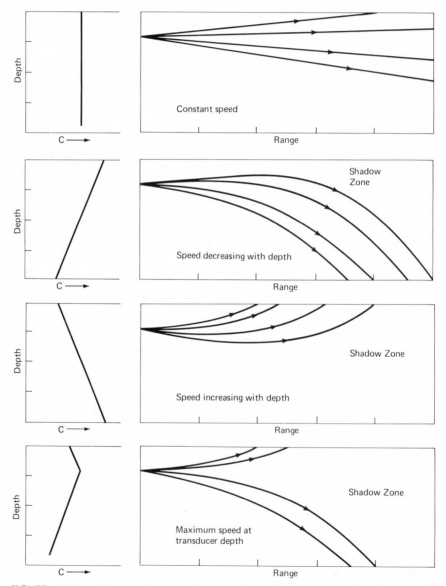

FIGURE 4.11 If the sound velocity is constant, the sound travels in straight lines; if the sound velocity changes, the sound rays are refracted. The direction is always toward lower velocities. (After McLellan, 1965.)

creases with increasing temperature, salinity, and pressure. The velocity of sound in water, however, is independent of the pitch or frequency. The high-frequency "ping" of a porpoise travels at the same speed as the deep, low-frequency "song" of the humpback whale.

Because the velocity of sound is not constant in the ocean, sound rays do not travel in straight lines but are bent toward lower velocity (Figure 4.11). In the case of a typical thermocline, sound rays can be refracted so that there is a shadow zone in which no direct rays pass. In World War II, submarines hovered under a shallow thermocline in an attempt to avoid detection from surface ships. Since the laws of physics have not changed in the intervening years, this tactic is probably still valid. However, more powerful sonars which bounce sound off the bottom make hiding under a thermocline less of a sure thing.

The effect of pressure tends to increase sound velocity with depth. Except for a few polar areas where the temperature is nearly isothermal from top to bottom,

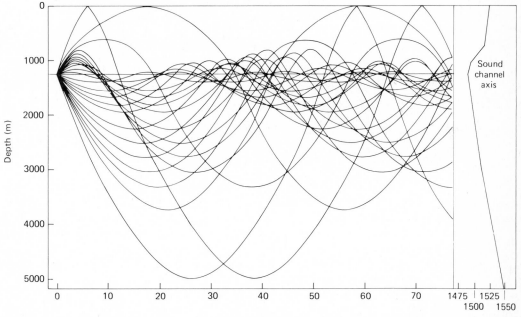

FIGURE 4.12 Because sound rays are always bent toward the lower velocity, a sound minimum tends to trap sound waves and to "channel" them. A sound source in the deep sound channel can often be heard hundreds and even thousands of miles away. The sounds that travel those long distances are of low frequency (the order of 100 to 1,000 c/sec) because low-frequency sound is less attenuated than high frequency. Although the ocean is reasonably "transparent" to sound waves, it is not completely so, and over a wide range of frequency, attenuation increases as the square of the frequency; thus a doubling of the frequency increases the attenuation by four times, which means that long-range sound transmission is best at low frequencies. Frequencies as low as 100 c/sec (with a corresponding wavelength of about 15 m) are commonly used for very long-range sound transmissions. Fresh water is even more transparent than salt water to the transmission of sound; for a given frequency the attenuation is 100 times less in fresh water. The difference is caused by a chemical interaction of the sound waves with one component of the sea salt, the magnesium sulphate ion.

the effect of the temperature structure of the ocean is to decrease sound velocity with depth; the effect is much more pronounced near the surface than at depth. The combination of temperature and pressure effects typically results in a sound-velocity curve with a minimum value at mid-depth. This minimum is called the *deep sound channel*; because sound rays are always refracted toward the lower velocity, a sound-velocity minimum serves to channel the sound (Figure 4.12).

The deep sound channel is a continuous feature in the ocean. It extends from near the surface in polar latitudes to a depth of nearly 2000 m off the coast of Portugal, with a typical depth being about 700 m. In the case of the very long transmissions referred to in the beginning of this section, both the sound source and the hydrophones were near the deep sound channel axis. Because whale sounds have been recorded over long distances, it has been speculated that they, too, make use of the excellent transmission qualities of the sound channel.

USE OF UNDERWATER SOUND

Because the ocean is nearly impervious to sunlight, radio waves, radar, and other forms of electromagnetic radiation, sound-energy technology has been developed as a substitute. We must use sound as our own underwater "eye," just as the bats have learned to do in air. In many ways sound is a poor substitute, and it is unlikely that echo ranging can ever be developed to give the range and discrimination of modern radar, or that sound transmission will ever have the communication capacity and range of present-day radio. By way of further comparison, the size and power requirements of underwater sound equipment are usually much larger than those for analogous electromagnetic equipment.

To those who worry about strategic military planning, the opacity of the ocean is critical, since the present nuclear detente with the U.S.S.R. is predicated in part on the valid assumption that neither side can be exactly sure where the other side's missile-launching submarines are at any given moment. Thus, neither the U.S. nor the U.S.S.R. can make a "first strike" with its own land-, air-, or sea-based weapons and hope to wipe out the other side's nuclear forces. Although satellite photographs may pinpoint land-based missiles, the submarines remain hidden, ready to launch a retaliating strike if required.

Because of its obvious requirements, the military has been responsible for most of the primary developments in underwater sound; however, just as World War II development of radar led to development of the smaller but very effective radars that one sees on all commercial ships and on a growing number of pleasure boats, so has the development of sonar echo ranging led to the development of "fish finders" that can be found on most modern fishing boats. Successful echo ranging for fish requires skill of modern electronics combined with the experience of the fisherman. With even the most discriminating sonar, it is difficult to distinguish one type of fish from another, but the density and spacing of the targets and their location with respect to water depth gives the experienced fisherman reasonable assurance about what type of fish he is "seeing."

SUMMARY

1. The ocean is cold, dark, salty, and stratified.

2. More than half the ocean has a temperature of less than 2.3°C. The warm water is confined to a relatively thin layer near the surface above the thermocline.

3. Because the density increases with depth, water can be mixed horizontally much more easily than it can be vertically.

4. More than 75 percent of the ocean has a salinity within 1 percent of the median, which is 34.69‰. About 85 percent of the salt in the ocean consists of sodium chloride.

5. Almost all the sunlight is absorbed in the top 100 m. Below a few hundred meters, the brightest light is that given off by bioluminescing organisms.

6. The ocean is comparatively "transparent" for sound radiation, as distinguished from light and other electromagnetic radiation. Sound devices are used as a substitute for radio devices in a number of applications.

Satellite photograph of Hurricane Gladys (1975) off the east coast of Florida and north of Cuba. (NOAA photo.)

The Earth as
a Heat Engine

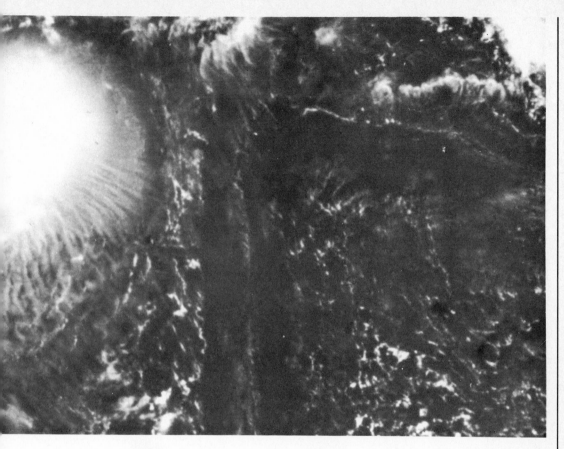

5

Well over 99.9 percent of the energy that determines our weather, our changes in climate, and the currents of the ocean comes from the sun. A very small amount of heat energy (a millionth of the average amount of the sun's energy that reaches the outer atmosphere) comes from the interior of the Earth. The atmosphere-ocean-land-ice system receives the sun's energy as radiant heat. This energy is sorted in various parts of the system, is transferred from one part to

CHAPTER

another, and is ultimately released to outer space. The details of this heat engine are awesomely complex, but the general principles are fairly straightforward.

RADIATION ENERGY BALANCE

Every object radiates energy over a relatively narrow range of wavelengths. The higher the temperature, the more energy radiated and the shorter the wavelength of maximum radiation. If a body is hot enough, such as a lighted match, an incandescent light bulb, or the sun, a significant amount of the energy is in the visible range. In the case of the sun, almost 50 percent is radiated in the visible spectrum between 400 and 700 nm. As Figure 5.1 shows, however,

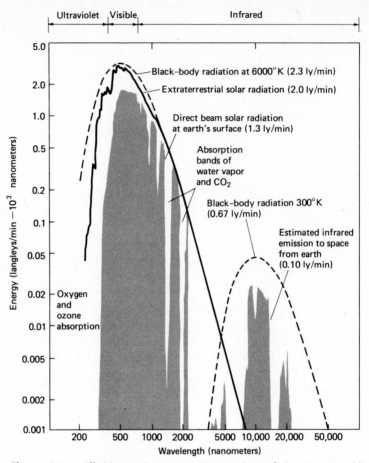

Figure 5.1   All objects radiate energy in proportion to their temperature. Most objects, including the sun, the ocean, the atmosphere, and the Earth, are sufficiently close to blackbody radiation that the details of the radiation energy flux can be determined from the temperature. The amount of energy radiated by a blackbody is proportional to the fourth power of the absolute temperature (Stefan-Boltzman law) and the wavelength of maximum radiation is inversely proportional to the absolute temperature (Wien's law). Thus the radiation power of a square centimeter of the sun's surface with a temperature of 6000°K is 160,000 times that of a square centimeter of sea surface with an absolute temperature of 300°K (27°C). Most of the sun's energy is in the visible part of the spectrum. All of the Earth's energy is in the infrared. The amount of energy reaching the Earth's surface is about half that which is available at the outer edge of the atmosphere. Note that both the short ultraviolet and the long infrared wavelengths are selectively absorbed. (After Sellers, 1965.)

for anything as cool as the ocean, land, or atmosphere none of the energy is in the visible part of the spectrum; all is in the infrared. You can photograph the Earth in the dark if you have infrared-sensitive film, but your eyes, which are sensitive to radiation only between 400 and 700 nm, cannot "see" the Earth in the absence of visible light.

The total amount of the sun's energy which enters the upper atmosphere is about 17×10^{20} cal/day. (A calorie is a unit of energy. It requires 1 calorie to heat 1 gram of water 1 degree Celsius.) At any given location, the amount varies with the time of day (being essentially zero at night) and with season (higher in summer than in winter). Averaged over a year, an equal amount must be reradiated into space from the ocean-land-ice-atmosphere system. If it is less, the Earth's surface will warm up; if more, the system will cool off. The extent to which the system is not in balance results in climatic change.

HEAT TRANSFER PROCESSES

As the sun's energy penetrates the atmosphere, a number of processes begin to occur. Some energy is reflected or scattered; some is absorbed either by clouds or by the air itself; and some penetrates the atmosphere, where it is either reflected by the ocean-land-ice surface or absorbed. A number of these processes are wavelength dependent. For example, much of the ultraviolet part of the energy spectrum is absorbed in the upper atmosphere in photochemical processes that produce ozone. Molecular or Rayleigh scattering increases rapidly as the wavelength decreases. Precisely, the scattering is inversely proportional to the fourth power of the wavelength, which means that the short blue wavelength at 400 nm is ten times more likely to be scattered than the longer red wavelength at 700 nm. This Rayleigh scattering law holds only when the particles doing the scattering are small compared to the wavelength of the light. When the particles are of the same size as or larger than the wavelength, the scattering is less dependent upon the wavelength. A person looking skyward and not directly at the sun perceives primarily scattered sunlight. A blue sky is a consequence of Rayleigh scattering; the deeper the blue, the greater percentage of the skylight resulting from Rayleigh scattering. Aerosols and dust and water particles cause more uniform scattering and a grayer, less blue sky.

Difference in cloud cover causes the greatest variation in the amount of radiation reaching the Earth's surface. A dense cover of low stratus clouds can absorb or reflect back to space more than 80 percent of the radiation from the sun. On the average the clouds cut off about 24 percent of the incoming energy. Back scatter, absorption by clouds and water vapor, and photochemical reactions account for almost half the sun's radiation energy. It is estimated that 51 percent reaches the Earth's surface. Of that reaching the surface of the sea, about 6 percent is reflected from the ocean surface (more at low sun angles than when the sun is directly overhead) and the remainder enters the ocean.

As part of the global radiation balance, the average amount of heat energy entering the world's ocean must be equal to that leaving; otherwise, the oceans will cool off or warm up. The heat losses from the ocean to the atmosphere are of three kinds: long-wave radiation, conduction, and evaporation.

NET LONG-WAVE RADIATION

Knowing the sea surface temperature, the energy radiated from the sea surface can be calculated by the Stefan-Boltzman law, as noted in Figure 5.1. If this calculation is made, it appears that the sea surface loses more energy than it is receiving from the sun. The solution to this apparent paradox is that much of the radiation from the sea surface is absorbed by the clouds and water vapor in the air and reradiated back to the surface. All this radiation is in the infrared part of the spectrum, that is, long-wave radiation. We must consider the *effective back radiation,* which is defined as the *net* long wave radiation loss from the sea surface. This term varies with the water vapor in the air, a fact that can be at least qualitatively verified by noting how much cooler it can become on a clear, dry night than on a cloudy one. On clear evenings most of the long-wave radiation escapes into space. On a cloudy night, or one with a high relative humidity, much of the long-wave radiation is absorbed by water vapor and is reradiated back to the earth's surface. Of the three processes which return heat from the ocean to the atmosphere (long-wave radiation, evaporation, and conduction), effective back radiation accounts for about one-third the heat transfer.

EVAPORATION

It takes about 600 calories to evaporate a cubic centimeter of seawater, and since it is estimated that the equivalent of a layer of water about 1.1 m thick is evaporated from the surface of the ocean each year, the equivalent heat loss can be readily determined. (For example, evaporating 1 m of water requires 60,000 cal/sq cm.) Ocean heat loss by evaporation is the largest of the three contributors to the ocean heat loss budget, and accounts for about 60 percent of the heat transfer. Details of the process leading to evaporation are difficult to determine, however, and charts of annual evaporation rates are subject to much uncertainty.

SENSIBLE HEAT LOSS

The third way in which heat is transferred from the ocean to the atmosphere is by conduction and convection. Heat energy "flows" from high temperature to low temperature and, because the average temperature of the ocean is higher than the average air temperature immediately above it, heat is transferred from the ocean to the atmosphere. The average air-sea temperature difference is generally less than two degrees, and the heat transfer resulting from this conduction-convection process is about 7 percent of the total. Figure 5.2 summarizes the different ways in which heat is transferred between outer space, the atmosphere, and the surface of the Earth.

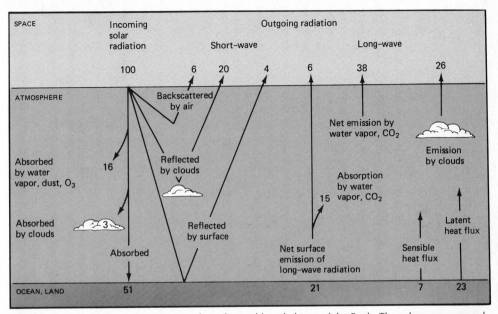

FIGURE 5.2 A schematic attempt to show the total heat balance of the Earth. The values are averaged over latitude and season as well as over land, ocean, and ice. Consideration of these averaged values is instructive because they show the relative importance of different processes in maintaining the heat balance of the Earth. The numbers are based on 100 units of incoming radiation; the 100 units are roughly equivalent to 0.5 cal/cm² minute. Of the 100 units of radiation energy entering the atmosphere, 16 are absorbed in the stratosphere and troposphere by water vapor, aerosols, and the air molecules themselves and 6 are back-scattered to space. About 23 units of the radiation react with clouds, 20 are back-scattered or reflected back to outer space, and 3 are absorbed by the clouds. At the Earth's surface about 4 units of radiation energy are reflected and 51 units are absorbed by the Earth. The outgoing radiation must equal the incoming. Thirty units are reflected shore wave radiation, which is referred to as the Earth's albedo. It is the reflected radiation from clouds, sky, water, and Earth that the astronauts see when they look down on the Earth from space, just as it is the albedo of the moon that we see from Earth. The 70 units of long-wave radiation come from the Earth's surface (6), from the atmosphere (38), and from clouds (26). (From U.S. Committee for the Global Atmospheric Research Program, 1975.)

LOCAL HEAT BALANCE

As discussed in Chapter 4, the ocean is comparatively opaque to radiation at all wavelengths. Even in the clearest of ocean water all but 1 percent of the incoming radiation is absorbed in 100 m, and for more turbid coastal waters the same amount is absorbed in 10 m (see Table 5.1). Note that in either case more than 50 percent of the energy is absorbed in the first meter. This is because all the infrared and ultraviolet are absorbed immediately, letting only the visible get deeper (see Figure 4.10 for transmittance as a function of wavelength). The absorbed energy heats the water, and the surface layer of the ocean undergoes a temperature cycle similar to that in the atmosphere. The surface water is warmer than average in summer and colder in winter. Like the atmosphere, the seasonal temperature range is generally smaller in the tropics than in higher latitudes. For much of the ocean the observed seasonal temperature cycle can be approximated by considering only the seasonal variation in the sun's radiation balanced against the heat losses of effective back radiation, evaporation, and sensible heat transfer. To see how this might be applied in a real situation, let us assume that the heat transfers for the Northern Hemisphere are as shown in Figure 5.3.

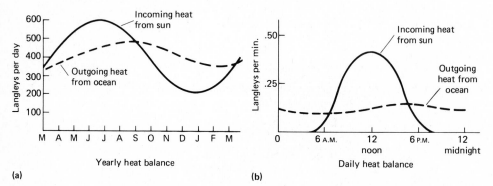

FIGURE 5.3   A schematic attempt to demonstrate the seasonal and daily heat balance. (a) From March to August more heat enters the ocean than is lost and the surface layer of the ocean warms up. From September to February the ocean loses more heat than it gains and the surface layer cools off. (b) On a daily cycle there is a net gain of heat during the day and a net loss at night. Note that the ocean is losing heat 24 hours a day. Whether there is a net gain or a net loss at the end of a 24-hour period is primarily a function of season.

TABLE   5.1   Percentage of Sun's Radiation Reaching Different Depths of Oceanic and Coastal Waters. (See Figure 4.10 for details concerning absorption by wavelength.) (After Jerlov, 1968.)

Depth (m)	Oceanic Water					Coastal Water				
	I	IA	IB	II	III	1	3	5	7	9
0	100	100	100	100	100	100	100	100	100	100
1	44.5	44.1	42.9	42.0	39.4	36.9	33.0	27.8	22.6	17.6
2	38.5	37.9	36.0	34.7	30.3	27.1	22.5	16.4	11.3	7.5
5	30.2	29.0	25.8	23.4	16.8	14.2	9.3	4.6	2.1	1.0
10	22.2	20.8	16.9	14.2	7.6	5.9	2.7	0.69	0.17	0.052
20						1.3	0.29	0.020		
25	13.2	11.1	7.7	4.2	0.97					
50	5.3	3.3	1.8	0.70	0.041	0.022				
75	1.68	0.95	0.42	0.124	0.0018					
100	0.53	0.28	0.10	0.0228						
150	0.056			0.00080						
200	0.0062									

There is a net gain of heat to the surface layer during the months of mid-February to mid-August and a net loss during the other six months. From mid-February to mid-August the average amount of incoming heat is larger than the amount of outgoing heat. During the remaining six months the situation will be reversed (Figure 5.3a). Furthermore, the incoming radiation is cyclic during a 24-hour period, whereas the outgoing heat is approximately steady (Figure 5.3b).

The incoming radiation is absorbed throughout several tens of meters while all the processes responsible for heat loss take place at the surface. Thus, on the average, and perhaps nearly all the time over the entire ocean, there is a net loss of heat to the atmosphere in the top millimeter of the ocean (Figure 5.4).

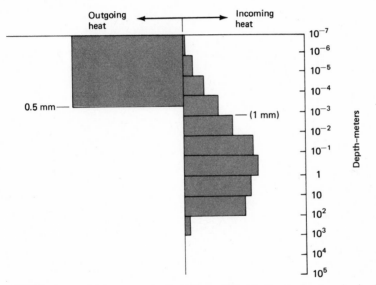

FIGURE 5.4   In the ocean the incoming radiation is absorbed over a depth of 10–100 m. All of the outgoing heat is transmitted from the surface skin of the ocean. (Note the log scale in the figure.) As a consequence of the net heat loss at the surface, the temperature of the top fraction of a millimeter is depressed, which results in a negative temperature gradient in the surface skin of the ocean. On a calm night the temperature of the surface of the ocean, which is radiating heat into space, can be several tenths of a degree colder than the water a millimeter below the surface.

From Figures 5.3 and 5.4 a picture of the local variation in the temperature structure emerges. The minimum surface temperature in the northern ocean occurs in mid-February. As the incoming heat exceeds the outgoing heat, the ocean begins to warm. The maximum heat content of the ocean occurs in mid-August, at which point the daily heat losses exceed the heat gains.

Figure 5.5 is drawn from actual average temperature records. A seasonal thermocline develops during the summer. The depth of the thermocline can be explained at least approximately by considering the depth of penetration of the incoming radiation and the methods by which the heat can be mixed downward.

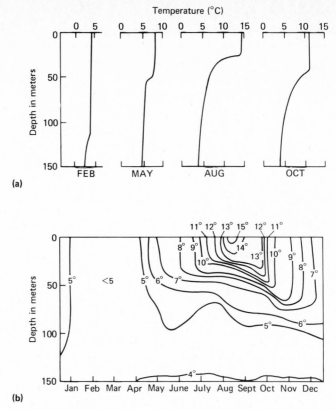

(a)

(b)

FIGURE 5.5   (a) Typical seasonal changes in temperature structure in the top 100 m at a Northern Hemisphere mid-latitude site. (b) Comparable data in a time versus depth plot. The build-up and destruction of the summer thermocline can be clearly seen.

Most of the mixing energy comes from the wind. The thermocline is deeper in spring than in summer because the average winds are stronger in spring than in summer and because, as the thermocline becomes stronger, more energy is required to mix the heat deeper. Thus, as summer progresses, the thermocline becomes both shallower and stronger, but the total heat content of the surface layer increases. In fall the thermocline weakens as the daily heat loss exceeds the heat gain. The combination of a weaker thermocline with higher winds and increased convection (generated by sinking in the surface layer as the surface water cools) drives the thermocline deeper. By February the seasonal themocline has disappeared and the process begins again.

During the spring, when there is a slight excess of incoming heat over heat loss, it is often possible to observe diurnal thermoclines. As can be seen in Figure 5.3b, there is a net heat gain during the day and a net heat loss at night. In some manner the excess heat remaining from these shallow diurnal thermoclines is mixed downward to form the seasonal thermocline.

GEOGRAPHIC DISTRIBUTION OF HEAT AND COLD

We must now look to the differences in heating and cooling that occur from place to place on the Earth's surface. Because the Earth is roughly spherical, radiation entering the top of the atmosphere is more concentrated at the equator and spread out toward the poles. The relationship is further complicated on a seasonal basis, because the axis of the Earth's rotation is tilted 23.5° from a line perpendicular to a plane defined by the Earth's orbit around the sun. As a result, from about March 21 to September 22 (vernal equinox to autumnal equinox) the Northern Hemisphere receives more radiation than the Southern Hemisphere, and during the other six months the Southern Hemisphere receives more radiation. The yearly average, however, is a simple pattern displayed in the upper curve of Figure 5.6a. Note that

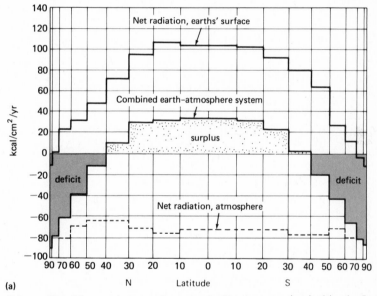

(a)

FIGURE 5.6a   The difference between the amount of heat energy absorbed by the Earth and that radiated to space from the Earth and atmosphere. The difference is the net surplus or deficit as a function of latitude. For example, between 10°N and 10°S the Earth receives about 105 kcal/cm²/yr, and the net radiation loss over the same latitude band is about 72 kcal/cm²/yr, thus resulting in a net surplus of about 33 kcal/cm²/yr. (Data from Sellers, 1965.)

the horizontal scale of this diagram is adjusted so that the spacing between each degree of latitude is proportional to the actual area represented on the globe. Whereas the incoming radiation absorbed by the Earth is much higher at tropical latitudes than at polar latitudes, the net radiation loss shows little change with latitude. As shown in Figure 5.6a there is a net surplus of incoming heat energy at low latitudes and a net surplus of outgoing heat energy at high latitudes. The oceans and the atmosphere must transport heat from the tropics to the polar regions (Figure 5.6b). If such a transport were not to occur, the poles would not only be cold, they would also be *cooling* and the tropics would also be *heating*.

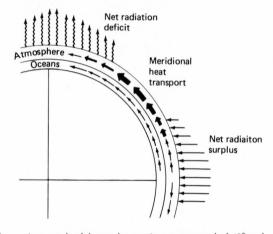

(b)

FIGURE 5.6b   As a result of the net heat gain equatorward of 40° and the net heat loss poleward of 40° there is a meridional transfer of heat by the atmosphere and the oceans from low to high latitudes. (From Strahler, 1971.)

The requirements of poleward energy transport are met by the global pattern of atmospheric winds and oceanic currents. These great systems of flowing fluid, with their intricate details of cyclonic storms and meandering loops, are but the response to the requirement for a poleward flux of heat in an attempt to maintain an equilibrium tememperature over the entire Earth. This energy flux is carried on by three basic mechanisms, one in the ocean and two in the atmosphere. Within the oceans, the currents carry heat in sensible form, as cold currents moving toward the equator on the surface and at depth, and as warm surface currents moving poleward. In the atmosphere, there is (1) a transport of latent heat in the form of water vapor and (2) sensible heat in the form of warm air masses. Viewed in this light, tropical storms and hurricanes which move poleward are more than annoyances and hazards; they are part of a great global process of transporting warm air and latent heat to high latitudes to maintain the thermal balance of the globe.

Latent heat transfer is accomplished as follows. About 600 cal are required to evaporate a gram of seawater and about 540 cal of heat are released when the pure water vapor is condensed and becomes rain. Thus every gram of water evaporated in the tropics, carried poleward, and falling as rain at higher latitudes is transporting 540 cal/g. Estimates for energy flux are subject to great uncertainty, but suggest that both the atmosphere and ocean currents play important roles in transporting heat.

GLOBAL WATER BALANCE

In addition to a global heat balance, there must be a global water balance. Water is evaporated from the ocean and lakes, rains on land and water, and runs from rivers into the ocean. Ninety-seven percent of the water is in the oceans and most of the rest is in the form of permanent ice (Table 3.4).

If the glaciers were to melt, sea level would rise about 60 m, or 200 feet, inundating most coastal cities and markedly changing the coastlines of the whole world. During the last ice age, some 18,000 years ago, the ice sheets expanded and increased in volume by a factor of three. Sea level was nearly 120 m lower and the shoreline in many places was at the outer edge of our present continental shelf.

TABLE 5.2 Water Budget by Ocean (Adapted from Budyko, 1974.)
The average precipitation and evaporation for the different ocean basins in centimeters per year. Also shown is the volume of water involved in the exchange. Note that water flows from the Pacific and the Arctic Oceans into the Atlantic and the Indian Oceans.

	Atlantic	Pacific	Indian	Arctic	Global
Precipitation					
cm/yr	78	121	101	24	103
100 km³/yr	74	214	76	2	367
Evaporation					
cm/yr	104	114	138	12	113
1000 km³/yr	99	202	103	1	405
Runoff					
1000 km³/yr	19	11	5	3	38
Exchange					
1000 km³/yr	−6	23	−22	5	—

An evaporation rate of 113 cm/yr over the oceans, as shown in Table 5.2, is equivalent to removing 405×10^{12} m³/yr, or .03 percent of the total ocean volume each year. Approximately an equal amount of water reaches the ocean by rainfall and river runoff. It is estimated that 10 percent comes by way of rivers and the remainder by rainfall. There is evidence from worldwide tide gauges and salt marshes that sea level may have been rising for the preceding 100 years at a rate of 0.2 cm/yr. This is less than a 0.2 percent difference from a balanced water regime, yet its effect can be detected. Table 5.2 estimates water balance by ocean as well as on a worldwide basis. The units are in centimeters per year as well as in volume per year. Note there is a net flow of water from the Pacific and Arctic Oceans to the Atlantic and Indian Oceans.

Although there is a global balance, there is, of course, no local balance. When ocean water evaporates, it leaves the salt behind; thus the remaining surface water becomes saltier. Charts of the surface salinity of the ocean demonstrate the effect of the local imbalance. In central ocean regions where evaporation exceeds precipitation, the surface salinity is higher than average; it is less than average in those areas where the reverse is true. In general, coastal regions have lower salinities than the open ocean because of the influence of river runoff. It has been possible to show at least an approximate quantitative relationship between observed salinity and the estimated differences between evaporation and precipitation.

That surface salinity can be approximately predicted by only considering the exchange of fresh water suggests that the salt balance regime is of smaller magnitude than the water balance regime, which we have seen requires a turnover of about .03 percent of the ocean volume each year. It is estimated that the rivers bring in about 4×10^{12} kg of dissolved solids per year (see Table 13.1). The total amount of salt in the ocean is about 5×10^{19} kg, which means that the ocean salt balance each year involves less than 10^{-7} of the total salt content of the ocean. Even if all the salt coming into the ocean were to be dissolved in seawater, which we know is not the case, it would be about 500 years before we could detect an increase in the average salinity, using our present techniques for measuring it.

LONG-RANGE WEATHER FORECASTING AND CLIMATIC CHANGE

The atmosphere derives its heat from the sun, but only about 20 percent of atmospheric energy is derived directly from the sun (see Figure 5.2). Most of the rest comes from the ocean in the form of effective back radiation, evaporation, and sensible heat transfer. The ocean transfers heat to the atmosphere in the night as well as in the day; it stores heat during the summer and releases it during the winter. The ocean's ameliorating effect on the climate can be seen by comparing the seasonal daily temperature ranges at coastal and inland stations. Inland stations at the same latitude show a much greater seasonal range in temperature than do coastal or island stations.

It is well known that there are year-to-year differences in the atmospheric weather; for example, some winters are colder than others and some springs are wetter. There is reason to believe that these year-to-year changes in seasons are related to small changes in the ocean circulation (or ocean weather). The evidence for year-to-year changes in the temperatures and currents in the ocean is of more recent origin and as yet fragmentary. However, it seems likely that the changes in year-to-year weather patterns are related to changes in the surface temperature of the ocean, even if the causal relations connecting the two are not yet known.

Of particular interest is the extent to which changing ocean conditions facilitate the formation of tropical cyclones (hurricanes in the Atlantic, typhoons in the Indian and Pacific Oceans). It has been known for some time that the ocean is the major energy source of the hurricane. A hurricane loses its force as it moves over land or even over cold water. It is the water evaporated from the warm ocean surface that provides the energy to drive the hurricane. As the water-vapor-laden air rises and cools, the condensation process releases enormous amounts of latent heat in a small space and in a short time. For some time it has also been known that there are certain tropical regions where hurricanes are most likely to form and that more form in some years than in others. What is less well understood is the exact combination of atmospheric and oceanic conditions that facilitate hurricane formation and, once they are formed, make them grow.

By "climate" we mean an average state of the air-sea-ice-land system. To make the term meaningful in a particular context, we must therefore specify the portion of the system over which observations are being averaged, and the interval of time. One level of climatic change is identified in the amelioration of climate since the last prehistoric ice age. Here we specify differences in the system averaged over thousands of years, but meteorologists also refer to climatic change in terms of changes of a few years—sometimes locally and sometimes worldwide. The drought of recent years in North Africa, for example, is a change in the previous average rainfall that occurred in that region and can be considered a climatic change. Other climatic changes have been documented from historical and natural records over hundreds of years and large areas of the Earth.

It is clear that the factors that control climate are enormously complex. The complexity stems first from the fact that there are interactions between and among all four elements in the climate system: air, sea, land, and ice. Within each of these subsystems, the physical laws governing their responses to changes external to them are complicated and are not completely understood. Second, there are numerous complex feedback relationships between elements in the system. For example, a small change in one part—say the amount of cloud cover in the southern ocean—will generate responses in adjacent parts of the system which will be positive in some cases (that is, amplify the effect) and negative in others (that is, tend to damp out the initial positive impetus). For example, if increased cloud cover reduces the surface temperature, which in turn reduces evaporation, which thus reduces the moisture content of the air and thus reduces the cloud cover, this would be a negative feedback. If, on the other hand, it could be shown that the primary effect of increased cloud cover in the southern ocean was a reduction of the surface temperature which resulted in a southward shift of the major storm systems and thus in an increase in the cloud cover, which resulted in a further decrease in the surface temperature and a further southward shift in the storm tracks, and so on, this would be a positive feedback. When such feedbacks are considered throughout the entire system, the task of imagining or predicting the end result of what at first glance seems a simple question proves to be extraordinarily difficult.

For these reasons, meteorologists and oceanographers have for a decade been working on large, numerical models of the climate system (Figures 5.7 and 5.8). These models embody the key physical equations governing the response of the system or at least as many as can be accommodated within the limited computational capacity of the largest existing digital computer. The job of modeling completely by equations is an impossible goal, however, because the Earth is so large and because the scales of many significant processes are so small.

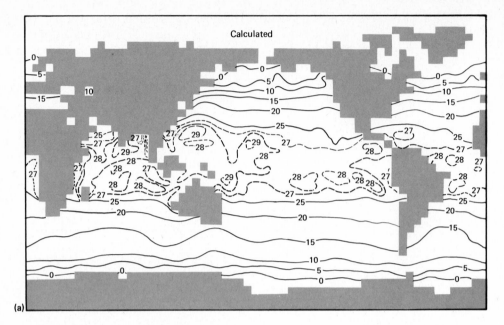

(a)

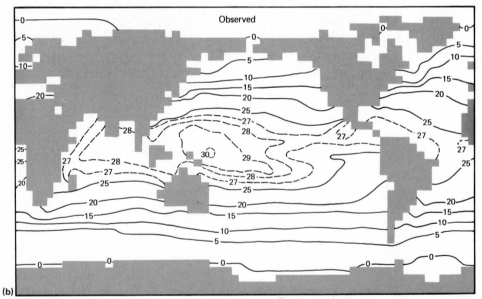

(b)

FIGURE 5.7 (a) Surface temperatures of the ocean, calculated by a numerical model of the atmosphere-ocean system at the Geophysical Fluids Dynamics Laboratory. The rectangular shorelines reflect the computational grid. Values plotted are average annual temperatures (°C). (b) Observed average annual surface temperatures of the ocean, °C, plotted on the same grid as (a). (From Bryan et al., 1975.)

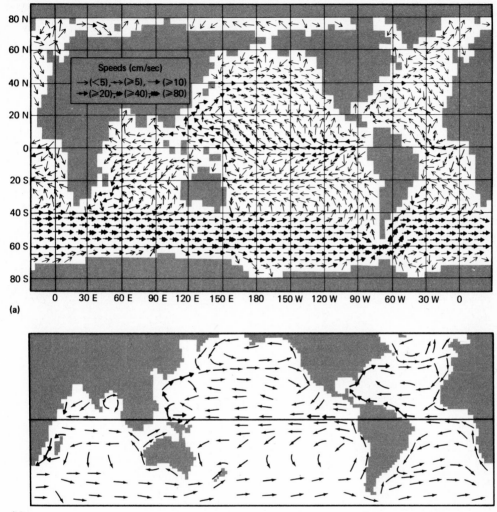

FIGURE 5.8 (a) Surface currents in March, calculated by a numerical model of the ocean at the RAND Corporation. (From Alexander, 1974.) (b) Observed ocean currents averaged over February and March. (From Bryan et al., 1975.)

SUMMARY

1. Although there is a global balance between the sun's energy entering the atmosphere and the ocean and that energy which is released to space, there need not be a local balance.

2. During spring and summer more energy enters the ocean at mid-latitudes than is lost from the ocean by evaporation, effective back radiation, and convection; consequently, a seasonal thermocline develops. During fall and winter there is a net heat loss from the ocean and the thermocline dissipates.

3. Since the ocean transports heat from low to high latitudes, more heat enters the ocean at low latitudes than is lost to space on a yearly basis, and similarly, more heat escapes from the Earth at high latitudes than comes from the sun. The winds of the atmosphere and the currents of the ocean are part of a complex heat engine which absorbs, transports, and releases the sun's energy.

4. There is also a water balance in the ocean. Each year about three-tenths of 1 percent of the ocean is evaporated. Most of the evaporated water falls on the ocean as rain. That which falls on land is eventually returned to the ocean by rivers.

Research vessel *Vema* rounding Cape Horn. (Drake photo.)

Principles of Ocean Circulation

6

"There is a river in the ocean. In the severest droughts it never fails and in the mightiest floods it never overflows. Its banks and its bottom are of cold water, while its current is of warm." Thus was the opening description of the Gulf Stream by Matthew Fontaine Maury in 1855 in his classic book *The Physical Geography of the Sea and Its Meteorology*. A little more than a hundred years later Henry Stommel characterized it more accurately but less poetically in his

CHAPTER

classic *The Gulf Stream* as a boundary flow between the warm saline water of the Sargasso Sea and the cold dense water of the continental slope.

Rivers or boundary currents, or something more complex?—the answer cannot be found only by gathering more observational data. We must also improve our description of the dynamics that control the ocean circulation. The mathematical equations which govern the dynamics can be solved only approximately (they are of a type mathematicians call nonlinear differential equations) and in many cases the approximate solutions available, even with the most modern computers, are not sufficient to determine the dynamical controls of the ocean circulation process under study.

It should not be assumed, however, that

oceanographers have no understanding of the principles of ocean circulation. For example, we know that the Gulf Stream is not a river, "its banks and bottom . . . of cold water." Rivers flow downhill—generally, the steeper the slope, the faster the flow. By such an analogy the Gulf Stream flows *around* the hill, the top of the hill being the warm water of the Sargasso Sea. The dynamics controlling the Gulf Stream are completely different from those which apply to channeled rivers, as we shall see later. This chapter is a brief description of the processes and principles that govern ocean circulation; Chapter 7 provides a description of some of the most important large-scale features of ocean movement such as the Gulf Stream and coastal upwelling.

WHAT DRIVES THE OCEAN— THE WINDS

Figure 6.1 is a chart of the mean surface currents of the world's oceans. The figure shows the Gulf Stream, the Peru (Humboldt) Current, the Antarctic Circumpolar Current, and many other currents that may sound at least vaguely familiar. Figure 6.2 is a stylized sketch of the surface currents in an ideal ocean. In this sketch the surface currents are characterized by two large gyres separated by a counterflow near the equator. The Northern Hemisphere gyre goes clockwise, that in the Southern

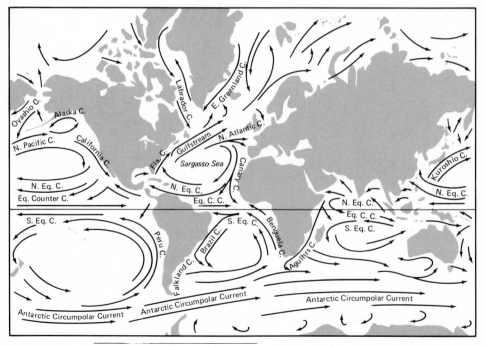

FIGURE   6.1   Major surface currents of the ocean.

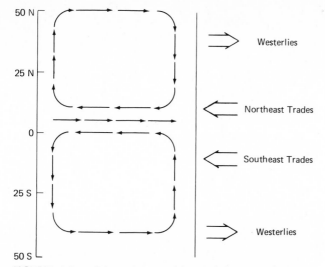

FIGURE 6.2 Schematic view of the wind-driven ocean circulation. The winds apply a clockwise force to the surface in the Northern Hemisphere and a counterclockwise force in the Southern Hemisphere which set up the two current gyres. A countercurrent separates the two gyres and is found slightly north of the equator between the region of the Northeast and Southwest Trades.

Hemisphere counterclockwise. Comparison of the idealized sketch of Figure 6.2 with the real ocean of Figure 6.1 shows many similarities in the Atlantic and the Pacific. The real ocean has a more complex set of countercurrents along the boundaries, such as the Labrador Current in the North Atlantic and a return flow in the Gulf of Alaska, and the currents on western sides of the oceans appear to be swifter than those on the eastern sides, but the schematic sketch of Figure 6.2 is a good first approximation. Figure 6.2 also shows an idealized surface wind distribution. Again, the actual mean wind field is a bit more complex, but to a fair degree of approximation the winds over the oceans can be divided into the *westerlies* of the mid-latitudes, which flow from west to east, and the Northeast and Southeast Trades of the tropics, separated by a region of light winds variously called the doldrums or the intertropical convergence.

Qualitatively, it is easy to imagine how winds of Figure 6.2 blowing on the surface of the sea can establish the major current gyres of the Northern and Southern Hemispheres. The Northern Hemisphere winds apply a clockwise *torque* to the ocean surface and the Southern Hemisphere winds apply a counterclockwise torque; the major ocean current gyres are the results. It is important to note that there is not a simple one-to-one relation between the wind and the current. We do not need stronger winds over the Gulf Stream, for example, to maintain that swift flow; rather there is an ocean-wide balance between the mean wind torque and the resulting ocean currents. Furthermore, an immense amount of energy is stored in these ocean currents; thus a shift in the mean wind field does not mean an automatic shift in the major ocean gyres. These ocean current gyres can be thought of as enormous flywheels, and the winds as an erratic engine that keeps pumping energy into the flywheel. The engine may change speed rather abruptly, but the flywheel responds very slowly to such changes. One way of thinking about the amounts of energy stored in these wind-driven ocean gyres is to consider the question: How long could these gyres maintain themselves if the winds should stop and no more energy would be pumped into the flywheel? The answer is probably at least six months and perhaps as long as two years.

WHAT DRIVES THE OCEAN— THE SUN

Superimposed on the great wind-driven gyres is another circulation "driven" by the sun, the so-called *thermohaline* circulation. *Thermo* comes from "temperature" and *haline* from "salinity." Together, the temperature and salinity of the water determine the density as discussed in Chapter 4. The thermohaline circulation is shown schematically in Figure 6.3. As discussed in Chapter 5 the ocean transports heat from tropical to polar latitudes. This is done in part by the major warm-water currents such as the Gulf Stream, but there is also a return flow of cold water to the tropics and much of this "flow" occurs at depths below the wind-driven gyres. Actually both systems are part of a single ocean circulation regime and the wind-driven system interacts with the thermohaline system. For example, although the thermohaline system is primarily a convection flow (the cold heavy water of the

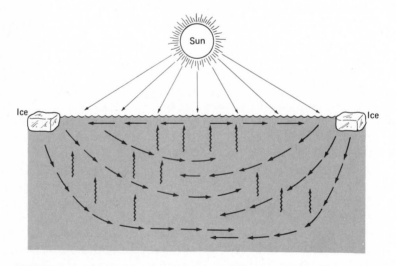

FIGURE 6.3   Schematic view of the thermohaline circulation. The sun provides more heat to the tropical latitudes than to polar latitudes. The cold, dense polar water sinks and is replaced by warm surface water. A weak upward advection of cold water occurs in the interior of the ocean.

polar latitude sinking and moving toward the tropics), we believe it is the winds that spread apart (cause a *divergence* of) the surface waters and in effect "pump" the cold water back to the surface to complete the cycle.

Although our conceptual picture of the thermohaline circulation is less complete than that of the wind-driven circulation, we believe we understand at least some parts of the process. For example, we believe the formation of sea ice in the Weddell Sea off Antarctica and in the Norwegian Sea off Greenland plays an important role in the formation of the cold dense water that fills the bottom of the North Atlantic. In both regions relatively more saline water is brought into the area and in winter is cooled to the freezing point. As the water freezes, much of the salt is left behind (see Chapter 4), thus increasing the salinity and the density of the remaining unfrozen water. This heavy water finds its way to the bottom, where it is generally referred to as Antarctic Bottom Water and North Atlantic Deep Water. We use "believe" advisedly. The actual process has never been observed, and some of the observational evidence is subject to more than one interpretation.

A second important feature of the thermohaline circulation relates to the density stratification of the ocean and the role of mixing. As can be seen in Figure 6.4,

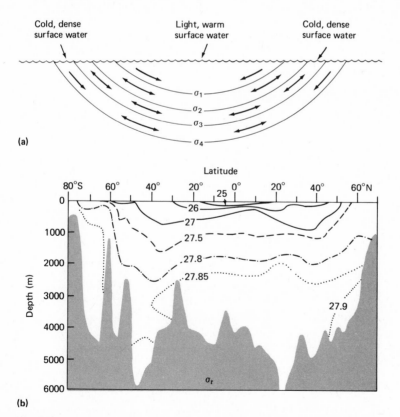

FIGURE 6.4 (a) Schematic view of the density distribution of the ocean. Water flows and/or mixes along lines of constant density. (b) For comparison, the actual density distribution in the Atlantic. (After Pickard, 1975.)

the density of the ocean increases with depth, and lines of constant density are nearly horizontal. It is considerably easier to mix properties along lines of constant density than across them. (See Density in Chapter 4.)

The third and final point concerning the thermohaline circulation is that there is considerable uncertainty as to how to characterize it. Does the water flow in simple currents or is it by diffusion? Consider the case of the Mediterranean outflow of Figure 4.7. Two extreme types of circulation can be imagined. The first is that a current flows along the core of high salinity, as in Figure 6.5a. The observed salinity distribution is simply a reflection of the current, and the decrease in salinity as one moves away from the Strait of Gibraltar is a measure of the loss of identity by mixing through vertical diffusion. The second extreme model (Figure 6.5b) can best be understood by considering the "thought experiment" of Figure 6.6.

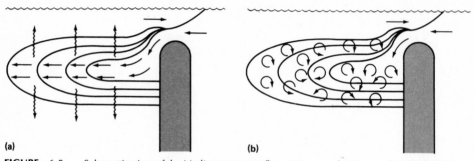

(a)   **(b)**

FIGURE 6.5   Schematic view of the Mediterranean outflow: (a) as an advection and vertical diffusion process; (b) as a horizontal diffusion process.

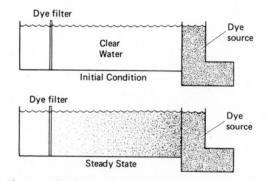

FIGURE 6.6   Consider a trough a few feet long filled with fresh water. At one end blue dye is slowly added. At the other end is a chemical filter which immediately absorbs any of the dye that comes in contact with it. Thus we have a mechanism for adding dye at one end of the trough and removing it at the other without disturbing the water. The dye will slowly diffuse through the water, and in time a color gradient will develop in the water ranging from nearly clear water at the filter end to deep blue at the other. This color gradient is maintained by diffusion processes. There is no net flow or current from one end of the trough to another. In the experiment there is no need for a *net* current flow to account for the distribution of dye. By a similar argument it is possible to account for the salinity distribution of the Mediterranean outflow, and similar features of the thermohaline circulation, without calling on ocean currents as the mechanism. Instead it is suggested that the observed distribution of temperature and salinity in the deep ocean is a result of large-scale eddies which move randomly.

In fact, both advection (the movement of water by ocean currents) and diffusion would appear to play important roles in the thermohaline circulation; determining the relative importance of the two in any given area or situation is one of the major tasks of physical oceanography. The reader may wish to make his own judgment in the following example. Figure 6.7 shows the paths of a number of floats at depths of about 2,000 m tracked over a period of 26 months. The floats

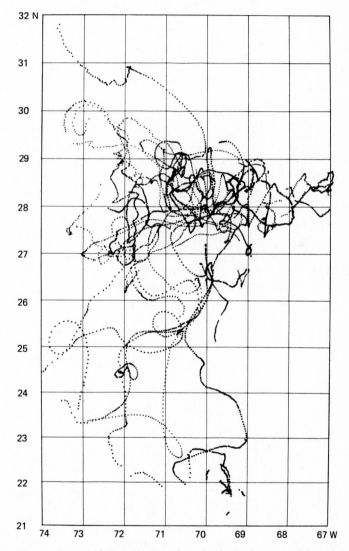

FIGURE 6.7 The movement of floats in the central North Atlantic, far removed from any major ocean currents. The floats were at depth of about 2,000 m and were originally dropped in the vicinity 28°N and 70°W. The 48 floats "diffused" from the center. Each "dot" represents a successive 24-hour position. The floats moved in a variety of directions and speeds, sometimes two or more moving together; sometimes independently; sometimes turning, executing large loops; sometimes moving only a few kilometers over a several-day period; occasionally traveling 10 km a day. (After Rossby, Voorhis, and Webb, 1975.)

can be considered as water particles. If the deep ocean was primarily a series of slowly moving currents, the floats could be expected to move with the current. If, instead, the deep ocean is best characterized as a region of weak currents and large-scale diffusion, then the movement might best be described in terms of the swirls and eddies that can be observed by watching the movement of cigarette smoke in a still room. With no wind there are swirls and eddies, which result in diffusion of the smoke, but little net drift (advection) of the smoke. Figure 6.7 suggests that diffusion may play a much larger role in the thermohaline circulation than many oceanographers had previously believed.

STEADY-STATE CIRCULATION
—THE CORIOLIS FORCE

It is not surprising that Maury thought of the Gulf Stream as a river in the sea, flowing steadily and constantly. The Gulf Stream is an example of a *steady-state* flow; rivers are our most common example of steady-state flow. By ''steady state'' we mean that the flow is constant; it neither speeds up, slows down, nor changes direction. Examples of true steady-state flow in nature are rare. Rivers curve and their speed changes as the width between banks widens or narrows, but the flow is approximately steady state and so is that of the Gulf Stream.

The physical requirement of steady-state flow is that there be an exact balance of the forces that control the flow. In a river this balance is provided by a pressure gradient force caused by the river flowing downhill, which in turn is balanced by a frictional force whose value is a function of the speed and depth of the river and the roughness of the river bottom. This balance can be written as

$$\text{pressure gradient} = \text{friction}$$
$$g \times i = R \times u \qquad (6.1)$$

where g is gravity, i is the slope of the river, R is the frictional constant related to the river depth and bottom roughness, and u is the velocity of the river. If it is assumed that R and g are constant, then the river velocity must change as the slope changes. Equation 6.1 makes the rather obvious statement that the steeper the slope, the faster the river flow. It should be noted that this ''obvious statement'' does not always agree with the real facts. As rivers broaden and deepen, the frictional coefficient, R, decreases, with the result that rivers speed up even though the slope may decrease.

The Gulf Stream is also a steady-state current, but the balance of forces is completely different from that for a river. In the ocean the friction term is very, very small. One consequence of this is that in the ocean (and the atmosphere) another ''force'' must be considered, a force that can mostly be ignored in rivers and in all situations where the other forces are large. This ''force'' is the *Coriolis force* which results from the rotation of the earth; quotation marks are used because the force is not a true force in the sense of friction, gravity, pressure gradient, and so on. Rather it is a "force of convenience." We spend our lives on a rotating sphere, but we mostly ignore this fact when we make observations of motion on this earth and, in a simple physics laboratory experiment, calculate the forces acting on a body.

Consider a pendulum suspended at the North Pole and free to swing in any direction. Assume that at twelve noon it is set in motion such that it is swinging along the 90°E–90°W longitude axis (see Figure 6.8). We know from Newton's first law that in the *absence of other forces* the pendulum will continue to swing in the same direction as the Earth rotates under it. Looking down on the North Pole from space, the Earth rotates counterclockwise 15° per hour but the pendulum does not rotate. To an observer on Earth standing near the pole, the pendulum appears to rotate *clockwise* 15° per hour. In 12 hours the pendulum will have rotated such that it is again swinging along the 90°E-90°W axis. Thus to an observer, it would

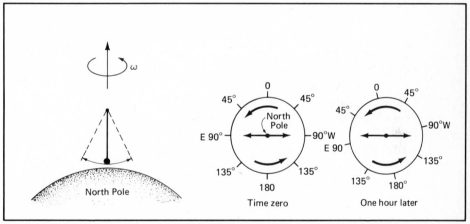

Pendulum at Pole

FIGURE 6.8   In the absence of other forces, a pendulum set in motion at the North Pole will continue to swing in its original plane as the Earth rotates under it. Looking down on the North Pole the Earth rotates counterclockwise 15°/hr. To an observer on the Earth the pendulum appears to rotate clockwise 15°/hr, thus taking 12 hours to come back to its original position. Try the same thought experiment for the South Pole and you can show that the pendulum will rotate counterclockwise 15°/hr.

Imagine now a similar pendulum swinging along the east-west axis at the equator. As the Earth rotates under it, the pendulum will continue to swing along the east-west axis. At the equator there is no rotation of the pendulum. There is no rotational force here at the equator. It can be shown that the period of rotation of a pendulum is

$$T = \frac{12 \text{ hours}}{\sin \phi}$$

where ϕ is latitude. At 90° latitude the period is 12 hours; at the equator the period is infinity. Such a pendulum is called a Foucault pendulum, after Jean Bernard Foucault, who demonstrated such a pendulum in Paris in 1851. Foucault pendulums are a part of many science museums. As the Foucault pendulum example suggests, the effect on the Earth's rotation changes with latitude. It is strongest at the poles and is zero at the equator. It reverses direction in the Southern Hemisphere.

appear that some small force has been applied to the pendulum to make its axis rotate clockwise 15° per hour. This "force," which causes the apparent clockwise or counterclockwise rotation of the pendulum, is called the *Coriolis force*.

Because we are not going to prove the mathematical relationship, we now require an act of faith as we describe the effect of the Coriolis force on a moving air or water particle. The effect of the Earth's rotation is such that every particle in motion appears to have a force applied to it which would make it curve to the right in the Northern Hemisphere. In the Southern Hemisphere the force is applied in the opposite direction and makes its curve to the left. The magnitude of the force is directly proportional to the velocity of the particle. The magnitude of the force also changes with latitude, being highest at the poles and zero at the equator. Figure 6.9 presents a partial explanation for one component of the Coriolis force.

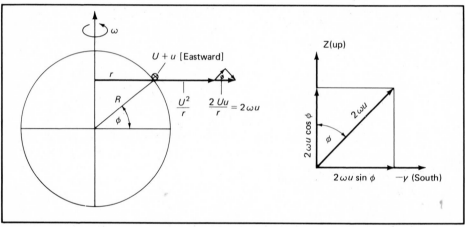

Coriolis Force

FIGURE 6.9   A complete derivation of the Coriolis force is best done with vector algebra; however, some insight can be gained by considering the centrifugal acceleration of a particle on the surface of the Earth. The centrifugal acceleration of a particle resting on the Earth is U^2/r, where $U = \omega r$, where ω is the Earth's angular velocity. If the particles were moving in an eastward direction, with a velocity u, the centrifugal acceleration is

$$\frac{(U + u)^2}{r} = \frac{U^2}{r} + \frac{2Uu}{r} + \frac{u^2}{r} = \omega^2 r + 2\omega u + \frac{u^2}{r}$$

Since U is the order of a thousand times greater than u, the first term is the order of a thousand times larger than the second, which is of the order of a thousand times larger than the third. This last term is small enough to be ignored. The second term is the Coriolis acceleration.

The Coriolis acceleration vector can be resolved into two components, one normal to the plane of the Earth, and the other parallel to the plane of the Earth. The latter has the value $2\omega u \sin \phi$ and is the horizontal component of the Coriolis acceleration which applies to east-west motion. Note that if the particle were moving westward, the same absolute value would apply, but the second term in the above equation would be negative. The vector would be pointed toward the Earth's axis and the component in the plane of the Earth would be pointing north.

GEOSTROPHIC CURRENTS

By analogy with our steady-state river flow we can write an equation which balances the pressure gradient force and the Coriolis force.

$$\text{pressure gradient} = \text{Coriolis force}$$
$$g \times i = f \times v \qquad\qquad (6.2)$$

where f is the Coriolis term whose magnitude is approximately $14 \times 10^{-5} \sin \phi$ sec^{-1}, where ϕ is the latitude ($\sin \phi$ will vary from 0 at the equator to 0.5 at 30°N and -0.5 at 30°S, and to 1.0 at 90°N and -1.0 at 90°S), and where the current direction v is perpendicular to the slope i.

Currents that obey the relationship shown in Equation 6.2 are called *geostrophic currents*. The major currents in the ocean such as the Gulf Stream (and all those shown in Figure 6.1) are very nearly geostrophic currents.

The most extraordinary feature of Equation 6.2 is that the current does not flow down the slope but flows *parallel* to the slope, because the Coriolis force acts at *right angles* to the pressure gradient. Thus geostrophic currents do not flow from high pressure to low pressure but flow parallel to the isobars (lines of constant pressure). The winds in the atmosphere also obey the geostrophic relationship, or nearly so. Through television weather reports most of us have become familiar with high- and low-pressure cells and with the wind patterns that result from the application of Equation 6.2. Because the winds are approximately geostrophic, the air does not flow from high pressure to low pressure as might be expected; rather, the winds blow parallel to the isobars, flowing clockwise around high-pressure cells and counterclockwise around low-pressure cells in the Northern Hemisphere and in the opposite direction in the Southern Hemisphere (Figure 6.10).

Returning now to the Gulf Stream as a geostrophic current, we can see how the Gulf Stream is a "river" that does not run down the hill but around the hill. In the region of the Gulf Stream the sea surface slopes downward from the Sargasso Sea and the Gulf Stream runs parallel to that slope. Only very small slopes are required to maintain currents of 0.5–1 m/sec (1–2 knots) (Figure 6.11).

If the major ocean currents are geostrophic currents, then according to Figures 6.1 and 6.2 the centers of the ocean gyres must be high-pressure cells and the currents flow around these high-pressure cells. Borrowing the terminology of meteorologists, this is an *anticyclonic* circulation. Flow in the opposite direction around low-pressure cells is *cyclonic*. Again, note that anticyclonic circulation is clockwise in the Northern Hemisphere and counterclockwise in the Southern Hemisphere.

MORE CORIOLIS EFFECTS

Entering the world of the Coriolis force is a bit like going through the looking glass with Alice. Physical intuition is of little avail as you attempt to work through the logical consequences of adding a force that acts at right angles to the direction of

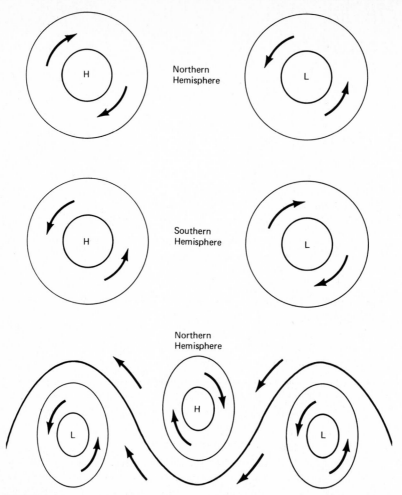

FIGURE 6.10 Under conditions of pure geostrophic flow, currents in the Northern Hemisphere move clockwise around high-pressure cells and counterclockwise around low-pressure cells. The directions are reversed in the Southern Hemisphere.

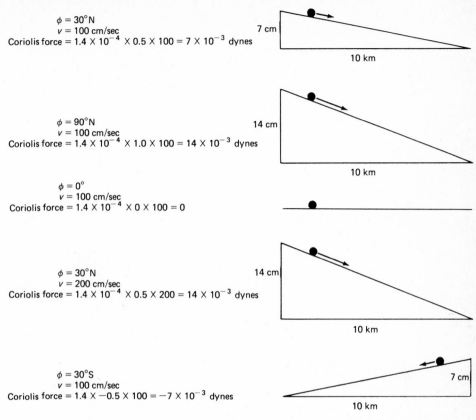

$\phi = 30°N$
$v = 100$ cm/sec
Coriolis force $= 1.4 \times 10^{-4} \times 0.5 \times 100 = 7 \times 10^{-3}$ dynes

7 cm

10 km

$\phi = 90°N$
$v = 100$ cm/sec
Coriolis force $= 1.4 \times 10^{-4} \times 1.0 \times 100 = 14 \times 10^{-3}$ dynes

14 cm

10 km

$\phi = 0°$
$v = 100$ cm/sec
Coriolis force $= 1.4 \times 10^{-4} \times 0 \times 100 = 0$

$\phi = 30°N$
$v = 200$ cm/sec
Coriolis force $= 1.4 \times 10^{-4} \times 0.5 \times 200 = 14 \times 10^{-3}$ dynes

14 cm

10 km

$\phi = 30°S$
$v = 100$ cm/sec
Coriolis force $= 1.4 \times -0.5 \times 100 = -7 \times 10^{-3}$ dynes

7 cm

10 km

Vertical exaggeration is 10,000

FIGURE 6.11 The numerical value of the horizontal Coriolis force is ($4\pi/24$ hours) times sine (latitude) times current velocity. The first term converts to 1.4×10^{-4}/sec. For a numerical example take the latitude as 30°N (sin 30° = 0.5) and the current is 100 cm/sec. The Coriolis force is $1.4 \times 10^{-4} \times 0.5 \times 100 = 7 \times 10^{-3}$ dynes, which is a very small force. It is approximately equivalent to the gravitational force applied to a ball bearing rolling down an inclined plane with a slope of seven parts in a million, which is the equivalent of a 7-cm change in elevation in 10 km. The horizontal Coriolis force is stronger at higher latitudes (as the sine of the latitude increases to its maximum value of 1.0 at 90°) and is weaker at lower latitudes, reaching zero at the equator. The Coriolis force also increases as the current strengthens and decreases as the current weakens. It changes direction in the Southern Hemisphere. To give some measure of the magnitude of the Coriolis force under differing conditions, the force is shown in the equivalent value to which a ball bearing would be subject on a frictionless inclined plane.

movement. We have already seen that in the world of the Coriolis force, water does not flow downhill but around the hill, and here we will look at three other situations.

Consider first what occurs if there are *no* other forces acting on a particle. Newton's first law states that a particle once in motion will continue in motion at the same speed and in the same direction until acted upon by another force. In the absence of other forces the motion is in a straight line at constant velocity. However, on a rotating Earth, a consequence of this law is that the motion is not straight, but circular. A particle in motion is acted upon by the Coriolis force. As can be seen in Figure 6.12, the motion is circular, clockwise in the Northern Hemisphere, counterclockwise in the Southern Hemisphere. The period is the same as that of a Foucault pendulum, 12 hours at the poles, 24 hours at 30°N, and infinity at the equator. The particle will continue its circular motion at the same speed, and at any given latitude the radius of these *inertial circles* is a simple function of the initial velocity.

Next consider what occurs when the wind blows on the water. As the wind blows, the water is set in motion; the wind-induced surface current is about 3 per-

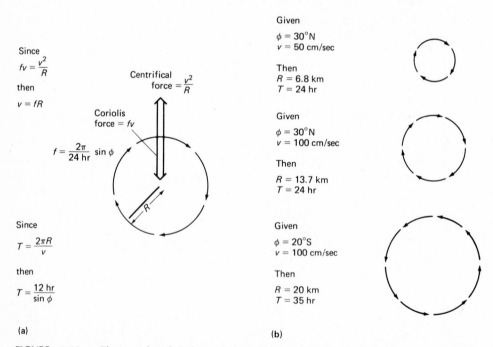

Since

$$fv = \frac{v^2}{R}$$

then

$$v = fR$$

Since

$$T = \frac{2\pi R}{v}$$

then

$$T = \frac{12 \text{ hr}}{\sin \phi}$$

Centrifical force $= \dfrac{v^2}{R}$

Coriolis force $= fv$

$$f = \frac{2\pi}{24 \text{ hr}} \sin \phi$$

R

Given

$\phi = 30°\text{N}$
$v = 50 \text{ cm/sec}$

Then

$R = 6.8 \text{ km}$
$T = 24 \text{ hr}$

Given

$\phi = 30°\text{N}$
$v = 100 \text{ cm/sec}$

Then

$R = 13.7 \text{ km}$
$T = 24 \text{ hr}$

Given

$\phi = 20°\text{S}$
$v = 100 \text{ cm/sec}$

Then

$R = 20 \text{ km}$
$T = 35 \text{ hr}$

(a)   (b)

FIGURE 6.12   The inertial circle is the equivalent of Newton's first law for bodies on a spinning Earth. The circular motion is a result of the balance between the Coriolis force and the centrifugal force (v^2/R) resulting from the circular motion. By simple manipulation of the balance terms it is easy to show that the velocity is equal to the *Coriolis parameter* [$4\pi/24$ hr times sine (latitude)] times the radius of the circle; that is; $v = fR$, where f represents the Coriolis parameter. Remembering that the circumference of a circle is $2\pi R$ it can readily be shown that the time required to complete a circle is independent of the circle size or the velocity, but does vary with latitude.

cent of the wind itself; that is, a 10-knot wind will generate a 0.3-knot surface current. However, the current will not flow directly downwind but will flow to the right of the wind in the Northern Hemisphere and to the left in the Southern Hemisphere. The simplest relationship worked out by V. W. Ekman in 1902 suggested that the current on the surface should be at 45° to the wind. The evidence suggests that the surface current is nearer to 10°–15° to the direction of the wind (the right or left depending upon whether the current flows in the Northern Hemisphere or Southern Hemisphere).

Of more consequence, in many ways, is what occurs to the total wind-driven layer which may extend several tens of meters deep. The Ekman theory suggests that the effect of the Coriolis force is that the *mean flow,* averaged over the entire depth, should be at right angles to the direction of the wind. Figure 6.13 indicates several consequences of such ideal Ekman flow. Perhaps the most important is that if Figure 6.13 is compared with the mean wind of Figure 6.2, it can be seen how water can be piled up in the center of the gyre to maintain the high-pressure cell and the required steady-state geostrophic currents. Using the analogy of the flywheel once again, the winds are the engine that intermittently pushes water into the center of the gyre; the potential energy associated with this distribution of mass is the energy of the flywheel. The steady-state geostrophic currents that flow parallel to the slope of the sea surface are a consequence of this distribution of mass, and the fact that the flow is parallel to the slope and not down the slope means that these major ocean currents are not extracting any energy from the flywheel.

Another consequence of Ekman motion is that in order to get *upwelling* along the coast, the wind should be parallel to the coast rather than blowing offshore. (Upwelling is the process by which water is brought up from beneath the surface.) When the wind blows along the coast, water can be transported to the right of the wind in the Northern Hemisphere and is replaced by water from below (Figure 6.13). The upwelled colder water is rich in nutrients and can support a large biological population. The cold water and high productivity found off the coasts of Peru and Chile are related less to the cold water brought up from the Antarctic than they are to the fact that there is wind-induced upwelling along the coast as the surface waters are moved to the left away from the continent (since it is the Southern Hemisphere). Similarly, the narrow band of cold surface water off the Pacific Coast of the United States is caused by upwelling (Figure 6.13). An examination of mean wind charts allows the prediction of areas of upwelling. Upwelling is nearly always accompanied by high biological productivity. One interesting example can be found in the Arabian Sea, where the seasonal wind pattern shifts with the monsoon. Marked upwelling occurs in January in the northeastern Arabian Sea off Pakistan and in July in the western Arabian Sea south of the Gulf of Aden.

Another example can be found along the equator where the trades cause an Ekman drift to the north in the Northern Hemisphere and to the south in the Southern Hemisphere. Where the easterly component of the trades is well developed, the surface temperature along the equator in the central Atlantic and the Pacific is often two degrees cooler than it is 100 km on either side of the equator. There may be another explanation for the lowered surface temperature found along

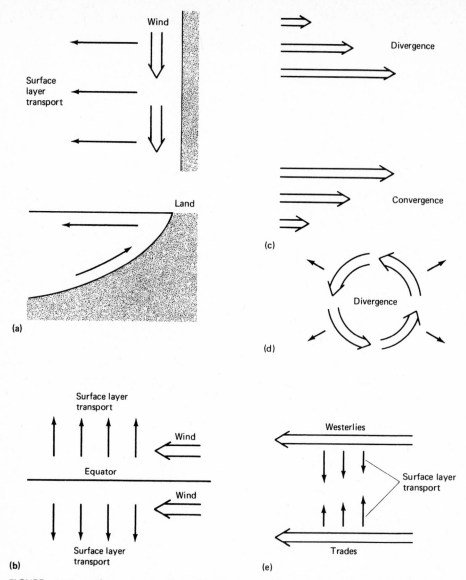

FIGURE 6.13 There are a number of interesting and important consequences of Ekman motion where the wind moves the water at right angles to the wind. For example, a wind parallel to the shore either piles up water on the shore or moves it offshore. (a) In the latter case it results in nutrient rich water being "upwelled" from below to replace the surface water. (b) Because of the change in direction as one crosses the equator, there is typically a divergence along the equator resulting from the southeasterly trade winds which results in cooler and nutrient rich water being brought to the surface. (c) Special changes in the horizontal wind speed (wind shear) can cause either divergence or convergence of the waters in the surface layer. (d) Extreme examples are hurricanes and typhoons which result in a divergence of the surface waters. This is true in both Northern and Southern Hemispheres, because the winds rotate in the opposite direction in a Southern Hemisphere typhoon. (e) Similarly the large anticyclonic wind patterns over the central oceans result in a piling up of water in the center of the ocean.

the equator, but the evidence to date suggests that it is at least in part the result of a divergence of the surface waters resulting from an Ekman transport very close to the equator.

SUMMARY

1. The deep ocean circulation is generally classified as thermohaline. Cold water at high latitudes sinks and slides equatorward under the warmer, less dense surface water.

2. The salinity of water under sea ice is increased by the salt brine, which does not freeze. This water is very dense and fills the bottom layers of the major ocean basins.

3. The surface currents of the ocean are primarily wind-driven.

4. There is a rough correlation between the large anticyclonic torques of the major wind system and the anticyclonic ocean current gyres of the major ocean basins.

5. Many features of the ocean are strongly influenced by the effect of the Earth's rotation (the Coriolis force). As a consequence of the Coriolis force, pendulums rotate (Foucault pendulum), major ocean currents run around the hill rather than downhill (geostrophic flow), and the wind drives water at right angles to the direction it blows (Ekman motion).

Preparing to release a buoy which will float at 2000 m depth. For results see Figure 6.7. (Photo courtesy H.T. Rossby.)

Major Ocean Currents

7

CHAPTER

Having discussed the principles of ocean circulation in the previous chapter, we now turn to a discussion of the major ocean currents themselves. There are two major subdivisions, the currents of the deep ocean, which are expressions of the thermohaline system, and the surface currents, which are driven primarily by the surface wind system.

THE DEEP OCEAN

All the deep water in the ocean is cold, and the origin of this water can be traced to various sources in Antarctica, in the Norwegian Sea, and off Greenland. As noted in Chapter 6, the deep water takes on its characteristic temperature and salinity at the surface, the heaviest water forming under ice, where the freezing seawater allows the addition of salt to the water already cooled to near the freezing point. The water sinks approximately along lines of constant density (specifically lines of constant σ_θ, (see Chapter 4), mixes with the surrounding water, and fills the deep ocean basins. As also noted in Chapter 6, the relative role of simple current flows versus eddy diffusion and mixing is difficult to judge. Probably both mechanisms are important. In particular, evidence has accumulated in recent years for significant current flow in deep water along the western boundaries of oceans. There is fair to excellent evidence now for western boundary currents in the North Atlantic and in the South Atlantic and South Pacific. All three western boundary currents flow equatorward. The evidence is of several kinds. A careful study of temperature and salinity distributions indicates a tendency for tongues of water to push down the western boundaries of ocean basins. This can be seen in tongues of high-salinity water overflowing from the Norwegian Sea

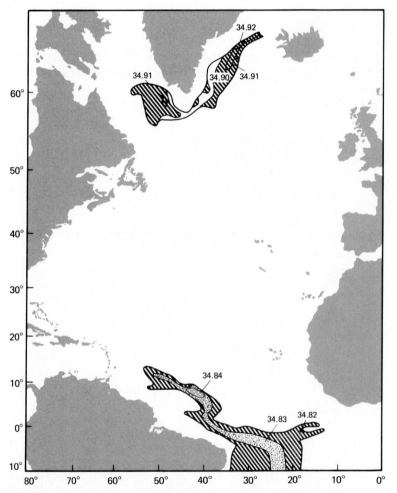

FIGURE 7.1   Cold, salty tongues of water creeping equatorward along the western sides of the North and South Atlantic. The southern tongue originates in the Weddel Sea and is Antarctic Bottom Water. The northern tongue is from the Norwegian Sea and is often referred to as either North Atlantic Deep Water or North Atlantic Bottom Water. This figure shows the high-salinity tongues (plotted on the 1.3°C potential temperature surface) of the North Atlantic Deep Water and the Antarctic Bottom Water. (From Worthington and Wright, 1970.)

through the Denmark Strait over sills in the Faeroe-Iceland ridge and up the western South Atlantic from Antarctica (Figure 7.1) and in the core of cold 1°-water hugging the New Zealand coast, as shown in the cross section of Figure 7.2. The slope of the isotherms in Figure 7.2 implies a horizontal pressure gradient from which it is possible to calculate the strength of the expected geostrophic flow. The calculation suggests a northward flow of several centimeters per second close to the bottom, and a water transport of about 15×10^6 m³/sec. Indeed all the named currents in the surface current chart of Figure 6.1 have transports of the order of 10–100 million cubic meters per second, and there is some evidence that the Gulf Stream and Antarctic Circumpolar Current are even larger than that. By comparison, the flow of the Mississippi is 20×10^3 m³/sec.

A primary difference between the deep circulation in the Atlantic and that in the Pacific is that the former is filled with cold water from both the Northern and Southern polar regions. Both sources are clearly evident in Figure 7.1. Moreover the high-salinity Mediterranean outflow contributes to the characteristics of the

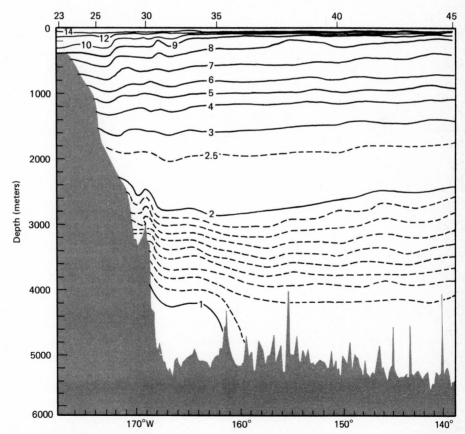

FIGURE 7.2   This temperature cross section at 43°S off New Zealand in the South Pacific shows clearly the cold western boundary current hugging the western edge of the South Pacific Basin. (After Warren, 1970.)

deep water in the Atlantic (Figure 4.7). In contrast all the deep water in the Pacific Ocean originates in the Antarctic Ocean. One result of these varying sources is that the deep water of the Atlantic is less homogeneous than that of the Pacific. For example, the salinity of nearly all the deep water in the Pacific is between 34.65‰ and 34.75‰, whereas the salinity range of the deep water in the Atlantic is several times that, even though the volume of the Atlantic Ocean is only half that of the Pacific.

The water masses of the deep oceans are defined by their temperature and salinity characteristics and are named. The tongue of water shown in Figure 7.1 pushing equatorward from the south is called Antarctic Bottom Water; that coming down from the north is referred to as North Atlantic Deep Water. In the areas where they meet, the North Atlantic Deep Water is slightly less dense and overrides the Antarctic Bottom Water. All the Pacific below 2000 m is filled with a highly homogeneous water mass, often referred to as Pacific Common Water.

If you were to tag a series of water molecules as they slid beneath the surface to form one of the deep water masses, how long would it be before currents, mixing, and other processes moved them back to the surface or from one ocean to another? We can only estimate the time. We know the deep water does not turn over every few months, and we are also certain that the deep water of the oceans did not originate in the last ice age some 15,000–20,000 years ago. Estimates of residence time can be inferred in a number of ways; one example for the deep Pacific is described in Chapter 9. We believe that the residence time for deep water in the Pacific is much longer than for the Atlantic; the order of 1000 years for the Pacific, perhaps 200–300 years for the Atlantic; a very long time by some measures, but much too short for those who have considered the waters of the deep oceans as a place to dispose of long-lived soluble radioactive and toxic chemical wastes.

MAJOR WIND-DRIVEN CURRENTS

The gross features of the surface circulation of the ocean can be explained as wind-driven currents. There are many similarities in the circulation patterns of the Atlantic and Pacific (Figure 6.1). In both the circulation is dominated by two large anticyclonic gyres (an anticyclone rotates clockwise in the Northern Hemisphere and counterclockwise in the Southern Hemisphere). The two gyres are separated by an equatorial countercurrent. In both oceans there are strong western boundary currents in the Northern Hemisphere (the Gulf Stream in the Atlantic and the Kuroshio or Japanese Current in the Pacific) and somewhat weaker ones in the Southern Hemisphere (the Brazil Current and West Australia Current). In both the North Atlantic and North Pacific there is a cold flow from the north along the western basin, the Oyashio in the Pacific, the Labrador and Greenland currents in the North Atlantic. There is also a small cyclonic gyre, north of the main gyre in the eastern basins.

Some differences between oceans are due to differences in geometry. The Atlantic, Pacific, and Indian Oceans are not the same shape. But some of the differences are clearly related to different wind patterns. This difference is most clearly seen in the Indian Ocean. The circulation pattern in the southern Indian Ocean is similar in its gross features to that in the South Atlantic and South Pacific; however, the northern Indian Ocean circulation is clearly dominated by the monsoon. In parts of the northern Indian Ocean there is a complete reversal in the surface currents between the summer and winter monsoons (Figures 7.3 and 7.4).

For several reasons, the closer to land, the greater the deviations from the gross circulation features. Permanent or semipermanent eddies can be established as a

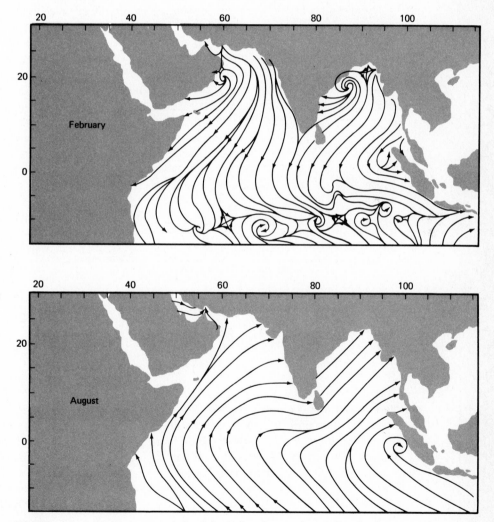

FIGURE 7.3   The monsoon winds of the Indian Ocean change direction between February and August.

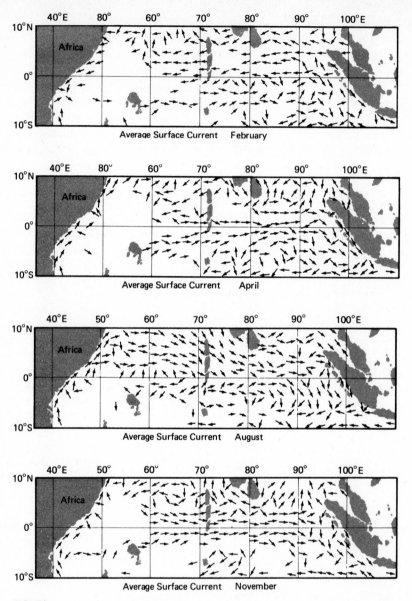

FIGURE 7.4   As a result of the monsoon winds, the equatorial current of the Indian Ocean also changes direction seasonally. In February the currents north of the equator flow westward; in August they flow to the east.

result of the interaction of major currents with the boundaries of the ocean. Similarly, local wind patterns are more likely to deviate from the mean close to shore than in the central ocean basins. In some areas the effects of river runoff and tides are important factors.

Finally, it should be noted that surface current charts such as Figure 6.1 are average-value charts. Precise navigation cannot be expected on the basis of such charts. At any given time the position of the current and its strength may be different from the average. Moreover, ocean currents are turbulent. Thus the instantaneous velocity may be quite different from the mean. This is particularly true in the central region of the ocean, where the average currents are weak.

WESTERN BOUNDARY CURRENTS—
THE GULF STREAM

The western boundary currents in the Northern Hemisphere (the Gulf Stream and the Kuroshio) appear to be better developed than their counterparts in the Southern Hemisphere for reasons that have yet to be satisfactorily explained. The Gulf Stream and the Kuroshio are similar in many respects; of the two we probably know more about the Gulf Stream.

If the Gulf Stream is considered as part of a continuous anticyclonic gyre, there is a problem in defining its beginning and its end. A strong flow can be seen through the Yucatan Channel between Mexico and Cuba, and this current often loops into the Gulf of Mexico before flowing through the Florida Straits. The Gulf Stream hugs the coast for some 1200 km, from Key West, Florida, to Cape Hatteras, North Carolina, and any meandering from the coast is rare. After leaving Cape Hatteras, the Gulf Stream ceases to follow a steady path. It cuts across the North Atlantic, passing below the Grand Banks, but the path varies and sizable wavelike meanders develop in the Gulf Stream. The Gulf Stream has been observed to 45°W some 2500 km from Cape Hatteras. Somewhere between the Southeast Newfoundland Rise and the Mid-Atlantic Ridge the Gulf Stream ceases to be a single current.

The width of the surface Gulf Stream is between 125 and 175 km. The left-hand edge (as one faces downstream) of the Gulf Stream is easily identifiable because of a horizontal temperature gradient (easier to see a few tens of meters below the surface) and a counterflow (Figure 7.5). The right-hand edge of the Gulf Stream is difficult to distinguish in terms of temperature, but there is often a discernable countercurrent. Surface velocities in the Gulf Stream can reach 250 cm/sec, better than 5 knots. The highest velocities are to the left of the center of the current, and the width of the stream increases slightly downstream (Figure 7.6). The transport appears to increase downstream from about 30×10^6 m³/sec off Florida to more than 100×10^6 m³/sec south of the Grand Banks.

Figure 7.7 is a composite picture of actual paths of the Gulf Stream observed over several months as defined by the left-hand edge of the stream. The individual paths show little relation to one another. However, a synoptic picture of the Gulf Stream taken every few days would show that the individual meanders or waves

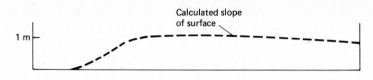

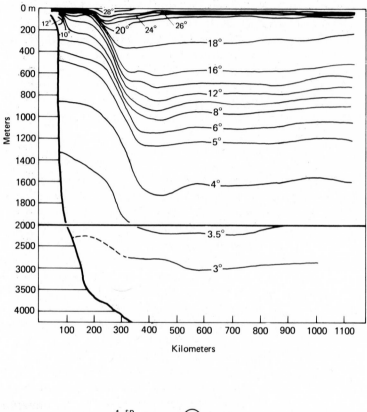

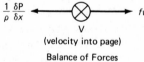

FIGURE 7.5   The sharp slope of the isotherm (the "cold wall") signals the Gulf Stream. The slope can be observed all the way to the bottom of the ocean. Shown above is the calculated slope of the sea surface in the region of the Gulf Stream. Since the Gulf Stream is a geostrophic current, the current flows parallel to the slope. The temperature section runs from Chesapeake Bay to Bermuda. (After Iselin, 1936.)

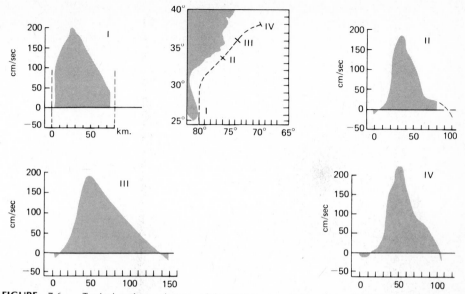

FIGURE 7.6   Typical surface velocities of the Gulf Stream. Although the Gulf Stream may be more than 100 km wide, the high-speed region (speed greater than 3 knots or 150 cm/sec) is often no more than 20 km wide. (From Knauss, 1969.)

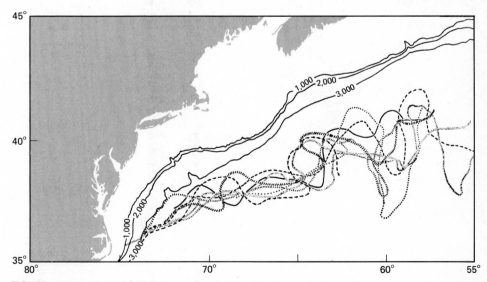

FIGURE 7.7   Once the Gulf Stream leaves Cape Hatteras its path can change from week to week. These typical paths of the Gulf Stream were observed over a several month period. The size of the meanders appears to increase downstream. (After Knauss, 1969.)

appear to progress downstream. Occasionally these meanders grow very large, and the path of the Gulf Stream becomes difficult to follow. Often the large meanders become unstable and break off to form either a large cyclonic eddy south of the Gulf Stream or an anticyclonic eddy north of the Gulf Stream (Figure 7.8). Apparently these eddies, or Gulf Stream rings as they have been called, last for some time, perhaps as long as three to five years (Figure 7.9).

Although the Kuroshio appears to be similar to the Gulf Stream in many aspects, it has at least one very fascinating difference. The Kuroshio can develop a larger meander than the Gulf Stream, which can remain stable for years. It would appear that the Kuroshio path is bimodal and can be stable in one of two positions: either with a marked meander centered at 130°E or without the meander (Figure 7.10).

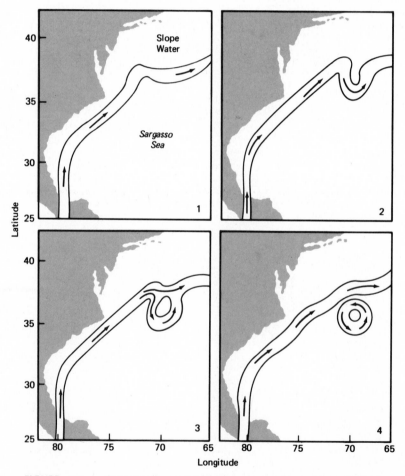

FIGURE 7.8   Occasionally a meander becomes unstable and throws off a ring. These can be thrown off either to the south, as shown here, or to the north of the Gulf Stream, where they will rotate in the opposite direction. It is believed that about five rings a year are thrown off to the south of the Gulf Stream.

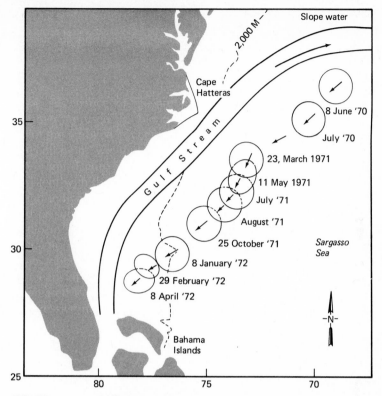

FIGURE 7.9   Gulf Stream rings decay slowly. This one was tracked for almost two years before breaking up in the shallow waters of the Blake Plateau. (After Richardson, Strong, and Knauss, 1973.)

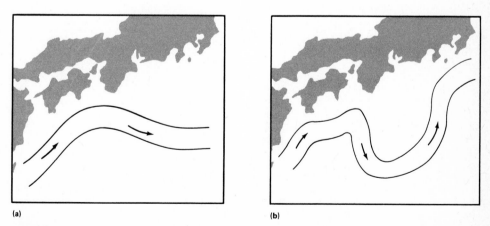

FIGURE 7.10   The Kuroshio, which is the Pacific equivalent of the Gulf Stream, has two stable positions. Either it flows in a more or less straight path (a) or it develops a large stable loop (b). It takes a few months to shift from one position to another and may stay in its new position for several years. No one has a very clear understanding of what causes the Kuroshio to shift from one stable mode to another.

CURRENTS ALONG THE EASTERN SIDES
OF OCEANS—UPWELLING

Although the major ocean circulation may be described by large anticyclonic gyres, as shown in Figures 6.1 and 6.2, the current systems that compose a single gyre differ markedly from one part of the gyre to another. The western boundary currents, such as the Gulf Stream and Kuroshio, are narrow, swift, deep flows with comparatively well-marked boundaries. The equatorward flows on the other side of the ocean basins, such as the California, Peru, or Benguela currents, are broad, weak, shallow flows whose boundaries are so poorly defined that some authors find it convenient to define different currents as one moves offshore.

The best observed of these eastern boundary currents is the California Current. The flow appears to be mostly confined to the upper 500 m. The transport is less than 15×10^6 m³/sec. A synoptic photograph of the flow at a single instant would not show a simple unidirectional flow, but rather a series of large eddies superimposed on a broad, weak equatorward movement (Figure 7.11). The speed and direction measured in the California Current at any given instance may be quite different from the average flow. A similar condition would appear to prevail in other eastern boundary currents. The coastal flow can be particularly complex and for descriptive purposes the coastal flow is often separated from the broader offshore currents by a different name.

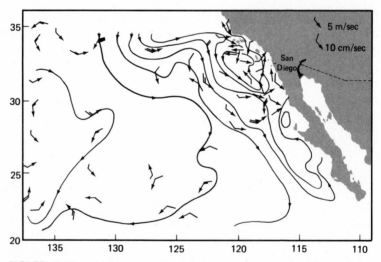

FIGURE 7.11   In contrast to the strong, narrow western boundary currents such as the Gulf Stream, the currents on the eastern side of the ocean basins are generally weak and diffuse. In this example of the California Current a large counterclockwise eddy can be seen off Southern California. The barbed arrows represent surface current observations; the lines represent the surface pressure field and indicate the direction of the geostrophic currents. (From Wooster and Reid, Jr., 1963.)

In many eastern boundary currents, coastal upwelling is a dominant factor in determining the surface distribution of temperature, salinity, and chemical properties (Figure 7.12; see Chapter 6 and especially Figure 6.13). Upwelling has major biological significance because the deep water carries nutrients to the surface layers which enhance photoplankton productivity (see Chapter 11). Regions of upwelling are among the richest biological areas in the world (see Chapter 12).

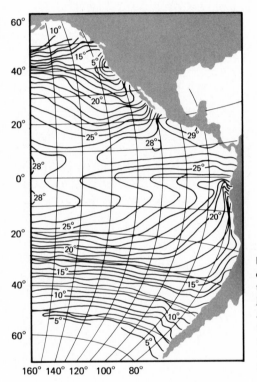

FIGURE 7.12   The effect of upwelling off the coasts of California, South America, and along the equator can be clearly seen in this chart of July surface temperatures for the eastern Pacific. As one moves toward shore the surface temperature rapidly drops because of cold water being upwelled along the coast. The decrease in surface temperature along the equator is also a result of upwelling.

EQUATORIAL CURRENTS

A simple way to consider the major currents of the tropics is to consider them counterpart of the trade-wind system. Over most of the Atlantic and Pacific there are the northeasterly trades of the Northern Hemisphere separated from the southeasterly trades, which are mostly in the Southern Hemisphere. The intertropical convergence (ITC) is an area of weak and variable winds between the two trade-wind systems, often referred to as the doldrums. The ITC separates the wind systems of the two hemispheres and may be thought of as the climatic equator. This climatic equator is generally found between 3°N and 10°N.

The major ocean currents of the tropics are a reflection of the wind system. Beneath the trades are the westward-flowing North and South Equatorial Currents, which are part of the main anticyclonic current gyres of the Northern and Southern Hemispheres (Figure 6.1). Between these two broad westward flows is a relatively narrow (300–500 km wide) eastward-flowing Equatorial Countercurrent (Figure 7.13). It should also be noted that the schematic picture of this figure is best realized in the central Atlantic and Pacific. Both the trade winds and the equatorial current system are more complex near the edges of the ocean basins.

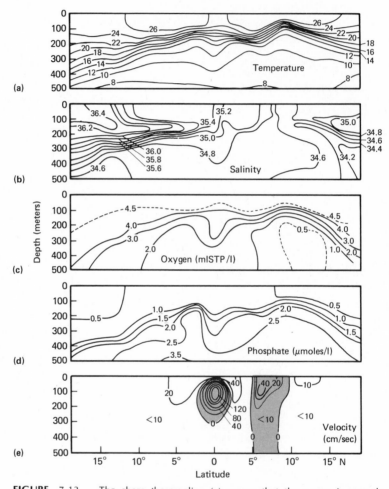

FIGURE 7.13   The sharp thermocline (a) means that the ocean is strongly stratified in this region. The thermocline is a barrier across which it is difficult to mix oxygen downward (c) and phosphate upward (d), or salinity (b) either up or down. The exception is the equator, where mixing across the thermocline appears to occur. (e) shows the major equatorial currents, the westward-flowing North and South Equatorial Currents, separated by an eastward-flowing Equatorial Countercurrent between 5°N–10°N. Also shown is the intense Equatorial Undercurrent which flows eastward along the equator. These cross sections are from the central Pacific. (After Knauss, 1963.)

The tropics are characterized by a well-mixed, warm surface layer and a sharp thermocline which separates the surface layer from the cold water beneath. This thermocline also serves as a lid separating the surface layer, which is high in dissolved oxygen and low in nutrients such as phosphate and nitrate, from the deeper waters, which are low in oxygen and relatively rich in nutrients. The equatorial currents are confined mostly to the surface layer with average speeds of 25–75 cm/sec. Below the thermocline, current speeds are much weaker than in the surface layer.

On the equator itself is a powerful eastward-flowing subsurface current in the region of the thermocline, the Equatorial Undercurrent, or Cromwell Current, as the Pacific Equatorial Undercurrent is often called. This current is like a thin ribbon, perhaps 200 m thick and 300 km wide, with peak speeds of up to 150 cm/sec (3 knots). The core of the current usually coincides with the thermocline and is generally centered on, or very close to, the equator (Figure 7.14). Occasionally the eastward flow has been observed at the surface, but usually these Equatorial Undercurrents are completely below the surface.

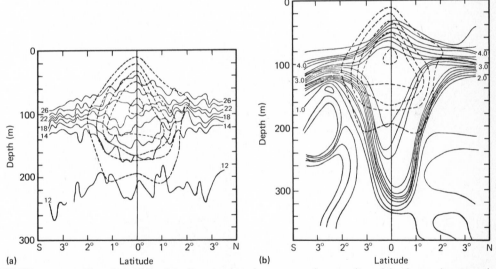

FIGURE 7.14   The relationship of the Equatorial Undercurrent to the spreading of the thermocline (a) and the downward mixing of oxygen (b) is clearly shown in this cross section from the central Pacific. The velocity contours are in dotted line at 25 cm/sec intervals and go from 150 cm/sec in the center to 25 cm/sec along the outside. The oxygen values are in ml/l. The transport of the Equatorial Undercurrent at this time was 40×10^6 m³/sec. (After Knauss, 1959.)

POLAR CIRCULATION

The circulation in the two polar regions is totally different. The Arctic Ocean is completely covered by pack ice. What little is known about the currents suggests a sluggish counterclockwise drift. The deep cold water of the Arctic is kept from mixing freely with that in the Atlantic and Pacific by two fairly shallow sills between the continental blocks. The shallow sill of the Bering Straits between the U.S.S.R. and Alaska is less than 100 m deep and effectively closes off flow between the Atlantic and Pacific via the Arctic.

The situation is totally different in the Southern Hemisphere. The Drake Passage between South America and the Antarctic continent is 300 miles wide and 3000 m deep, thus assuring water movement between the Atlantic and Pacific. The Antarctic Circumpolar Current (ACC) that flows eastward through the Drake Passage extends to the bottom and has a transport variously estimated between 100–200 × 10⁶ m³/sec, which makes it the largest current in the world.

The ACC is driven by the prevailing westerly winds. Its average speed and transport are determined by a balance between the surface stress applied by the winds and the frictional losses to the bottom. Current meters placed 300 m off the bottom in the Drake Passage recorded speeds of 3–9 cm/sec in a generally eastward direction, clear evidence that the ACC extends to the bottom. The apparent influence of bottom topography on the path of the ACC can be demonstrated by the fact that the current tends to move southward as the water deepens and northward as it shallows (Figure 7.15).

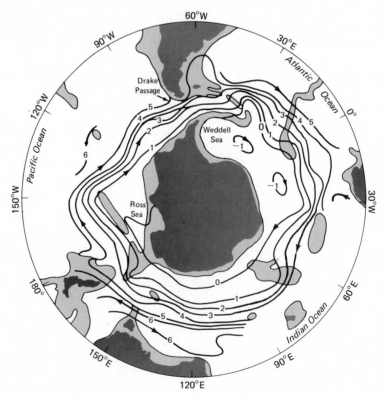

FIGURE 7.15   The Antarctic Circumpolar Current has the largest transport of any current in the ocean; approximately 200 × 10⁶ m³/sec. It extends to the bottom, and the meandering pattern partly reflects an adjustment to the bottom topography. The transport between any two lines is estimated at about 30 × 10⁶m³/sec. The lightly shaded areas are less than 3000 m deep.

SUMMARY

1. The major surface currents are part of the anticyclonic gyres in the Northern and Southern Hemispheres separated at the climatic equator by an eastward-flowing equatorial countercurrent.

2. The currents on the western sides of the gyres, such as the Gulf Stream and Kuroshio, are stronger and more intense than those found on the eastern sides of the ocean basins.

3. In the Indian Ocean there is a seasonal reversal of flow north of the equator with the changing monsoon winds.

4. In the Atlantic and Pacific Oceans a well-defined equatorial undercurrent flows eastward just below the surface.

5. The Antarctic circumpolar current, which circles Antarctica and flows eastward at all depths, is the largest of all currents in terms of transport, about 200×10^6 m³/sec compared to about 100×10^6 m³/sec for the Gulf Stream and $15–50 \times 10^6$ m³/sec for most other currents.

6. All the deep water in the Pacific comes from a single source in the Antarctic; thus the temperature and salinity of the deep Pacific water is much more homogeneous than that in the Atlantic, which has several sources.

7. The best-defined advection flows in the deep ocean are along the western boundaries of the basins.

Waves from a distant storm break on the shore. (Official U.S. Navy photo.)

Waves and Tides

8

Everyone has seen surface waves on water, the slight rippling of the surface of a lake by a breeze, whitecaps produced by a storm, or the breaking of the surf on a beach. The slow oscillation of sea level in a harbor or lake after the passage of a line squall is also a wave, as is the tide. Waves are periodic, and all have certain common characteristics of length, amplitude, and period. The speed or velocity of waves can also be characterized. As shown in Figure 8.1, the *wave-*

CHAPTER

length is the distance between crests or troughs; the *wave height* is the distance from trough to crest and is twice the *amplitude;* and the *period* is the time between passage of two successive crests (or troughs) past the same point.

Wave height and period are the simplest properties of waves to observe along the coast or in a boat at sea, although in photographs wavelength is generally the dominant observable feature. The range of these parameters is wide. The "cat's-paws," the waves caused by a gust of wind on a calm lake, have characteristic heights of a centimeter and periods of a second or less. Breaking surf often has wave heights measured in meters with periods of 4 to 12 seconds. Harbor oscillations have periods measured in minutes, and tides are measured in hours.

Ocean waves come in various shapes and forms. A wind that blows steadily for as long as an hour can establish a choppy *sea,* and if the wind is as strong as 10 knots the sea can become large enough to generate a few breaking waves or whitecaps. Occasionally in the ocean when the winds are strong, waves appear to be running in a number of different directions at one time, occasionally coming together to form a very high wave. At the other extreme, the slow undulating *swell* can be detected moving past a ship at sea on even the calmest days. Waves generated by the local wind are called *sea. Swell* refers to those waves which have moved out from the area of the wind or remain after the wind ceases to blow. The confused sea of a storm or the breaking surf on a beach seem to bear little resemblance to the idealized wave of Figure 8.1, but even the most complex set of ocean waves can be characterized by wave height and period.

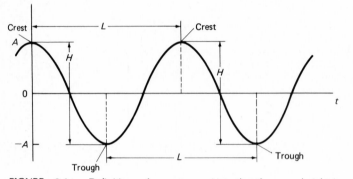

FIGURE 8.1 Definitions of wave terms. Note that the wave height is twice the amplitude. When seamen estimate wave height, they are measuring the distance between the crest and the trough.

Attempts have been made to summarize the average characteristics of the ocean surface. Figure 8.2 is one such attempt. The horizontal scale is wave frequency, $2\pi/T$, where T is the period. The vertical scale is the square of the wave amplitude, since this is a measure of the amount of energy in a wave. As can be seen in the figure, most of the energy is in the range of 4 to 12 seconds, a fact known to every surfboard rider. The only other periods with appreciable energy are the semidiurnal and diurnal tides. There is, however, at least some wave energy in all periods—from a second to more than a day.

Wave motion can be characterized in three ways. The first, and most simple, is the *phase velocity.* If a sinusoidal wave train such as shown in Figure 8.1 moves past a point, the phase velocity is simply the distance traveled divided by the time, or the wavelength divided by the period.

$$C = \frac{L}{T} \qquad (8.1)$$

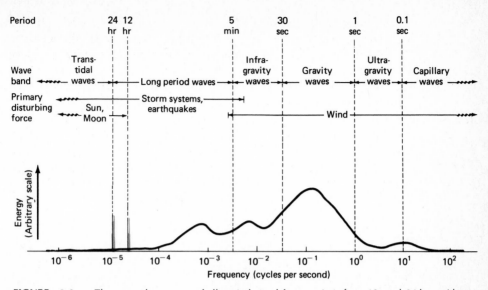

FIGURE 8.2 The ocean has waves of all periods (and frequencies), from 12- and 24-hour tides to the very small capillary waves that ruffle the surface when the wind blows. The energy of a wave is proportional to the square of the wave height. Most of the surface wave energy is in waves with periods from 4 to 12 seconds. (After Kinsman, 1965.)

The second kind of wave motion, which is the particle velocity, is a bit more subtle. You can take a length of rope and with a flick of your wrist start a wave moving down its length, as shown in Figure 8.3. The wave moves the entire length of the rope, but the rope has not moved any horizontal distance. Similarly with ocean waves; waves travel along the surface, the water stays put. A cork floating on the

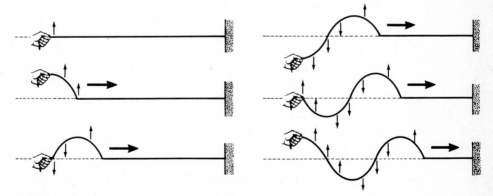

FIGURE 8.3 A wave is sent down a rope with a flick of the wrist. The wave travels horizontally, but the elements of the rope move vertically. Similarly with surface waves at sea. The waves travel across the ocean, but the water particles move in circles or ellipses (see Figure 8.6).

surface bobs up and down as the waves pass, but it does not travel along the surface with the speed of the wave. It is necessary to distinguish between the movement of the wave crest (phase velocity) and the movement of the water itself (particle velocity).

A third type of motion, which will be discussed later, is called *group velocity* and must be considered when the *phase velocity* of a wave is a function of the wave period.

PHASE VELOCITY

A large portion of the observed characteristics of ocean waves can be accounted for by assuming that ocean waves are a combination of simple sinusoidal waves, as shown in Figure 8.1. If it is assumed that the wavelength is much longer than the wave height (a quite reasonable assumption for the ocean), a relationship may be derived between phase velocity, wavelength, and depth of water (h). This general relationship can be further simplified to include nearly all the cases of surface waves of interest in oceanography.

$$C_s = (gh)^{1/2} = 3.1h^{1/2} \text{ m/sec} \qquad (8.2)$$

$$C_d = \frac{gT}{2\pi} = 1.55T \text{ m/sec} \qquad (8.3)$$

where h is measured in meters and T in seconds and where the subscripts s and d refer to *shallow-water waves* and *deep-water waves* respectively. The terms "shallow" and "deep" refer to the relation of the wavelength to the water depth and *not* to the absolute depth of the water. For the shallow-water equation to be applied, the wavelength must be at least twenty times the depth of water. For the deep-water equation to be applied, the wavelength must be less than four times the water depth. For wavelengths of intermediate size, a more general but complex relation must be applied. Fortunately, most of the waves of interest in the ocean can be characterized as either shallow- or deep-water waves (Figure 8.4).

Shallow- and deep-water waves are quite different from each other. In contrast, the phase velocity of shallow-water waves is controlled by the depth of water; the deeper the water, the faster the wave. The phase velocity of deep-water waves is independent of the depth. As long as the water is sufficiently deep, the wave speed and other characteristics of the wave are independent of the water depth. Deep-water waves "do not know the bottom is there." Shallow-water waves "feel bottom."

One way to visualize the relation of deep-water waves to shallow-water waves is to imagine what occurs to a deep-water wave if its length and period are increased. Let the ocean be 4000 m deep. Start with a 10-second wave. According to Equations 8.1 and 8.3 the wave would have a wavelength of about 155 m and a speed of 15.5 m/sec, clearly a deep-water wave. If the period was 20 seconds,

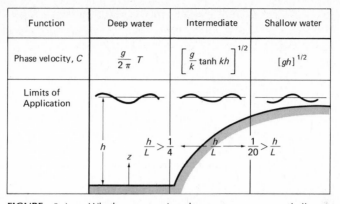

Function	Deep water	Intermediate	Shallow water
Phase velocity, C	$\dfrac{g}{2\pi}T$	$\left[\dfrac{g}{k}\tanh kh\right]^{1/2}$	$[gh]^{1/2}$
Limits of Application			

FIGURE 8.4   Whether a wave is a deep-water wave or a shallow-water wave depends upon the ratio of wavelength to water depth rather than the absolute depth of the water.

the wavelength would be 310 m and the speed 31 m/sec, still a deep-water wave. It would continue to be a deep-water wave until its period was about 100 seconds, at which time the wave speed would be 155 m/sec, and its length 15.5 km (about four times the water depth). By the time the period was 4 minutes, the wave would be a shallow-water wave. It would be traveling at about 198 m/sec and would have a wavelength of nearly 50 km. If the period was 10 minutes, the wavelength would be 120 km, but the speed would remain the same, 198 m/sec. The shallow-water wave speed is the *maximum* phase velocity of a surface wave for a given water depth. The relation between phase velocity, wavelength, and water depth is shown in Figure 8.5.

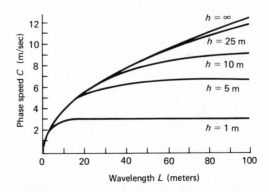

FIGURE 8.5   The speed of deep-water waves increases as the wavelength increases, until the wavelength becomes so long that the wave becomes a shallow-water wave. The speed of the shallow-water wave is determined by the depth of the water.

Deep- and shallow-water waves differ in characteristics other than their phase velocity. Figure 8.6 shows the characteristic particle motion for deep-water waves and for shallow-water waves.

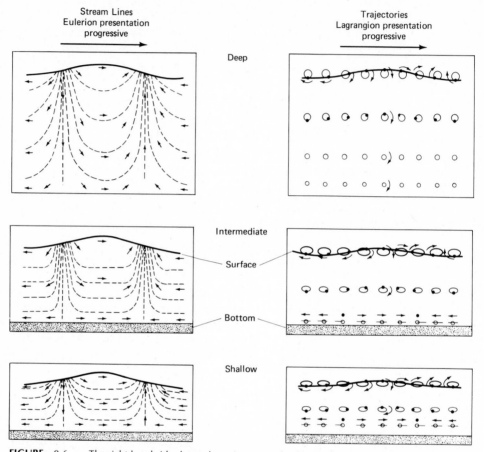

FIGURE 8.6   The right-hand side shows the trajectories of individual water particles. They are circles for deep-water waves and ellipses for intermediate and shallow-water waves. The left-hand side shows the movement of the water as the wave passes. The water is moving upward before the wave crest and downward after the crest passes. Water under the crest is moving backward, and water under the trough moves in the direction of the wave. For deep-water waves the particle movement decreases exponentially with depth. The diameter of the circle at the surface is equal to the wave height and is reduced to 4 percent of the height by the time the water depth is half a wavelength. For shallow-water waves the major axis of the ellipse is unchanged with depth. The minor axis decreases proportionately with depth, until the motion is simply a horizontal back and forth motion at the bottom. (After Kinsman, 1965.)

WAVE DISPERSION AND GROUP VELOCITY

Waves whose phase velocity is dependent on period or wavelength are called dispersive waves. According to Equations 8.2 and 8.3 deep-water waves are dispersive waves and shallow-water waves are not. Consider two wave trains of slightly different periods superimposed upon each other as in Figure 8.7a. The resulting envelope of Figure 8.7b shows regions where the waves are in phase (where the

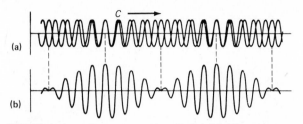

FIGURE 8.7 Two wave trains of slightly different wavelength produce beats. The waves amplitude add when they are in phase and cancel each other out when out of phase. The waves observed at sea are the "beats" in (b) and not the individual components of (a).

resulting wave height is largest) separated by regions where the waves are out of phase (where the wave height is minimum). If both wave trains traveled at the same phase velocity, so would the resulting envelope. The maxima and minima, corresponding to the regions of in phase and out of phase, would also move at the same velocity as the individual phase velocities. If the wave trains move at different phase velocities, however, the regions where the two waves are in phase and out of phase will change. The regions of maximum wave height in the envelope will travel at a velocity different from either of the individual wave trains. The speed of the envelope is called the group velocity (V), and for dispersive waves with the characteristics of deep-water waves it can be shown that

$$V = \frac{1}{2}\, C = \frac{gT}{4\pi} = 0.78T \text{ m/sec} \qquad (8.4)$$

For shallow waves the group velocity is the same as the phase velocity.

Those who have been to sea and have attempted to identify individual waves and watch their progress are continually frustrated. The waves seem to disappear. The reason, of course, is that what is being observed is not individual wave trains of identical periods, but the envelope made up of the addition of a series of wave trains of slightly different periods. Because the envelope is propagated more slowly than the individual waves, what is seen is one wave appearing to disappear slowly, while another behind it seems to build, as in Figure 8.8. On the other hand, by the time a wave gets in shallow water, the group velocity and the phase velocity are identical, since the waves are no longer dispersion waves. A surfer who picks a wave to ride need not worry that the wave will disappear while the one behind builds.

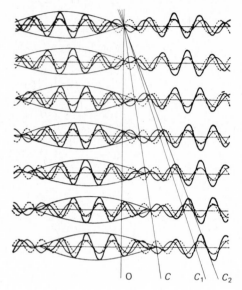

FIGURE 8.8   When two waves of different wavelength travel at different speeds, the beat travels at a speed different from both waves. As shown here, the minimum which can be identified on the first panel and can be tracked for successive times travels at a speed (C) that is half the phase speed of the individual waves C_1 and C_2. (After Defant, 1961.)

WAVE SPECTRUM AND FULLY DEVELOPED SEA

The waves generated by the wind bear little resemblance to the idealized form of Figure 8.7. The surface of the ocean appears as an almost random pattern of hills and valleys (Figure 8.9). The complex surfaces of Figure 8.9, however, can be constructed by assuming they are composed of a series of simple sinusoidal waves (as in Figure 8.7) of varying amplitude by a process called spectral analysis.

The state of sea and the resulting wave spectra vary depending upon the strength of the wind and upon how long and over what distance it has been blowing. As the wind blows, waves are formed. The more the wind blows, the higher the waves become. A wind of 12 knots blowing for a few hours will build waves that begin to break, and whitecaps begin to appear. At 50 knots the wind will blow the tops off waves, and at 100 knots it is difficult to distinguish the "air-sea interface." For any given wind speed there is an equilibrium point, a *fully developed sea,* when the energy imparted to the waves by the wind equals the energy lost by the waves through breaking (Figure 8.10). The fully developed seas do not occur every time the wind blows. The wind must blow for a long time and over a wide area. The region over which the wind blows is called the *fetch* area. It takes longer to build a fully developed sea for a strong wind than for a light wind. A strong wind also requires a longer fetch over which to blow than does a light wind, but, of

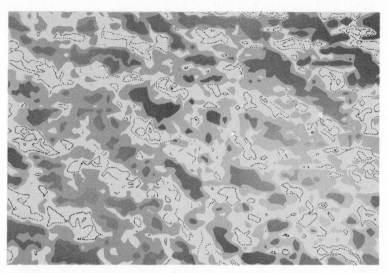

FIGURE 8.9 Contours of wave height taken from stereo pairs. Contour intervals are approximately ½ m; shaded intervals are below mean sea level; light is above. Area shown is approximately 810 × 540 m. (After Coté et al., 1960, from Neumann and Pierson, 1966.)

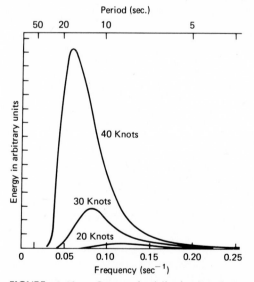

FIGURE 8.10 Energy of a fully developed sea for different wind fields. The energy is proportional to the square of the wave height. Thus a 3-m wave has 2.25 times the energy of a 2-m wave (2.25 = 9/4.) Note that the period of waves with maximum energy increases as the wind speed increases and that most of the energy is concentrated in a relatively narrow range of wave periods. (From Neumann, 1953.)

course, the waves generated are running through the fetch areas. If the *fetch* area is not long enough, the waves might run through and out of the generating area before the wind has had sufficient opportunity to make them fully developed.

Wave forecasting is a highly developed art. Figure 8.11 relates minimum time and fetch necessary to build a fully developed sea for different wind speeds. Of course, much of the time over much of the ocean, conditions for building a fully developed sea do not hold. In these cases it is expected that the total energy is less (which is obvious), that the maximum energy is found at a shorter period, and that the available energy is distributed over a broader frequency range than in the case of the fully developed sea.

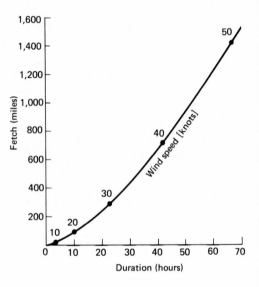

FIGURE 8.11 Minimum fetch and duration for a fully developed sea. For a 20-knot wind, the wind must blow for at least 10 hours over a fetch of 70 miles. A 32-knot wind requires 28 hours and a fetch of at least 400 miles. If one or the other conditions do not hold, the sea will not reach the equilibrium state of a fully developed sea. The highest seas in the world occur in the Antarctic Ocean, where very large, slow-moving storm systems provide the combined fetch and time requirements for fully developed seas at very high wind speeds.

WAVE PROPAGATION, REFRACTION AND BREAKING

Assume there is a simple storm 2000 km from the coast which generates a fully developed sea comparable for a 30-knot wind, as shown in Figure 8.10. All the significant wave components in Figure 8.10 have wavelengths that are short compared to the depth of the open ocean, which means that all the waves are dispersive waves and that the energy envelope is traveling at the group velocity. Thus according to Equation 8.4 a wave gauge near the shore would begin recording 18-second waves 40 hours after they left the generating area. The wave height would be measured in centimeters and it would require a very special wave recorder to see such small amplitudes superimposed on the usual background level of wave activity which might be the order of a meter. However, some 32 hours after these early "forerunners," major swell from the storm with a period of 10 seconds would begin breaking on the beach.

REFRACTION AND BREAKERS

Except for the slight shift in the frequency spectra caused by selective attenuation of the higher frequencies, there is little significant change in the swell until it reaches the shore. Eventually the water depth decreases sufficiently that the waves become first intermediate and then shallow and begin to *refract* (change direction).

Consider shallow-water waves moving onto a beach with both a submarine ridge and a submarine valley. Because the group velocity is now $(gh)^{1/2}$, the waves adjacent to the spit are slowed in relation to those in deeper water, while those over the valley move faster. The wave crests are no longer parallel to the beach. Furthermore, as the waves slow down, the distance between crests decreases. This occurs because the period of the waves does not change, and the wavelength must decrease as the velocity decreases according to Equation 8.1.

It is possible to draw rays which are everywhere perpendicular to the wave crests (as in Figure 8.12). Assume for a moment there are no frictional losses during the brief interval when the waves run toward shore. Under these circumstances the average amount of energy in the area between two crests and two rays (orthogonals) remains constant. As the waves slow down, the distance between crests decreases, and thus the original area decreases. This means the wave height must increase, since the average energy per unit area has now increased.

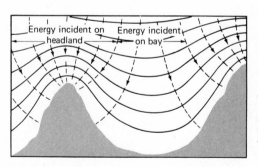

FIGURE 8.12   When waves "feel bottom" and are no longer deep-water waves, their speed is determined by the depth of water. Over a shallow ridge or at a headland, the waves slow down, and the wave rays, which are perpendicular to the wave crests, converge. As a result there is a concentration of wave energy over the ridge, which means the waves are higher. The opposite effect occurs over an underwater valley or in a bay. Here the rays diverge and the average wave height is less. It is easier to launch a small boat through the surf in the vicinity of an underwater valley than near a headland. (After Gordon, 1972.)

When the near-shore ocean bottom is not uniform, the distance between the wave orthogonals also changes because of wave refraction. The orthogonals move closer together over an underwater spit and separate over an underwater valley. As a result there is an increase in the wave height and a decrease over the valley (Figure 8.12).

As the waves move closer and closer to shore, the wavelength decreases and the wave height increases. When the wave height is greater than about seven-tenths of the water depth, the wave becomes unstable and breaks. Knowing the position and strength of the storm as well as the details of the local topography permits the prediction of the arrival and approximate strength of the storm waves. Local topography plays an important role in determining the effect of a storm.

WAVE FORECASTING

Forecasting wave conditions is dependent upon good weather maps. If one knows what the winds are over the ocean one can not only determine the local waves, but can also forecast the arrival time and intensity of high surf conditions on the beach and the possibility of bad wave conditions in the open ocean some distance from the generating area of the storms. Wave forecasting was developed during World War II to facilitate amphibious craft landings on the beaches of West Africa, Normandy, and the Pacific Islands. Since then, techniques have been improved and forecasts are routinely made for such activities as ship routing on transocean voyages, offshore oil drilling operations, and recreational surfing.

TSUNAMI WAVES

If you drop a rock in a quiet pond, you can observe waves radiating in all directions away from the source. Similarly underwater earthquakes generate *tsunami waves,* which radiate from the site of the earthquake. Because of their long period, these waves behave like shallow-water waves in the deep ocean. The fact that tsunami are caused by earthquakes has long been known, as has the additional puzzling, but happy fact that not all underwater earthquakes cause tsunami. In popular literature, a tsunami wave is often referred to as a tidal wave; it was also so referred to in the scientific literature before 1950. "Tsunami" was suggested as an alternative to "tidal wave" to avoid confusion between the two terms. Tides and tsunami have little in common except that they are both shallow-water waves. The effect of tsunami waves can be felt at great distances, and the damage and loss of life caused by waves more than 6 m high have been considerable. In 1896 a tsunami killed an estimated 27,000 people on the coast of Japan. Tsunami warning systems exist, but perhaps only one in ten large underwater earthquakes causes noticeable damage and the chances of destructive damage occurring in a given location are much less than that. Even the timid hate to leave their homes when the false-alarm rate is that high. Thus before sounding the alarm, those in charge of the alarm systems usually require more evidence than mere notice of an earthquake.

Present evidence suggests that tsunami are directly coupled to crustal movement. A shallow focus earthquake (see Chapter 16) that causes major crustal movement of the ocean bottom will cause a tsunami; an equally strong earthquake that does not result in large crustal movement will not. A somewhat more remote possibility is that tsunami are triggered by major turbidity flows resulting from earthquakes.

At its origin the tsunami may appear as a single pulse, the leading edge of which travels with the speed of a shallow-water wave. No one has successfully measured the wave shape in deep water remote from shore, but it is inferred that the initial pulse can be focused such that energy need not be propagated equally in all directions. Thus even knowing that there is probable crustal movement is not sufficient to predict the region of major damage.

With knowledge of the location of the earthquake, which is routinely provided within a few minutes through a worldwide seismic network, and of the depth of the ocean between the source and any possible target, the arrival time is easy to predict. In fact, the average depth of the Pacific was inferred in 1855 by A. D. Bache by noting the time of arrival of tsunami on various tide gauges.

The average depth of the oceans is about 4000 m, which means that as a shallow-water wave a tsunami travels just under 200 m/sec (about 450 mph or 400 knots). It takes less than five hours for a tsunami generated in the Gulf of Alaska to reach Hawaii. As it moves into shallow water, the tsunami slows down. At a depth of 100 m, its speed is 31 m/sec; at 50 m, it is reduced to 22 m/sec.

As the wave slows down, the wave height increases. It can be shown by a simple conservation of energy calculation that the ratio of wave height to water depth for a shallow water tsunami wave is

$$\frac{h_1}{h_2} = \left(\frac{H_2}{H_1}\right)^4 \qquad (8.5)$$

According to Equation 8.5, if the wave height H_2 was 20 m in a water depth h_2 of 20 m, then the tsunami wave would be about 5.3 m high in the open ocean, where h_1 is 4000 m. If a characteristic period of 10 minutes is assumed, the wavelength in deep water is about 120 km, resulting in a slope of the sea surface of 1 in 30,000. It is little wonder that tsunami are not seen by ships at sea.

It is less clear why tsunami cause more damage in some localities than others. Distance from the source, local refraction effects, and focusing of the source pulse are all important, but it is also thought that a wide continental shelf serves both as a wave reflector (sending much of the energy back across the ocean) and as an absorber of tsunami energy through friction along the bottom. Shallow shelves can also trap wave energy. Whatever the reasons, incidents of significant tsunami damage are rare in regions with wide continental shelves.

SEICHES AND OTHER TRAPPED WAVES

It is possible to set up wavelike motion in closed or partially closed basins such as lakes or harbors. In many ways these waves are identical to the sloshing motion generated in a bathtub or a pan of water. In both cases the period of the slosh is determined by the depth of the water and the length of the basin. This type of motion is called a *seiche* and is a form of *standing wave* (Figure 8.13). There are a variety of ways in which a seiche may be excited. One of the most common in a lake is the passage of a storm in which the wind piles up water at one end of the

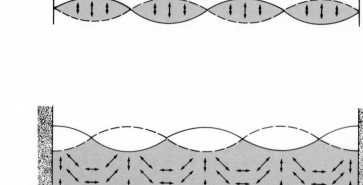

(a)

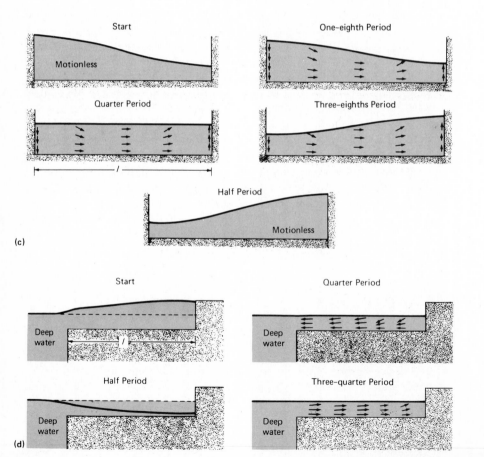

(b)

(c)

(d)

lake. When the wind dies and the stress of the wind is removed, the lake level begins to oscillate. Occasionally small boat harbors have been built which have a seiche period close to that of the prevailing swell or other natural forcing functions, with the result that the harbor is in a continual state of oscillation and is not of much use for anchoring boats.

Seiche-like oscillations have been observed between islands and in other - semienclosed areas, but not all such changes in sea level can be explained by seiches. Waves can be trapped on the continental shelf and can run along the coastline. These are called *edge waves;* their characteristics are determined by the slope and width of the shelf. Hurricanes and large storm systems often trigger storm surges, which are shallow-water waves and sometimes a form of edge wave. If the crest of one of these waves should occur at high tide, the damage can be greater than that caused by the wind waves associated with the passage of the storm.

TIDES

For those of us raised inland and whose swimming and exposure to beaches is confined to lakes, often the most striking experience of our first day at the ocean shore is not the surf pounding on the beach, but the slow rise and fall of sea level and the movement of the shoreline. Rocks and beaches are first covered, then exposed, and one finds an extraordinary community of animals and plants thriving in this intertidal region which one might imagine as one of the most demanding environments where they are alternately covered with seawater and then exposed to the air.

The fact that the tides are somehow related to the moon has been known for centuries to all people who reside by the sea, but the basis for a true explanation was not available until after Sir Isaac Newton's *Principia Mathematica* (1684). After Newton the tidal prediction problem engaged the interests of some of the greatest

FIGURE 8.13   In a standing wave there are nodes (points of no vertical movement) and antinodes (points that are alternately crests and troughs). (a) A vibrating violin string is a well-known example of a standing wave. The points of suspension are nodes and there can be one or more nodes along the string. (b) For a standing wave or seiche in a water-filled channel, the end boundaries are antinodes and the water particles have both horizontal and vertical movement. At the antinodes the motion is purely vertical, and at the nodes it is horizontal. (c) The most common seiche for a closed basin is one where the wavelength is half the basin length. This means when there is a "crest" at one end of the basin there is a trough at the other end and vice versa. (d) For an open basin such as a bay or estuary opening on the ocean, the wavelength is a quarter of the length of the basin. The closed end is alternately a trough or crest, but at the open end there is always a wave node (otherwise there would be a waterfall at the joining of the basin and the ocean as the standing wave oscillated up and down).

For a closed basin such as a lake, the period (T_c) of the seiche is given by

$$T_c = \frac{2l}{\sqrt{gh}}$$

where l is the length of the lake and h is the depth of the water. In the case of a harbor, or basin which opens on a larger body of water the period (T_o) is twice as long for a closed basin of the same dimension

$$T_o = \frac{4l}{\sqrt{gh}}$$

The relationship of period to length in the two cases is somewhat analogous to the calculation of the pitch for an open or closed organ pipe.

mathematicians of the eighteenth century, including Euler, D. Bernoulli, d'Alembert, and Laplace. It is Laplace who built the foundation on which all subsequent work has followed. The greatest practical contribution to tidal prediction, however, occurred around 1870, when Lord Kelvin suggested that tidal prediction for a given location could be made on the basis of sorting the various tidal components at a given location in terms of the known movements of the sun and moon, and who proceeded to build what may have been the world's first analogue computer. The Kelvin tide-predicting computer was used for nearly a century to produce the critically important tidal-prediction tables of the world, and only recently has its wonderful arrangement of pulleys and gears been relegated to museums to be replaced by the modern, high-speed digital computers.

TIDE-PRODUCING FORCES

The tide-producing forces are the gravitational attraction of the sun and the moon. The magnitude of this force is proportional to the mass of the sun (or moon) and inversely proportional to the *cube* of the distance of the sun (moon) from the Earth. Furthermore the force is approximately equal on both sides of the Earth. Based on the mass of the sun and moon and their respective distances from the Earth, the tide-producing force of the moon is about twice that of the sun. The interaction of these forces is complex because of the different positions of the sun and moon with respect to each other and with rotation of the Earth, as shown diagramatically in Figure 8.14. When the moon and sun are nearly in line with the Earth, as they are at full moon and new moon, the tides are the highest (these are the so-called spring tides). Because the planes of rotation of the sun and the moon are not parallel, the

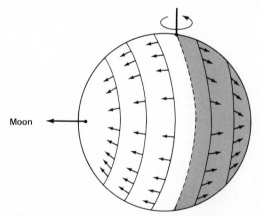

Moon

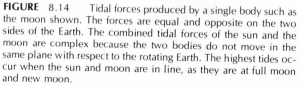

FIGURE 8.14   Tidal forces produced by a single body such as the moon shown. The forces are equal and opposite on the two sides of the Earth. The combined tidal forces of the sun and the moon are complex because the two bodies do not move in the same plane with respect to the rotating Earth. The highest tides occur when the sun and moon are in line, as they are at full moon and new moon.

effect of the force of the moon and sun varies with the seasons as well as with the stage of the moon. The point of the Kelvin tide machine is that it takes these various orbits and breaks them down in a series of periodic forces of different amplitude, which can then interact with one another to produce the exact combination of forces for a given location.

TIDAL WAVES

Although tides are produced by the gravitational attraction of the sun and the moon, the relationship is not one of a simple gravitation equilibrium. If it were thus, the high tides caused by the sun would always be at solar noon and midnight and similarly with the moon. Furthermore, the line of high tide would proceed from east to west, at about 15° of longitude per hour as the Earth rotates on its axis. Knowing the tidal forces, we can calculate the amplitude of this *equilibrium tidal wave*, and its maximum range would be 55 cm for the lunar tide and 25 cm for the solar tide, which would combine to produce a maximum tidal range of 80 cm at full moon and new moon when the two are nearly in line with the Earth. The tidal range varies enormously (Figure 8.15), and sometimes the high tide advances from west to east, opposite to the direction of rotation. The explanation is simply that the tidal forces produce a tidal wave of the proper period, but that the path of the wave and its amplitude is determined by the shape of the ocean basin. Generally these tides can be classified as either progressive tides, as in Figure 8.16, or rotary tides, as in Figure 8.17. The semidiurnal lunar tide in the Atlantic is a combination of a progressive tidal wave in the South Atlantic and a rotary tidal wave in the North Atlantic (Figure 8.18).

The reason for the difference in height of the different tidal waves as shown in Figure 8.15 is partly related to the *resonant period* of the basin. As discussed under seiches, a closed or open basin has a resonant period (or seiche period) related to its length and depth. Whenever the resonant period is close to the 12-hour semidiurnal tidal period, very high tides can occur. Such is the case for the Bay of Fundy, whose 22-m tides are the highest in the world. A useful analogy is to consider how a swing is pushed to make it go. The swing has a natural period determined by the length of rope, and the period is nearly the same whether the arc of the swing is long or short. The person who pushes must push at exactly the same period as the swing. Once the swing is moving, relatively little effort is required to keep it moving. We instinctively push at the resonant period, but a blindfolded swing pusher who did not know the period would have little success in making the swing go high. The tidal force is the swing pusher, but in the case of the tidal forces, the period of the "push" is given. If a basin has a natural period close to the tidal forcing function, very high tides can be generated. Elsewhere the tidal range is less.

It is fairly easy to show that the requirement for a resonant tide along the equator on an ocean-covered earth is for the water depth to be 22 km. For the running of ports and for other commerce, it is probably just as well that resonance, such as occurs in the Bay of Fundy, is the exception and not the rule (Figure 8.19).

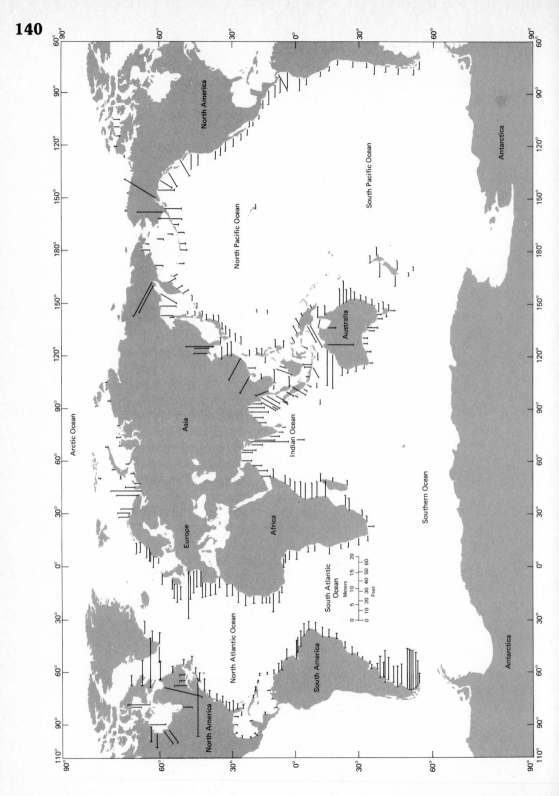

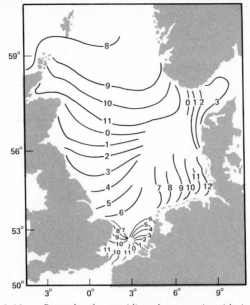

FIGURE 8.16   Example of a semidiurnal progressive tide in the North Sea. The numbers represent the hour of high tide relative to an arbitrary time. High and low tides occur at approximately the same time on opposite sides of the basin, as the tidal wave moves up and down the basin.

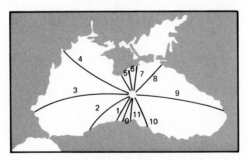

FIGURE 8.17   Example of a semidiurnal rotating tide in the Black Sea. The numbers represent the hour of high tide relative to an arbitrary time. The tidal wave moves counterclockwise around the basin. The first point is a region of no tide and is called an *amphidromic* point.

FIGURE 8.15   Tidal ranges vary considerably over the ocean. Generally the range is less on mid-ocean island stations than on the continents. (After Gordon, 1972.)

FIGURE 8.19 The largest tidal range in the world is in the Bay of Fundy. (From Defant, A., 1953, Ebbe und Flut des Meers der Atmosphare und der Erdfeste. Springer-Verlag, Berlin; with permission of the publisher.)

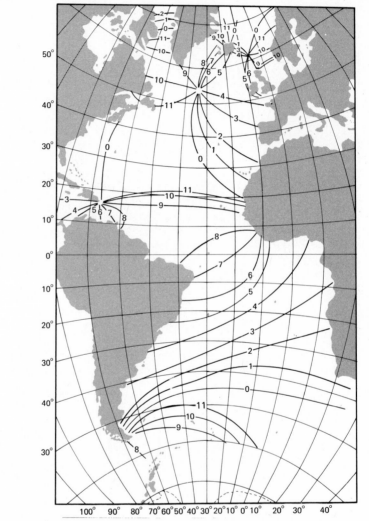

FIGURE 8.18 The semidiurnal tide of the Atlantic combines a progressive tide in the South Atlantic with a rotary tide in the North Atlantic.

SUMMARY

1. The stronger the wind, the higher the waves and the longer the period of the waves. The longer the period, the faster the wave travels.

2. Except when a wave is breaking, the water is not transported with the waves but describes circles as the wave passes.

3. As waves move into shallow water, their speed is controlled by the water depth; the speed decreases as the depth decreases.

4. The speed of very long waves such as tides and tsunamis is independent of the wave period. For these so-called shallow-water waves, the speed is controlled by the water depth.

5. Tsunamis result from movement of the ocean bottom caused by earthquakes.

6. Tides are generated by the changing gravitational field of the sun and moon as their distance and position from the Earth changes.

7. The tidal effect of the moon is about twice that of the sun.

8. The large differences in tidal range at different locations is primarily controlled by the shape of the ocean basins.

(a)   (b)

Sampling the ocean depths for water: (a) the cocked sampling rosette deployed. (b) The lowering of the sampler into the depths. (c) The computer console tracking water properties as sampler descends. Samplers will be closed automatically by an electronic signal from the scientist. (d) The returned samplers being tapped for water. (Turekian photos.)

Chemical Cycles Within the Sea

(c)

(d)

9

In dealing with the physical processes in the ocean, we have proceeded on the simple approximation that seawater composition is sensibly constant everywhere except for small but significant differences in the total concentration of dissolved substance (salinity). It has been assumed that the relative proportions of the elements remain constant from place to place. Indeed, this assumption is true for most of the major components of seawater, thus the composition of Table 4.1

CHAPTER

can be taken as a good picture of the gross chemistry of seawater.

The only major component shown to have a significant variation is calcium. The concentration of this element varies by about 2 percent in relation to the other major elements with low calcium concentrations in surface waters compared to the deep ocean. The reason is that calcium is removed from surface waters by calcium-carbonate-depositing organisms.

Thus we see that there are two major cycles involving the dissolved components of seawater. One cycle is the differential transports of water and total dissolved salt, yielding the variations of salinity observed in the ocean. The other cycle involves the transport of those chemicals in seawater that are associated with the activities of life in the ocean.

Elements can show variations in concentration in the oceans to the extent that they, like calcium, are involved in the biological cycle. The amount of variation depends on the fraction of the element reservoir involved in the on-going cycle. Although potassium and magnesium, for example, are clearly involved in the biological cycle, the oceanic reservoir is so large in relation to the amount cycled biologically per unit time, that whatever variations in concentration are generated, they are so small that they cannot be detected analytically.

CONVENTIONS ABOUT CONCENTRATIONS

In this and later chapters, the figures will show concentrations of chemical species as a function of depth or location in the oceans, or will relate the concentration of one chemical species to another. The choice of concentration units in the representations is determined by the most frequent use to which the data are put. We can always convert from one concentration unit to another, as the occasion requires.

The term "concentration" has been used in the definition of salinity as the grams of dissolved substance in 1000 g (or 1 kg) of seawater. One of the more useful ways of representing concentration, for purposes of chemical reactions, is the *mole per liter* or *mole per kilogram*. A *mole* is the number of grams of a component divided by its atomic or molecular weight. For example, 12 g of carbon is equal to 1 mole of carbon, because the atomic weight of carbon is 12. Similarly, 44 g of carbon dioxide (CO_2) is equal to 1 mole of carbon dioxide, because its molecular weight is 44 (1 carbon with atomic weight of 12 and 2 oxygens each with atomic weight of 16). Note that the amount of carbon is the same in one mole of carbon and one mole of carbon dioxide.

Seawater has a density about 3 percent greater than pure water (with a density of one at 25°C); thus for a given concentration there are 3 percent more moles per liter than moles per kilogram; this small difference will not normally affect our discussions.

BIOLOGICALLY CONTROLLED ELEMENTS

NUTRIENT ELEMENTS: PHOSPHORUS AND NITROGEN

The most dramatic variations in concentration are shown not by the major components of seawater but by the so-called nutrient elements, phosphorus and nitrogen. These elements, together with carbon, hydrogen, and oxygen, are the major building-block elements of life. They occur mainly as phosphate (PO_4^{-3}) and nitrate (NO_3^-) dissolved in seawater, but are converted into organic compounds by marine plants and reorganized in the bodies of marine animals.

The atomic proportion of carbon to nitrogen to phosphorus in most marine plants and animals is about 1 part phosphorus, 15 parts nitrogen, and 80 parts carbon. The concentration of phosphate and nitrate in seawater is much less than that of bicarbonate (and carbonate); therefore the limiting nutrient elements in the ocean for the sustenance of life are nitrogen and phosphorus, and not carbon. When these two are used up, plenty of carbon is still left.

If we plot the nitrogen concentration in ocean water samples against the phosphorus concentration, we find that the best line through the data points goes through the origin (Figure 9.1), and the slope of the line, $\Delta N/\Delta P$, is close to 15. This is consistent with the observed ratio of nitrogen to phosphorus (16) in marine organisms and indicates that the formation or destruction of biological tissue in the sea will add or subtract nitrogen and phosphorus in the ratio of 16 to 1: roughly the same ratio in seawater.

The extraction of phosphorus and nitrogen from seawater occurs in the ocean surface by phytoplankton (Chapter 11). All other organisms utilize this primary production by means of an elaborate food web. The net result is that the nutrient elements are released to the deeper water after extraction from surface waters as the result of progressive metabolism.

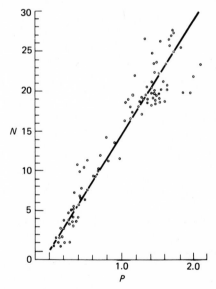

FIGURE 9.1   Correlation between nitrogen as nitrate and phosphorus as phosphate in waters of the western Atlantic. The concentration units are micromoles per liter. (After Redfield, Ketchum, and Richards, 1963.)

Figure 9.2 shows several vertical phosphorus (as phosphate) and nitrogen (as nitrate) profiles in the world oceans. It is evident that the deep Atlantic concentrations are lower by about a factor of 2 than the Pacific and Indian Ocean concentrations. This is a direct consequence of the way in which deep water is formed in the different oceans (Chapter 7). The North Atlantic Deep Water is formed essentially by the sinking of nutrient-depleted water, whereas the Pacific and Indian Deep Waters have their origins around Antarctica, where the water has already

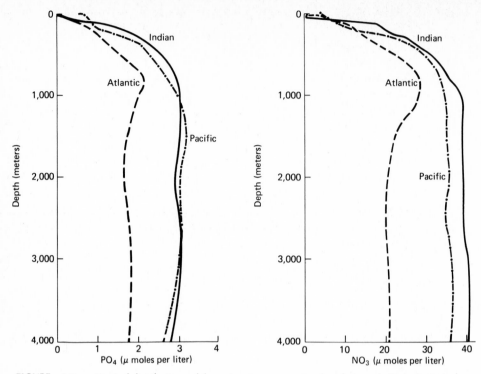

FIGURE 9.2 Vertical distributions of the nutrient components, phosphate and nitrate, in typical water columns in the Atlantic, Pacific, and Indian Oceans. (After Sverdrup, Johnson, and Fleming, 1942).

been imprinted with high concentrations of phosphate and nitrate from the supply of nutrient-rich intermediate depth waters. (Pacific Deep Water is also referred to as Pacific Common Water.)

OXYGEN

The oxygen level in the atmosphere is maintained by its manufacture by photosynthesis on the one hand and its biological utilization for metabolism of food on the other. Atmospheric gases (including oxygen) dissolve in the ocean. The lower the temperature of water, the larger the amount of atmospheric gas that will dissolve. The surface ocean layers are maintained at a high level (commonly above the saturation value) of oxygen not only by dissolving atmospheric oxygen but also by incorporating oxygen produced by marine plants. Thus all deep water of the ocean, which is formed in the high latitudes, starts out with values of dissolved oxygen around 100 percent saturation.

As this surface water moves from its source to supply the deep waters of the oceans, it progressively receives the burden of particulate organic matter resulting from surface productivity. This material is metabolized and the oxygen is used up bit by bit according to the chemical reactions shown in Table 9.1. In the world

TABLE 9.1   Metabolic Reactions Occurring in the Oceans

1. Oxidation by Oxygen
$$(CH_2O)_{106}(NH_3)_{16}H_3PO_4 + 138O_2 \rightarrow 106CO_2 + 122H_2O + 16HNO_3 + H_3PO_4$$
2. Denitrification
$$(CH_2O)_{106}(NH_3)_{16}H_3PO_4 + 94.4HNO_3 \rightarrow 106CO_2 + 55.2N_2 + 177.2H_2O + H_3PO_4$$
3. Sulfate reduction
$$(CH_2O)_{106}(NH_3)_{16}H_3PO_4 + 53H_2SO_4 \rightarrow 106CO_2 + 53H_2S + H_2O + 16NH_3$$

oceans the renewal time of deep water is so rapid that all the oxygen is never used up by this process before it is replaced by new water with high oxygen content. In special environments, such as the Black Sea and the Cariaco Trench, renewal times are so slow in relation to the flux of organic particles falling from the surface water, that all the oxygen in the deep-water column is used up, and the basin becomes *anoxic.* Futher metabolism of the organic matter then occurs by the reduction of sulfate (Table 9.1) in seawater with the production of hydrogen sulfide (the gas with the rotten-egg smell). Certain estuarine basins, such as parts of the Baltic Sea (Figure 9.3), periodically become anoxic if the renewal rates of oxygenated waters from the North Sea are not rapid enough to metabolize the organic matter flux.

In normal marine systems, where dissolved oxygen is used for metabolism, the decrease in oxygen is directly coupled to the increase in the concentration of the

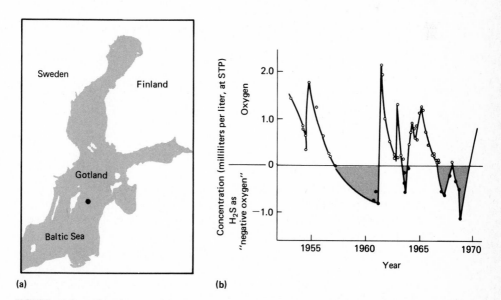

(a)   (b)

FIGURE 9.3   Certain parts of the Baltic Sea, such as the deep waters of the Gotland Deep (a), periodically become anoxic at depth. The variation (since 1953) in the oxygen and hydrogen sulfide concentration in water at 200 m depth is shown in (b). Only with the absence of dissolved oxygen is hydrogen sulfide produced by sulfate-reducing bacteria. Oxygen concentration is expressed in milliliters of oxygen gas at 25°C and 1 atmosphere pressure (that is, standard temperature and pressure, STP) per liter of seawater. Hydrogen sulfide (H_2S) is given as "negative oxygen" in the same units. (After Fonselius, 1969 and later data.)

nutrient elements. For every nitrogen atom transformed from organic nitrogen compounds (for example, amino acids) to nitrate, 17.25 atoms of oxygen are used up (Table 9.1). If the ratio of phosphorus to nitrogen uptake or release is 16, then for every phosphorus atom released from biological compounds, 276 atoms of oxygen are used up. In any one water mass these are the changes that are found. It must be remembered, however, that the oxygen content of water at the surface is primarily determined by the temperature of the water and is independent of the concentration of the nutrient elements, so, in principle, a deep-water mass having its origin at high latitudes will have a phosphate-to-oxygen ratio determined solely by the phosphate concentration of the sinking water. It is the *change* in oxygen relative to the *change* in phosphorus that yields a value of 276 and not the actual ratio in a particular water sample.

SHELL-FORMING COMPONENTS: SILICON DIOXIDE, CALCIUM CARBONATE AND STRONTIUM SULFATE

Marine organisms grow "shells" (or more properly *tests*), made of a number of different materials extractable from seawater. Aside from organophosphatic tests typical of certain organisms such as crustacea, most of the tests are of inorganic compounds (Figure 9.4, facing page). The microscopic plantlike plankton, for example, deposit tests of calcium carbonate (the coccoliths) and silicon oxide (diatoms), whereas the zooplankton inorganic tests include not only the above two compounds (foraminiferan tests and radiolarian tests respectively) but also strontium sulfate.

As these shells descend through the water column, they are subject to dissolution. Strontium sulfate is the most easily dissolved, thus no Acantharia have ever been found in deep-sea sediments. Calcium carbonate and silica (silicon dioxide) tests also dissolve, the site and efficiency of dissolution of each being different and complexly controlled (see Chapter 14). On dissolution of these compounds at

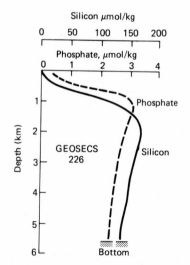

FIGURE 9.5   Profile of phosphorus and silicon in the North Pacific (30°N 150°W) from the GEOSECS expedition. Although both silicon and phosphorus increase with depth and show maxima at this station at depth, they do not completely follow each other.

FIGURE 9.4 (a) diatom (100 μm diameter); (b) radiolarian (150 μm long); (c) acantharia (50 μm diameter); (d) coccosphere (50 μm diameter); (e) foraminiferan (800 μm long). (Photos (a) (b) (c) (d) courtesy of S. Honjo; photo (e) courtesy of A.W.H. Bé, 1968.)

depth, the concentrations of their constituent ions increases. Thus the vertical distributions of silicon (Figure 9.5) and (to a much more subtle degree) calcium and strontium resemble the nutrient element profiles like those in Figure 9.2. Also like phosphorus and nitrogen, the concentrations of calcium, silicon, and strontium are higher in deep Pacific water in relation to Atlantic deep water.

TRACE METALS

Highly accurate data are now available on the distribution of several trace metals in oceanic profiles, including barium, copper, nickel, and cadmium (Figure 9.6). They all show a strong relationship to the nutrient element distributions, thus suggesting that they also take part in the biological cycle. The data as yet are too few to identify the biological carriers of each of these elements and the depth at which they each are dissolved.

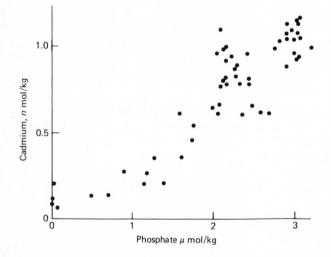

FIGURE 9.6 Cadmium distribution in the oceans seems to follow primarily the phosphorus distribution. The data plotted in this diagram are from a variety of depths in the oceans (n mole = nanomole = 10^{-9} mole). (Data mainly from Boyle, Sclater, and Edmund, 1976.)

TRACKING OCEAN CURRENTS WITH CHEMICAL TRACERS

We have seen that there are two major classes of chemicals in the oceans. One class comprises major elements that show little or no variation in relation to each other and whose total concentration (salinity) varies slightly. The second class comprises the nutrient elements, oxygen, the test-forming elements, and the biologically controlled trace elements. These elements show distinctive depth and geographic patterns, controlled both by biological activity and by ocean circulation patterns. As we shall see, we can learn something about the patterns and rates of circulation of the oceans by tracking these biologically sensitive tracers. Even more can be learned, if we consider, in addition, another class of chemical tracers, the natural radioactive isotopes such as carbon-14. We can then set limits on the rates of circulation by considering both the biologically controlled clock and the radioactively controlled clock.

NUTRIENT ELEMENTS AND OXYGEN AS OCEANIC TRACERS

The distributions of nitrogen, phosphorus and oxygen in the western Atlantic Ocean are shown in Figure 9.7. We can take two extreme positions regarding these observed distributions of the nutrient elements (and test-forming elements) in waters below 1000 m. (1) We can assume that any variation in concentration along a traverse is due to an *in situ* alteration of the abundance of the component. For example, the increase in dissolved phosphate along a traverse or with depth can be due to the oxidation of organic material supplied from the overlying surface waters. This would be matched by an appropriate decrease in the dissolved oxygen concentration (Figure 9.7). (2) We can assume that variations are the result of the mixing gradient between two water types having distinctively different nutrient element and oxygen concentrations.

If our first assumption is correct, then in principle we can measure how long deep water remains in a particular ocean basin, since the longer it remains, for a constant organic particulate flux from the surface, the higher the nutrient element concentration and the lower the oxygen concentration. We need to know the fraction of surface-produced organic material that is transported to the deep water. Unfortunately this value is not known with any certainty.

On the other hand, major changes in the concentrations of the biologically related elements may occur in highly circumscribed regions. The major biological productivity of the oceans, for example, occurs in regions of upwelling in the oceans (see Chapter 11). These are primarily along the eastern margins of the ocean basins and at the high latitudes (Figure 11.12). In these areas the total organic particulate transport to depth is very large. This results in a high rate of oxygen consumption at depth, so much so that the underlying sediments along these continental margins, and in places, even the water become anoxic. The waters thus are generally lower in oxygen and higher in nutrient elements than the surrounding waters. This nutrient-rich, oxygen-poor water is then transported by advection and diffusion to mix with water of a different, less biologically affected, history. Under these conditions there is no opportunity for telling the age of the water with this technique, because the chemical changes are ascribable to mixing processes alone.

NATURAL RADIOCARBON AS A TRACER

Even before nuclear bomb testing, the oceans were supplied with radioactive carbon-14 (or radiocarbon) by cosmic ray interactions with our atmosphere. The discovery and use of this isotope of carbon for dating archaeological and geological

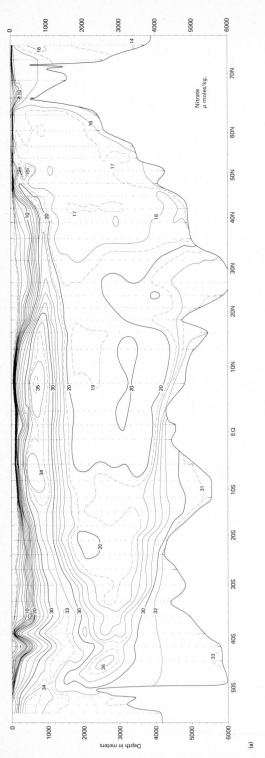

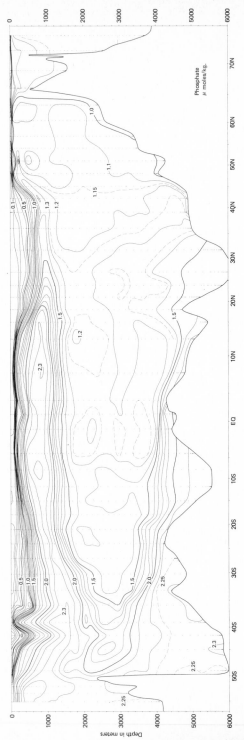

(a)

(b)

FIGURE 9.7 Nitrate (a), phosphate (b), and oxygen (c) distribution in the western Atlantic Ocean. These profiles were made as part of the GEOSECS program studying the distribution of oceanic properties for the purpose of understanding the patterns and rates of ocean circulation. Note that the section is made along the traverse shown at the left; therefore the spacings between latitudes is not uniform. (Courtesy of GEOSECS, to be published as part of the GEOSECS Atlas.)

deposits is one of the great scientific success stories of our times—earning its discoverer, Willard F. Libby, a Nobel Prize. Carbon-14 has a half-life of about 5700 years, which means that without addition of new carbon-14, the amount will decrease by radioactive decay by a factor of 2 every 5700 years. This provides us with a clock if we know how to use it. Its use in dating marine deposits is shown in Chapter 15; it can also be used in dating seawater movement.

Radiocarbon has been used to calculate the water renewal rate of the Pacific Ocean (Figure 9.8). In the eastern North Pacific there appears to be a single continuous water mass at depth. There are no obvious layers near the bottom and no intrusions of alien waters at intermediate depths. The vertical movement of tracers in this deep water is controlled by two processes, discussed in Chapter 6. One process is mathematically simulated by the equations describing diffusion; but as the transport is effected by small eddies in the bulk fluid, rather than by true molecular diffusion, it is sometimes called "eddy" diffusion. The other process is by bulk flow, or advection. As we are dealing here with upwelling of water supplied to the

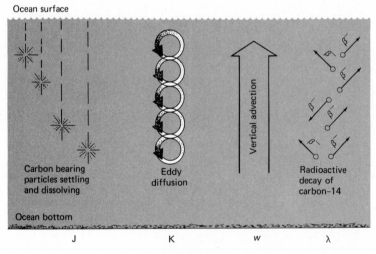

FIGURE 9.8   Model for determining the different rates affecting the distribution of radiocarbon (carbon-14) in a column of water in the eastern North Pacific as developed by Harmon Craig of Scripps Institution of Oceanography. It is based on the steady-state balance of input and output terms which yields the observed radiocarbon distribution as well as the parameters, salinity, temperature, and total dissolved carbon. At each depth for radiocarbon:

Flux downward due to eddy diffusion
+ flux downward due to particulate transport from above and dissolution
= flux upward by vertical advection
+ flux out of system by radioactive decay.

The ratios K/w and J/w are determined by curve fitting of the salinity profile and total dissolved carbon profile sequentially. With these ratios established, λ/w is determined by the radiocarbon profile; since λ is known (it is the characteristic constant of radioactive decay for carbon-14), w is determined and then also K and J.

ocean bottom from the Antarctic region, we will be concerned with *vertical advection*. As the tracer is moved vertically in the water column, some more of it is added from the surface by particle transport and subsequent dissolution. If the tracer is a radioactive isotope, it is also being lost by radioactive decay.

Radiocarbon is subject to all these processes, whereas ordinary carbon is subject to all except radioactive decay. The combination of measurements on the distribution of total carbon, carbon-14, salinity, and temperature allows an estimate of the dynamics of ocean circulation.

From either the salinity or temperature distribution we can get the ratio of the eddy diffusivity to vertical advection velocity. This ratio will apply to all dissolved species in seawater, including ordinary carbon and radiocarbon. We can fit the ordinary carbon data to the best-fit curve that will give us the ratio of carbon compound dissolution to vertical advection velocity.

Note that up to now we have only ratios; that is, we know the following:

$$\frac{\text{Eddy diffusivity}}{\text{Vertical advection velocity}} = \frac{K}{w} \tag{9.1}$$

$$\frac{\text{Carbon supply rate}}{\text{Vertical advection velocity}} = \frac{J}{w} \tag{9.2}$$

The value of w can be obtained only when we consider the behavior of radiocarbon. It follows ordinary carbon in the cycle, and the fraction releasable at depth by particle transport and dissolution is proportional to J through the radiocarbon-to-ordinary-carbon ratio in the particles. This can be determined independently. However, the fact that radiocarbon decays radioactively with a half-life of 5700 years (see Chapter 15 for a detailed discussion) permits one more piece of information.

$$\frac{\text{Radiocarbon decay constant}}{\text{Vertical advection velocity}} = \frac{\lambda}{w} \tag{9.3}$$

where λ is known ($0.693/5700$ yr^{-1} or 1.215×10^{-4} yr^{-1}). Knowing K/w from salinity or temperature distribution, and J/w for carbon (and therefore for radiocarbon as well) by curve fitting to the total carbon (formally noted as carbon dioxide) concentration variation with depth, we can obtain the value of λ/w by curve fitting of the total radiocarbon concentration profile. Since λ is known, w is determined.

Analysis of the data yields the result that water supplied to the deep ocean moves upward at a rate of about 3 to 4 m per year. Thus the traverse from the ocean bottom to the surface would take over 1000 years in the Pacific Ocean. Moreover, once we know the value of w (the vertical advection velocity), then the values of eddy diffusivity (K) and rate of supply of carbon (J) can be determined as well, since the ratios of these parameters to w are known.

The renewal time of deep water in the Pacific is several times longer than in the Atlantic (Chapter 7). If the vertical advection velocity in the Atlantic is the same as in the Pacific, the implication is that the horizontal transport of deep water out of the Atlantic to the Antarctic (or Southern) Ocean is faster than return to shallower depths by vertical transport.

SUMMARY

1. The relative proportion of the major ions making up the salt content of seawater is constant to a good precision.

2. The concentrations of the "nutrient elements," phosphorus, nitrogen, carbon, increase with depth in the oceans and vary among ocean basins.

3. Oxygen is continually supplied to the deep ocean by the sinking of cold high-latitude water, in which it is dissolved.

4. The biologically associated trace and minor elements vary in concentration sympathetically with the nutrient elements.

5. Ocean currents are tracked by the nutrient elements as well as by temperature and salinity variations.

6. By the use of carbon-14 and other chemical properties of the water column, the rate of upward motion of water in the oceans can be estimated to be about 4 m per year.

A diver (left) and a school of jacks (right) swim over the coral reef along the north coast of Jamaica. (Courtesy of W. Sacco.)

Life in the Sea

10

As discussed in Chapter 9, marine organisms significantly influence both the chemistry of ocean water and the composition of sediment on the sea floor. In these and other ways, marine life has an important impact on its environment. To understand the ocean, therefore, we must examine some of the biological processes which maintain life in the sea. In this chapter we first look at the diversity of life in the sea and then examine a set of processes which are life's unify-

CHAPTER

ing feature: the synthesis of organic compounds. These processes require that marine organisms, like all organisms, capture energy from, and exchange materials with, their environment. Such considerations lead to the conclusion that all marine organisms form part of an integrated global system, the marine ecosystem, which ultimately links every animal and every plant to each other and to the physical and chemical condition of the sea. The transfer of energy and materials within the marine ecosystem is thus the main theme of marine ecology. Each species plays a particular role in the ecosystem according to the energy source it taps, where it lives, and how it moves about or remains fixed. In Chapter 11 we will consider in more detail the structure and dynamics of the marine ecosystem, and pay particular attention to the factors influencing the rate at which the system as a whole produces living tissue. In Chapter 12 we will assess the potential of the ocean as a food source for man.

DIVERSITY OF MARINE LIFE

Like life on land, animals and plants in the sea exhibit an extraordinary diversity in size, form, and function. The range of size alone is very great: from one-celled plants, which can be seen only with powerful microscopes, to seaweeds 30 m long, and from microscopic bacteria to the great blue whale, the largest animal ever to live on earth. Far more impressive than this range in size, however, is the range of form and activity. Lifetimes spent in examining museum collections would be insufficient to know a fraction of the total number of marine species. One group of marine animals alone, the planktonic copepods, has several thousand species. Faced with such an exuberance of form, one natural question is, "How can marine biologists organize information about such a vast and complex array of subjects?" One answer is provided by the science of *taxonomy*, which aims to place every organism into a hierarchical set of named categories. Individual organisms classified as a *species* are judged to play a discrete or at least a definable role in the living world. For the majority of organisms that exchange genes through sexual processes, species so defined are homogenous genetic units—that is, they represent a group of organisms potentially capable of interbreeding and producing fertile offspring. The next step in the taxonomic hierarchy is formed by grouping as a *genus* those species judged to have a common ancestor. Genera are grouped into *families*, and so on, until all marine organisms are grouped into some 30 animal and plant *phyla*. One aim of such a classification is to record our current understanding of evolutionary history. Such classifications also have great practical value, because all members of a category (*taxon*) will share many basic anatomical and biochemical features. Once one has learned the basic structure and physiology of a clam (Phylum Mollusca, Class Pelecypoda), for example, this information serves as an easy introduction for examining any species of clam. Table 10.1 presents a taxonomic classification of the most important types of marine organisms.

The number of species is much greater on land than in the sea, in large part because terrestrial insects, birds, and mammals are so diverse. In all, the number of terrestrial species is at least an order of magnitude greater than those of the sea—one or two million species to the sea's hundreds of thousands. This is because of the greater variety of habitats and ecological situations on land. In the sea, however, there is a much greater diversity of major *structural types* of plants and animals. Two major marine plant groups, the brown and red algae, do not occur in fresh waters at all; and several animal phyla, notably the echinoderms, chaetognaths, and lower chordates, are completely restricted to the sea. The ocean supports a vast array of organisms such as clams and oysters that live by filtering microscopic organic particles from the water, and many others that feed by sweeping up detritus from the bottom. For each kind of earthworm that subsists on the rich material of moist soil, there are hundreds of marine worms that do the same thing.

TABLE   10.1   Classification of Principal Marine Organisms*

Phylum	Class	English Equivalent	Main Habitat as Adults — Plankton	Nekton	Benthos
Schizophyta		Bacteria	X		X
Mycophyta		Fungi	X		X
Cyanophyta		Blue-green algae	X		X
Chlorophyta		Green algae			X
Phaeophyta		Brown algae			X
Rhodophyta		Red algae			X
Chrysophyta		Diatoms	X		X
		Coccolithophorids	X		
Mastigophora		Flagellates	X		
Tracheophyta		Marine grass			X
Rhizopoda	Foraminifera	Foraminiferans	X		X
	Radiolaria	Radiolarians	X		
Cnidaria	Hydrozoa	Hydras	X		X
	Scyphozoa	Jellyfish	X		
	Anthozoa	Corals, etc.			X
Ctenophora		Comb jellies	X		
Platyhelminthes		Flatworms	X		X
Chaetognatha		Arrow worms	X		
Annelida		Segmented worms	X		X
Arthropoda		Arthropods	X	X	X
Branchiopoda		Brine shrimp	X		
Ostracoda		Ostracods	X		X
Copepoda		Copepods	X		X
Cirripedia		Barnacles			X
Euphausiacea	Crustacea	Krill, etc.	X		
Decapoda		Shrimp, crabs, lobsters, etc.	X	X	X
Amphipoda		Amphipods	X		X
	Arachnida	Horseshoe crabs, sea spiders			X
Cliophora		Ciliates	X		
Porifera		Sponges			X
Bryozoa		Moss animals			X
Brachiopoda		Brachiopods			X
Mollusca	Monoplacophora				X
	Amphineura	Chitons			X
	Scaphopoda	Tusk shells			X
	Pelecypoda	Clams, oysters		X	X
	Gastropoda	Snails	X		X
	Cephalopoda	Squids	X	X	X
Echinodermata	Crinoidea	Crinoids	X		X
	Asteroidea	Starfish			X
	Ophiuroidea	Brittle stars			X
	Echinoidea	Sea urchins			X
	Holothuroidea	Sea cucumbers			X
Protochordata		Tunicates, acorn worms	X		X
Chordata	Agnatha	Jawless fish		X	
	Elasmobranchii	Cartilagenous fish		X	X
	Osteichthyes	Bony fish		X	X
	Reptilia	Reptiles		X	
	Aves	Birds		X	
	Mammalia	Mammals		X	

* Modified from P. K. Weyl, 1970.

Although a taxonomic classification such as that in Table 10.1 is basic to the science of marine biology, its use exclusively for this purpose tends to obscure the study of ecosystem processes—because each major taxonomic category includes species that play quite different ecological roles. Further, different taxonomic categories include animals playing essentially the same role. Let us illustrate the problem by considering the classification of three taxonomic groups: fish, squid, and whales. Each group constitutes a natural evolutionary unit, for each originated from a single ancestral stock. Fish, for example, originated some 400 million years ago, but during their long history they have evolved considerably and now play many different roles in the ecosystem. Some, like the herring, feed on plankton from the surface waters of the ocean; others, like the tuna, are predators, feeding on other fish; still others, like the cod, live near the bottom of the ocean and feed mainly on clams and other bottom-living invertebrates. Thus, from the ecological viewpoint, each of these three types of fish is different and should be classified differently, whereas squid and certain species of whales are predators, and for many ecological purposes should be classified with the tuna.

Hence the biological units that we use in this book are natural ecological units: groups of species that share similar basic functions within the ecosystem. For example, animals are grouped according to what they eat and how they move— that is, how each species goes about its primary business of obtaining food.

BIOLOGICAL ACTIVITIES

BIOCHEMICAL SYNTHESIS

In spite of life's diversity, the processes of growth and reproduction are characteristic activities of all living things. Here we use the term "growth" to describe not only the process by which each individual becomes larger, but also the processes by which virtually all parts of an organism are continuously replaced, modified, and added to as part of an organic maintenance system. To carry out these activities, chemical compounds must be *synthesized*; that is, they must be formed from smaller and simpler components. Thus, biochemical synthesis can be regarded as the most pervasive feature of life.

Biochemical synthesis occurs by many different processes. Because chemical work must be performed, each process requires a source of energy. Two processes (photosynthesis and chemosynthesis) can be regarded as primary, in the sense that they are the only means by which simple, inorganic substances are incorporated into living organisms. Of these, *photosynthesis*, which utilizes the energy of sunlight and is responsible for the initial production of almost all organic compounds found in living things, is by far the most important. The process can be represented in the following much-simplified equation.

$$CO_2 + H_2O + \begin{matrix} \text{kinetic} \\ \text{energy of} \\ \text{sunlight} \end{matrix} = \text{sugar} + \text{oxygen}$$

carbon dioxide + water + sunlight = sugar + oxygen

Four aspects of photosynthesis are essential concepts in arriving at a fundamental understanding of life in the sea. (1) Only certain types of marine organisms are capable of carrying out photosynthesis. These include plants (algae, grasses, diatoms, coccolithophorids), and certain flagellates (Table 10.1). (2) Raw materials for the reaction are simple inorganic compounds (water and carbon dioxide). (3) Oxygen is produced. (4) Energy is stored in chemical form in the sugar molecule. As discussed below, both plants and animals use the potential energy stored in sugar to carry out most of life's essential activities.

In addition to photosynthesis and chemosynthesis (the other type of primary synthesis, to be discussed later) all living organisms carry out many different kinds of secondary synthesis; that is, they form larger and more complex organic molecules from simple units such as sugar molecules. One common synthetic process is that of *dehydration synthesis*. Here, two simple sugar molecules (*glucose*, the principal ingredient in corn syrup, and *fructose*, or fruit sugar) are formed into a molecule of *sucrose* (table sugar). In the process, a molecule of water is released. Using empirical chemical formulas, the process may be summed up this way.

$$C_6H_{12}O_6 + C_6H_{12}O_6 = C_{12}H_{22}O_{11} + H_2O$$
glucose + fructose = sucrose + water

Dehydration synthesis is one important way in which the materials of life are compounded. Sucrose is a larger and more complex molecule than either glucose or fructose, and therefore has more chemical energy stored in it. But it is by no means the most complex organic molecule. A common *fat* such as stearin, for example, has an empirical formula $C_{57}H_{110}O_6$ (a total of 173 atoms to make up each molecule). And even the simplest *proteins* are much larger still. Human hemoglobin, for example, has atoms of carbon, hydrogen, oxygen, nitrogen, sulphur, and iron combined in the proportions $C_{3032}H_{4816}O_{872}N_{780}S_8Fe_4$.

ACQUISITION OF ENERGY AND MATERIALS

All the processes of secondary biochemical synthesis must utilize some source of energy. This source is the potential energy stored in chemical form in molecules of various types: sugar, and other carbohydrates such as starch and cellulose; fats; and protein. This potential energy is released both by plants and by animals during respiration, an oxidation process typified by the following reaction.

$$C_6H_{12}O_6 + 6O_2 = 6CO_2 + 6H_2O + \text{kinetic energy}$$
sugar + oxygen = carbon dioxide + water + energy

Thus the solar energy initially captured by a green plant and stored as sugar can later be used by that plant or by some animal which has acquired that sugar molecule as food. For each molecule of glucose oxidized, 10×10^{-21} calories are released. All life on earth, including life in the sea, therefore depends upon a flow of energy which originates in the sun, enters the biosphere through the photosynthetic activities of green plants, and is transferred from one organism to another in chemical form as food.

Before discussing the consequences of this *energy* flow further, we must consider the origin and the fate of the *materials* making up life at sea. As the chemical reactions considered above indicate, the main building blocks of life are atoms of carbon, hydrogen, and oxygen. This is true of the compounds that make up the bulk of living tissue, including the carbohydrates (sugar, starch, cellulose), fats, proteins, and the nucleic acids DNA and RNA. But small amounts of other elements are also needed. Chlorophyll, the catalyst of photosynthesis, contains magnesium. Adenosine triphosphate (ATP), used in all energy transfers, contains phosphorus. Proteins contain nitrogen. In addition, iron, copper, cobalt, and other elements are required in trace amounts. Further, as discussed in Chapter 9, the nonliving, skeletal parts of marine organisms are compounds of silicon, calcium, strontium, and phosphorus. Clearly, the maintenance of life in the sea requires a continuous intake of material. Plants obtain directly from the sea the materials they need, whereas animals ingest some as food and obtain the rest directly from the sea.

These considerations lead us to the important conclusion that every animal and plant in the sea must acquire both energy and certain materials from its environment. Thus the capture and transfer of materials and energy can be regarded as the essential feature of marine ecology. But the flow of materials is not simply a one-way process—in which organisms acquire substances they need from the sea —but an *exchange* of materials between seawater and the biosphere. Therefore we can only really understand the chemistry and the biology of the ocean in terms of a *marine ecosystem*, that is, a system characterized by interactions between living and nonliving components.

PATTERNS OF ENERGY CAPTURE

In the following chapter we examine the marine ecosystem. But to understand how the marine ecosystem works, we must first analyze how various organisms obtain the energy they need—that is, what fundamental competitive strategy they each adopt in solving life's basic problem. We will examine three aspects of this competitive strategy: (1) what energy source is used; (2) what kind of activity the organism employs to obtain the energy it needs; and (3) where the organism lives.

ENERGY SOURCES

In Table 10.2, marine organisms are classified according to the energy source they employ. Two major types can be defined: *autotrophic organisms* (autotrophs) and

TABLE 10.2 Organisms Classified by Energy Source

Organism Type	Energy Source
AUTOTROPHIC	Chemical energy derived from inorganic surroundings
Photosynthetic	Sunlight
Chemotrophic	Chemical energy stored in simple chemical substrates (sulfides, sulfates, nitrates, and others)
HETEROTROPHIC	Chemical energy stored in complex organic compounds ("food")

heterotrophic organisms (heterotrophs). Autotrophs ("self-feeders") do not depend on other organisms for food. Instead they take in inorganic compounds such as CO_2, H_2O and H_2S. To synthesize carbohydrates, proteins, and other compounds a variety of energy sources are used. The vast majority of autotrophs utilize sunlight and carry out the process of photosynthesis as already discussed. The symbolic reaction rewritten here in more detail is

$$CO_2 + 2H_2O + energy = CH_2O + H_2O + O_2$$
carbon dioxide + water + sunlight = sugar + water + oxygen

But other autotrophic systems are known, such as one that synthesizes hydrogen sulfide (H_2S) and carbon dioxide to yield the same sugar.

$$CO_2 + 2H_2S + energy = CH_2O + H_2O + 2S$$
carbon dioxide + hydrogen sulfide + chemical energy =
sugar + water + sulfur

In typical bacterial systems which accomplish this, the energy source is not sunlight but chemical energy derived from the oxidation of some compound, for example, hydrogen sulfide.

$$2H_2S + O_2 = 2H_2O + 2S + energy$$
hydrogen sulfide + oxygen = water + sulfur + heat

Such autotrophs are therefore called *chemotrophic*. Other substrates which can be substituted for hydrogen sulfide in the reaction include nitrogen (N_2) and sulfate (SO_4).

Heterotrophs ("other-feeders") depend upon other organisms for food; to stay alive they must take in either the living or once-living tissues of other organisms. The organic matter in their food provides all the chemical energy required to carry out biochemical synthesis, and many of the essential materials as well.

MODES OF LIFE

Having classified marine organisms by their energy sources, we will now find it helpful to consider their mode of life, that is, where they live (their *habitats*) and how they move about (their *habits*). Broadly speaking, marine organisms occupy two major habitats: the sea bottom (benthic realm) and the water above (pelagic realm) (Table 10.3). Benthic organisms (collectively called *benthos*) may live either on the bottom (*epibionts*) or in the bottom (*endobionts*). Within these realms organisms display a great variety of activity patterns (habits) in their search for food and energy. Something of the range of habits is shown in Table 10.4, where mobile organisms are distinguished from immobile, or *sessile*, organisms, and these categories are further subdivided to reflect a range of activity types.

TABLE 10.3   Marine Habitats

Habitat	Organism Type	Description
Benthic realm	Benthos	Living on or in the seabed
	epibiont	On the seabed
	endobiont	In the seabed
Pelagic realm	Pelagic organisms	Living in the water above the seabed

TABLE 10.4   Habits of Marine Organisms

Habit	Organism	Description
Mobile		
1. Drifting	Plankton	Passive drifting
2. Swimming	Nekton	Active movement against currents
3. Crawling	Crawling animal	Propulsion over bottom
4. Burrowing	Burrower	Propulsion within sediment
5. Nectobenthic	Nectobenthos	Combinations of 2-4
Sessile		
6. Boring	Boring animal	Excavation in hard substrate
7. Cemented	Cemented animal	Chemical adhesion
8. Rooted	Rooted animal	Rootlike attachment
9. Reclining	Reclining animal	Living free on bottom

Species with the habits named above occur typically in the taxonomic categories listed below. See figures, pp. 170–5.

1. Flagellates (Fig. 10.1), diatoms (Fig. 10.2), foraminiferans (Fig. 10.3), radiolaria, ctenophores, pteropods, euphausids, amphipods, ciliates, and chaetognaths
2. Squid (Fig. 10.4), porpoises (Fig. 10.4), whales, sharks, and fish (Fig. 10.4)

Pelagic organisms in general may be classified into two types: the *nekton*, which actively swim and are capable of making headway against normal currents; and the *plankton*, which cannot swim against a normal current. The specific gravity of planktonic organisms is very close to that of seawater, so that generally they can be considered to float in the water column and drift with the ambient current. Plant plankton *(phytoplankton)* and many smaller individuals of the animal plankton *(zooplankton)* exhibit this purely passive, floating habit. Many larger planktonic animals, however, can change their vertical position, and some move up and down the water column each day.

Benthic organisms display a diversity of habits. Some sessile epibionts *cement* themselves to a hard substrate, such as a rock or the shell of another organism. These include many species of barnacles, corals, and other animals, as well as many types of algae. Other epibionts attach themselves by rootlike structures to the substrate. These include most of the large seaweeds, mussels, and many other organisms. *Reclining* epibionts simply lie unattached on the bottom. This habit, which holds many obvious risks, is adopted by the adult phases of many oyster species, for example. Many familiar marine invertebrate animals *crawl* or otherwise propel themselves along the surface of the seabed, or *burrow* beneath the surface.

Living in different habitats, eating different diets, and adopting different habits, marine organisms display a wide-range of life styles which may be realized by any individual organism at a particular moment in time. Individuals of some species occupy one habitat and assume one habit over their entire life span. Most species of phytoplankton do this. Many animal species, however, change their mode of life systematically during their life cycle, passing from a temporary larval stage as plankton and developing into adult forms which adopt nektonic or a wide variety of benthic habits. Others have immotile stages or have no larval stage at all. Furthermore, the adult members of many species may assume different habits at different times. Lobsters, for example, are *nektobenthic* animals, crawling along the surface of the sea bottom one moment and swimming short distances the next; and many crabs, on brief crawling or swimming expeditions, leave their protective burrows to forage for food. Although most fish are purely nektonic as adults, many are nektobenthic. Bottom-feeding fish such as the cod and the flounder, for example, spend most of their time swimming near or reclining on the seabed. These are known as *demersal* fish.

3. Acorn worms (Fig. 10.6), echinoids (Fig. 10.5), snails, crustacea
4. Clams (Fig. 10.7), crustacea (Fig. 10.7), brittle stars (Fig. 10.8), echinoids, annelids
5. Octopods, crustacea (lobsters, crabs, and so on), scallops
6. Sponges, clams, blue-green algae
7. Barnacles, corals, red algae
8. Mussels, sponges (Fig. 10.9), coelenterates, grasses, kelps, brachiopods
9. Oysters, corals

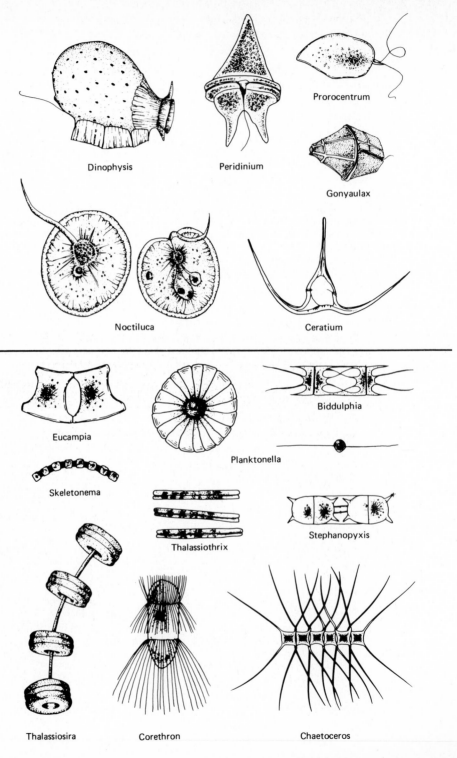

Dinophysis

Peridinium

Prorocentrum

Gonyaulax

Noctiluca

Ceratium

Eucampia

Planktonella

Biddulphia

Skeletonema

Thalassiothrix

Stephanopyxis

Thalassiosira

Corethron

Chaetoceros

FIGURE 10.1 (Facing page, top) Common species of marine flagellates. The genera shown are all dinoflagellates. These small planktonic forms (cell sizes typically from 25–500 microns) move about by moving one (sometimes two) flagellae. Under certain conditions dense concentrations called blooms occur in surface waters, which may color the water red, brown, or green. At night, *Noctiluca* is luminescent. Some forms (for example, *Gonyaulax*) produce toxins so that during bloom conditions animals feeding on them die. Shellfish ingesting these dinoflagellates may be poisonous to man. (From Sumich, J., *An Introduction to the Biology of Marine Life*. Dubuque, Iowa: Wm. C. Brown Co. Publishers. Used with permission.)

FIGURE 10.2 (Facing page, bottom) Common species of planktonic diatoms. Among the most abundant members of the phytoplankton, diatoms range in size from about 15 microns to 1 mm, and produce a frustule of opaline silica. (From Sumich, J., *An Introduction to the Biology of Marine Life*. Dubuque, Iowa: Wm. C. Brown Co. Publishers. Used with permission.)

FIGURE 10.3 (Below) Living *Hastigerina pelagica*, a common species of planktonic foraminifera. Like other members of this group, this species secretes a test of calcium carbonate which can be seen as the coiled shell at the center, bubblelike capsule made up of numerous transparent, gas-filled compartments. This flotation structure is 2 mm across. In it can be seen several species of dinoflagellates which live symbiotically with *Hastigerina*, exchanging nutrients, carbon dioxide, and oxygen with their host. A copepod, located inside the bubblelike capsule to the lower left of the shell, has fallen prey to and is being digested by the foraminifer. (Photo by Allan W.H. Bé, Lamont-Doherty Geological Observatory.)

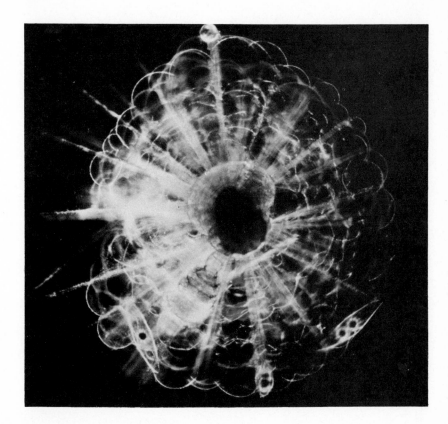

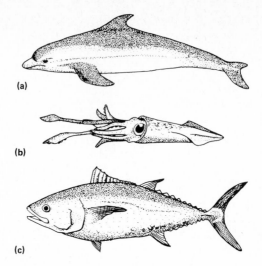

(a)

(b)

(c)

FIGURE 10.4 Typical streamlined, swimming animals of the nekton: (a) bottle-nosed porpoise, *Tursiops;* (b) squid, *Loligo;* (c) tuna, *Thunnus.* (From Sumich, J., *An Introduction to the Biology of Marine Life.* Dubuque, Iowa: Wm. C. Brown Co. Publishers. Used with permission.)

FIGURE 10.5 A spiny echinoid *Diadema* scavenges for food on the bottom, near an attached, filter-feeding black sponge. Photograph taken at a depth of 2 m, near Frazer's Hog Cay, B.W.I. (Imbrie photo.)

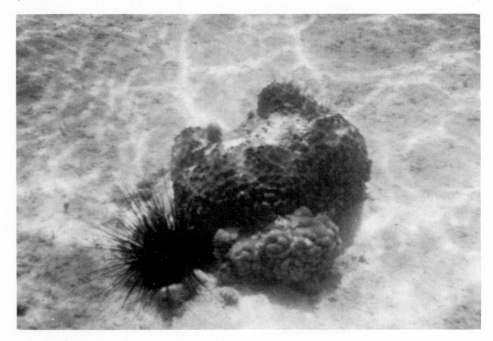

FIGURE 10.6   An abyssal acorn worm crawling along the bottom at a depth of 4871 m on the side of the Kermadec Trench, southwest Pacific. The animal's body ends shortly past the first bend, where the long fecal coil begins. A similar feces is seen at upper right. (From Heezen and Hollister, 1971.)

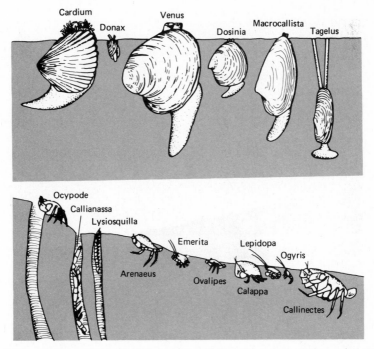

FIGURE 10.7   Species of clams (*top*) and crustaceans (*lower*) burrowing in the sand of a North Carolina beach. (Adapted by Thorson, 1971, from Pearse, 1942.)

FIGURE   10.9   A glass sponge attached to the bottom by a rootlike structure at a depth of 5378 m
along the western flank of the Bermuda Rise. The sponge is filtering microscopic food suspended in the
Antarctic Bottom Current, which is sweeping northward. (Courtesy of E. Schneider.)

FIGURE   10.8   Ophiuroids burrowing in the ooze at a depth of 1503 m off Cape Hatteras. (Courtesy of
R. Pratt.)

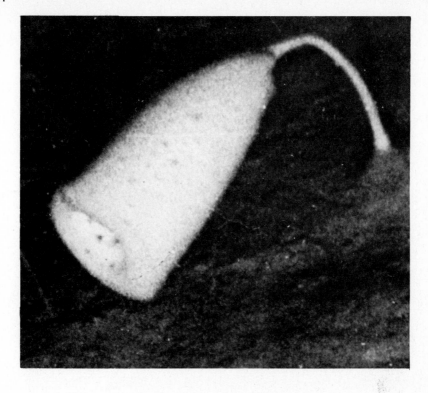

SUMMARY

1. To grow and to sustain life, every animal and plant must acquire energy and materials from its environment. How different species capture the energy and materials they need determines their basic role in the marine ecosystem.

2. Photosynthetic organisms use the energy of sunlight to fashion organic compounds from inorganic materials in seawater (chiefly water, carbon dioxide, nitrogen, and phosphorus).

3. A part of the sun's energy is stored in chemical form in the tissue of photosynthetic organisms, where it can be used later by that individual or acquired as *food* by *heterotrophic* organisms.

4. Marine organisms exhibit a wide range of adaptive strategies as they go about their primary business of energy capture. *Pelagic* organisms live above the seabed, and get about either by drifting (the *plankton*) or by swimming (the *nekton*).

5. *Benthic* organisms live on or in the seabed. They may *burrow* into a soft bottom, *bore* into a hard bottom, *crawl* over the surface, *attach* themselves by rootlike structures, or *lie free*.

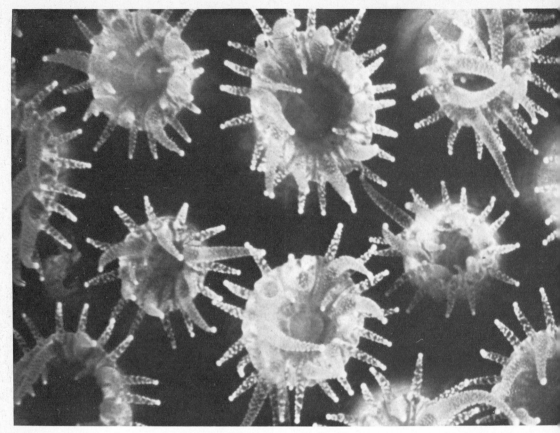

Polyps of the reef-building coral *Montastrea cavernosa* feeding along the south coast of Curaçao. The shape of the coral reefs is determined by a balance between biological processes which construct the reef and physical processes which tend to destroy it. (Courtesy of W. Sacco.)

The Marine Ecosystem

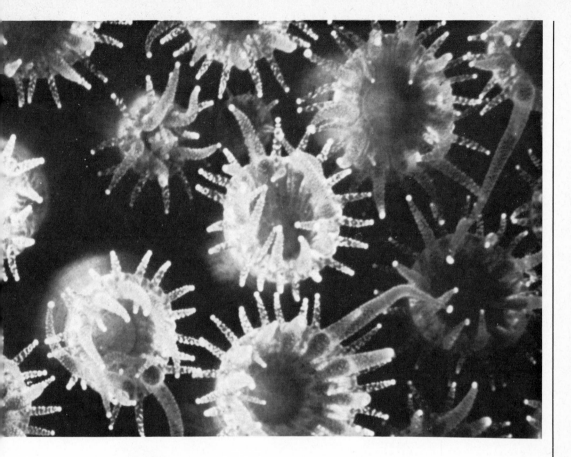

11

CHAPTER

Organisms do not simply *occur* in the sea; they *live* there. Each individual is linked by a network of interactions with other organisms and with the physical and chemical properties of the sea itself. This system of interactions is called the *marine ecosystem*. In the foregoing chapter we identified the transfer of energy and materials as an important aspect of the marine ecosystem. In this chapter we analyze these transfers in more detail, consider the geometry of the transfer paths, treat the ecosystem as a machine for producing organic tissue, and find out what controls the rate of organic production. In the final section we examine particular associations of organisms, and study how the nature of these biotic communities is shaped by their interactions with the sea and with the sea floor.

SIMPLE ECOSYSTEM MODELS

MAIN PATHS OF ENERGY AND MATERIAL FLOW

To understand a system as vast and complex as the marine ecosystem, we need models that simplify without distortion. One much-simplified model is given in Figure 11.1, in which the main paths of energy and material flow are idealized. Solar energy is absorbed by the plants and is transmitted to animals and bacteria as potential energy along the *main food chain*. These consumer groups exchange carbon dioxide, inorganic nutrients, and oxygen with plants. The flow of organic substances is therefore basically cyclic and conservative, in that the same materials pass back and forth between the biotic components of the system, either directly or through the ocean reservoir. Although in detail the flow of energy is also cyclic, there is on balance a one-way flow in which all the incoming energy is eventually dissipated as heat by the mechanical and chemical activities of the biosphere.

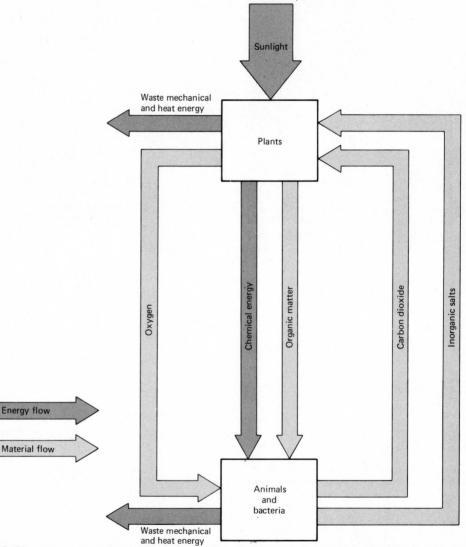

FIGURE 11.1   A simple model of the flow of energy and materials in the marine ecosystem. (From Sumich, J., *An Introduction to the Biology of Marine Life*. Dubuque, Iowa: Wm. C. Brown Co. Publishers. Used with permission.)

COMPONENTS OF THE ECOSYSTEM

A more detailed model of ecosystem components is given in Table 11.1, which lists the essential plant nutrients (see Chapter 9) and emphasizes that the organic side of the ecosystem includes both living and nonliving tissue. The latter is progressively broken down into smaller and smaller particles by physical attrition and bacterial decay. Ranging in size from the body of a whole dead whale to organic molecules, this *biogenic debris* makes up perhaps one-half the marine biosphere. Suspended in seawater, buried in the bottom sediment, and adhering to all exposed surfaces, this material represents a vast food reservoir. Some animals in the pelagic realm feed on it exclusively; for many others it forms part of their diet, along with living plankton. However, most of the feeders on biogenic detritus are bottom dwellers.

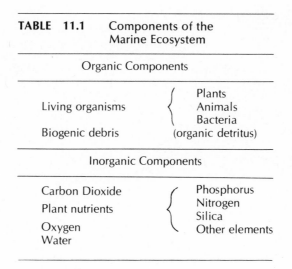

TABLE 11.1   Components of the Marine Ecosystem

Organic Components	
Living organisms	Plants / Animals / Bacteria
Biogenic debris	(organic detritus)

Inorganic Components	
Carbon Dioxide	Phosphorus
Plant nutrients	Nitrogen / Silica
Oxygen	Other elements
Water	

ECOSYSTEM PROCESSES

Energy and materials flow into and through the marine ecosystem by five major biological processes (Table 11.2). The total amount of plant tissue formed by *photosynthesis* over a specified time is called the *gross primary production*, and the plants that do this are called primary producers. In offshore areas, virtually all organic materials are derived from these producers. In nearshore areas, however, organic matter derived from terrestrial plants may make substantial contributions to the biosphere.

Five kinds of feeding can be distinguished: grazing, predation, scavenging, deposit feeding, and decomposition. Herbivores graze directly on plants. Most grazing occurs in the pelagic realm, where hosts of small animals eat diatoms and other phytoplankton. In shallow waters, seaweeds attached to the bottom and blue-green

TABLE 11.2   Basic Ecosystem Processes

1. Photosynthesis
2. Feeding
    a. Grazing
    b. Predation
    c. Scavenging
    d. Deposit feeding
    e. Decomposition
3. Death
4. Nutrient regeneration
    a. Indirect (bacterial decomposition)
    b. Direct (metabolic regeneration)
5. Respiration

FIGURE 11.3   The carnivorous moray eel awaits its prey in the reef off Curaçao. (Courtesy of W. Sacco.)

FIGURE 11.2   The sea anemone *Cerianthus* feeding at night on plankton at a depth of 10 m along the south shore of Curacao. (Courtesy of W. Sacco.)

algae adhering to or boring in the bottom are both eaten by grazing animals. Higher in the food chain, predators (carnivores) prey on live animals that are generally smaller than themselves (Figures 11.2 and 11.3). Baleen whales are among the most notable predators, engulfing entire schools of small animals. Scavenging animals feed on larger particles of dead organisms, often supplementing their diet with other fare. Scavengers include many invertebrate animals such as echinoids, snails and crabs (Figure 10.5). Biogenic debris and small living animals in sediments on the sea floor make up the diet of many deposit-feeding animals (Figures 10.6 and 11.4). Finally, decay organisms, mainly bacteria, decompose dead organic matter and thereby release inorganic compounds back to the sea.

Ordinarily we consider *death* an unfortunate event terminating the life of a single individual. In the context of the ecosystem as a whole, however, we come to see death as an essential ecosystem process. From the ecological point of view there are two kinds of death. When death occurs by predation, chemical compounds in the body of the prey are directly ingested by the predator and there is a virtually instantaneous transfer of materials within the biosphere. When death arises from other ("natural") causes, a substantial time lag may occur before the activity of scavenging animals or decay organisms can recycle the material.

Nutrient regeneration is an important set of processes by which organisms re-

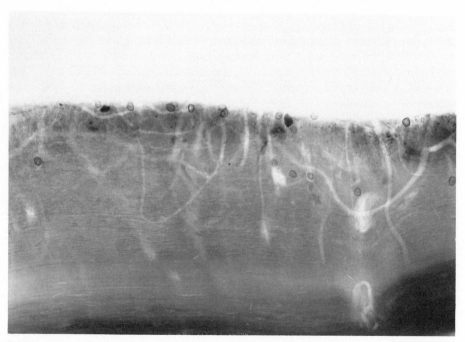

FIGURE 11.4   X-radiograph of mud bottom in New York bight shows U-shaped burrows of deposit-feeding clam, *Nucula,* and its 4-mm-long shells. (Courtesy of R.C. Aller.)

duce complex organic compounds into chemically simpler forms and return them to the sea as inorganic plant nutrients. As noted, much of this work is accomplished by decay organisms, chiefly the bacteria. This may occur at times and places far removed from the life habitat of the organism undergoing decay; hence the process is called *indirect regeneration*. The chemical products of metabolism in plants and animals also return simple compounds to the sea. This is called *direct regeneration*. Both animals and plants use up oxygen in *respiration*.

TROPHIC LEVELS

Among the most important biological interactions are those involving food. These *trophic* relationships are shown in Figure 11.5. Here the main steps in the food chain are identified, beginning with the gross primary production by phytoplankton. These plants are defined as the first *trophic level*. The second trophic level constitutes the herbivorous zooplankton, and the production there is defined as

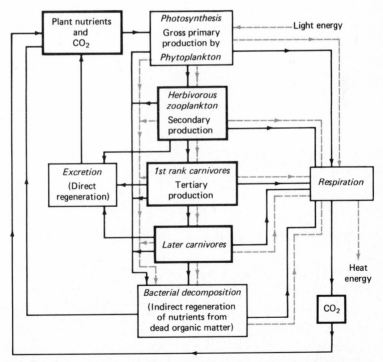

FIGURE 11.5   The flow of energy (dashed lines) and carbon-bearing compounds (solid lines) in the marine ecosystem. Reservoirs of material are indicated by boxes with heavy lines; processes by boxes with light lines. (From Tait and De Santo, 1972.)

secondary. Carnivores at successively higher trophic levels are spoken of as first-rank carnivores, second-rank carnivores, and so on.

As a generalized diagram of the flow of energy in the ecosystem, this model is valid. However, the chain model distorts many aspects of the real system, which is rather more appropriately spoken of as a *food web*. Figure 11.6 illustrates this model, which takes into account not only the biogenic debris (organic detritus), but the planktonic bacteria as well. The nutritive value of the particulate organic matter is substantially increased by bacteria in the process of feeding on it. Many small particles suspended in the sea or comprising part of the sea bed sediment are *fecal pellets*. At the first trophic level different species of plants are symbolized $P_1, P_2 \ldots P_i$. The second trophic level consists of herbivorous species (such as Z_3), feeding exclusively on plants, on bacterial aggregates only (Z_1), or on both sources of food (Z_2). The total biomass synthesized per unit time at this second trophic level is the secondary production of the system. At the third level lie various species that are either carnivores of the first rank (such as Z_3 or Z_2) or omnivores (such as Z_1). The total production at this level is referred to as tertiary production. At the fourth trophic level stand second-rank carnivores that feed on lower trophic levels. Finally, at the fifth trophic level are carnivores of the third rank.

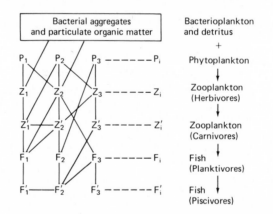

FIGURE 11.6   Pelagic food web. (See text for discussion.) (From Parsons and Takahashi, 1973.)

The concept of trophic level makes possible an analysis of the *efficiency* of the total system. Here we adopt the view that the biosphere is a machine, powered by the sun, which is supplying energy at a fixed *rate* (calories per unit time). Energy is supplied to each trophic level from some source: either from the sun or from food. A substantial fraction of the energy supplied to each level is dissipated there and cannot be passed on to higher levels. The losses include all the physical and chemical work performed by organisms in order to maintain themselves. In addition, animals living at higher trophic levels consume only a portion of the crop available at lower levels. Some plants and animalsthus die a natural death. As a result, the energy extracted from a trophic level by organisms higher in the food web is much less than the energy supplied to that level. The ratio of these quantities is known as the *ecological efficiency*, *E*, of a level.

$$E = \frac{\text{Energy extracted from a trophic level in unit time}}{\text{Energy supplied to a trophic level in unit time}} \qquad (11.1)$$

The values of *E* are rather low, generally estimated to be between 0.1 and 0.2.

In general, the production (*P*) at level *n* can be estimated from an equation of the form

$$P = B \times E^{(n-1)} \qquad (11.2)$$

where *B* is the primary production. Given a production of phytoplankton of 1.9×10^{10} metric tons per year, the production of fish at the fifth trophic level would be only 1.9×10^6 metric tons per year for *E* = 10 percent. If *E* were increased to 20 percent, *P* would be 30.4×10^6 metric tons per year, a fifteenfold increase.

ECOSYSTEM GEOMETRY

Models of the marine ecosystem discussed so far have omitted its three-dimensional structure. The diagram given in Figure 11.7 shows in simplified form the location of major paths of energy and nutrient exchange as they occur in the real ocean. One important fact brought out by this diagram is that in the open ocean a substantial distance separates the sunlit surface zone, where photosynthesis occurs, and the deep-water areas, where photosynthesis does not occur. Energy in chemical form is imported into deep water. This transfer of chemical energy occurs within the food web and by gravitational settling of organic detritus, especially as fecal pellets.

The constant downward settling of fecal pellets and of unpelleted dead organic tissues represents a substantial and continual drain on the supply of nutrients in surface waters. Thus the processes of death, feeding, defecation, and gravitational sinking together constitute an ecological pump, draining surface waters of essential nutrients. Unless counterprocesses were at work to replace the lost material, surface waters of the ocean would lose all nutrients, and life there would cease. That this global disaster does not occur is due primarily to those physical processes known collectively as upwelling, by which deeper waters are brought to the surface at rates averaging about 3.5 m/yr. As has been explained in earlier chapters of this book, these processes are intensified to a depth of 300 m along certain coasts (coastal upwelling); in the turbulent zones of currents; in areas of equatorial upwelling; and in high latitudes during seasons when the seasonal thermocline is destroyed, and convective mixing affects the entire water column.

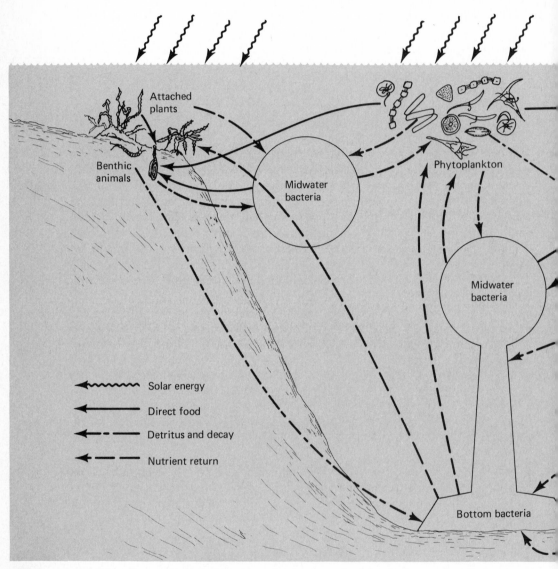

FIGURE 11.7   Schematic cross section of the ocean showing the main paths of energy and nutrient exchange. (© W. D. Russell-Hunter, 1970; *Aquatic Productivity*; Macmillan.)

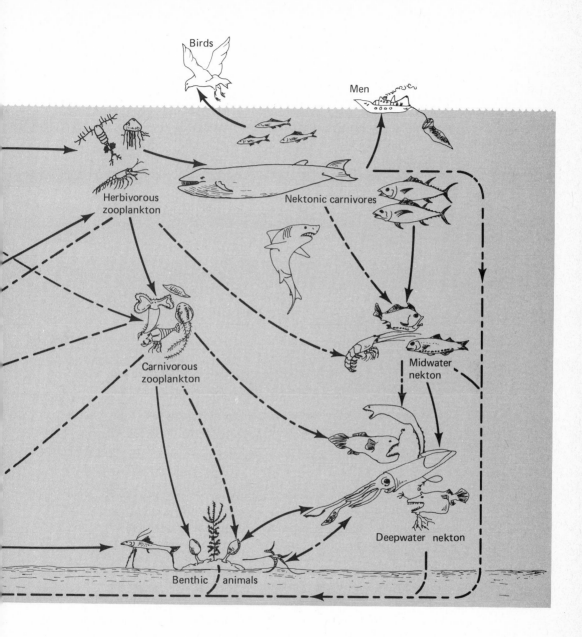

Birds

Men

Herbivorous
zooplankton

Nektonic carnivores

Carnivorous
zooplankton

Midwater
nekton

Deepwater nekton

Benthic animals

ECOSYSTEM DYNAMICS

ABUNDANCE OF ORGANISMS

The number of organisms living in the sea varies from place to place and from time to time. Plankton tows taken in the blue waters of the Sargasso Sea contain far fewer plankton per unit volume of water than the greenish waters off the New England coast. Why do these differences occur? And what controls the total number of organisms living in any particular part of the sea or in the global ocean itself? A natural first step in understanding organism abundance is to examine measurements made *at a particular instant*. By analogy with a cattleman counting the size of his herd, this measure is known as the *standing stock*. For a particular purpose we may wish to speak of the biomass of some particular group (such as phytoplankton or whales), or the total biomass. Standing stocks are reported variously as mass per unit area, mass per unit volume, or number of individuals per unit area or volume.

A little reflection will show that we really cannot understand biomass distributions without considering the rate at which organic matter is produced. To attempt to do so would be like estimating the success of a business by examining the contents of the cash register at a particular instant.

ORGANIC PRODUCTION

In any given region of the ocean, new organic matter is being formed at a certain rate, which is defined as the gross primary production. Because some of the organic matter produced in photosynthesis is later used by the plants themselves as an energy source, we must distinguish this gross production from the rate at which plant tissue *available for higher trophic levels* is formed. The latter quantity is the *net primary production*. The difference represents the energy required to keep the primary producers producing (Figure 11.8).

Several different units of measurement are employed in studies of organic production. Normally, the organic production is calculated for a given unit of time and over a given unit area of sea surface—because the sun's energy input is across this surface. Occasionally, for special purposes such as studies of production at different levels of the ocean, production may be defined per unit volume. The quantity of organic matter itself may be expressed in four different units: (1) as *wet weight* of organic tissue—for example, the production of pelagic fish in the English Channel has been estimated as 0.28 g/m²/yr of wet tissue weight; (2) as dry weight—for example, the dry weight of the pelagic fish just mentioned would be about 0.047 g/m²/yr; (3) as *organic carbon*—as a generalization, 1 g dry weight of organic matter contains about 0.44 g of carbon. The fish production mentioned above would then be recorded as 0.021 gC/m²/yr; and (4) as *energy content*—because we are interested in tracing the flow of energy through the system, it is often important to consider the energy content of a given portion of the system. Since the calorific

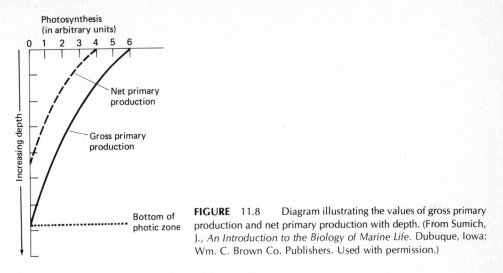

FIGURE 11.8 Diagram illustrating the values of gross primary production and net primary production with depth. (From Sumich, J., *An Introduction to the Biology of Marine Life*. Dubuque, Iowa: Wm. C. Brown Co. Publishers. Used with permission.)

value of fish is about 3.0 kcal/g dry weight, the energy content of the fish is estimated as 0.14 kcal/m²/yr.

Considerations of the energy content of marine biomass immediately spark the question: How much of the solar energy available in an area is represented by different trophic levels? Table 11.3 indicates the answer calculated for the North Sea. The losses at each step in this food chain are very large. Note that the total fisheries yield is only 0.039 kcal/m²/yr, or 0.066 percent of the gross primary production.

TABLE 11.3 Pelagic Production at Various Trophic Levels in the North Sea (expressed in units of energy)*

Level	kcal/m²/yr
Solar energy reaching sea surface	250,000
Gross primary production	500
Net primary production	350
Herbivores	56
Tertiary carnivores	5
Pelagic fisheries yield	0.14
Demersal fisheries yield	0.25

* From Tait and de Santo, 1976.

We can now turn to a consideration of the factors that control the standing stock, using the example of Table 11.3. The net primary production, expressed in terms of the energy content of phytoplankton tissues, is 350 kcal /m² /yr; expressed in terms of dry tissue weight, it equals 175 g /m² /yr. This does not mean, however, that we would at any particular time find 175 g of phytoplankton living beneath a square meter of the sea surface, because the life spans of the phytoplankton are short and the population is replaced many times over the year. By measuring the average standing stock, which is only 4 g /m², we can actually calculate the *turn-over rate constant*, which determines the number of times per year the population replaces itself.

$$\text{Turnover rate constant} = \frac{\text{net primary production}}{\text{standing stock}} : \frac{175 \text{ g/m}^2\text{/yr}}{4 \text{ g/m}^2} = 44\text{/yr}$$

(11.3)

Alternatively, we can calculate the standing stock if we know the net primary production and the turnover rate constant.

Standing stock = net primary production/turnover rate constant

Thus, the standing stock observed at any time reflects the length of the reproductive cycle as well as the net production.

FACTORS REGULATING ORGANIC PRODUCTION

Because the production of the marine ecosystem as a whole depends on the production at the first trophic level, it is important that we study the factors controlling this primary production. Four factors are involved: light intensity, temperature, nutrient supply, and grazing rates.

Light Because the process of photosynthesis is powered by sunlight, the amount of light available at a particular point in the ocean is an important limiting factor in organic production. In turn, the amount of light in the sea is controlled in the first place by geographic and meteorologic factors, especially the angle of the sun above the horizon and the cloudiness. Once sunlight enters the sea, its intensity diminishes very rapidly with depth (see Chapter 4). On the average, at a depth of 10 m, only 10 percent of the light entering the sea is available for photosynthesis; at 100 m, only 1 percent is available. This fact alone limits the zone of net primary production to the upper few hundred meters (Figure 11.9) of the sea. In coastal waters, which generally have more suspended matter than the open ocean, light penetration is further inhibited.

Temperature Although the intensity of sunlight is an important environmental factor controlling the rate of photosynthesis, and hence primary production, other factors must be considered as well. One of these is water temperature. Every algal species has a certain temperature range at which the rate of photosynthesis is a maximum, if all other conditions, such as light intensity, are held constant. Temperatures above or below this *optimum* range cause photosynthesis to decrease (Figure 11.10).

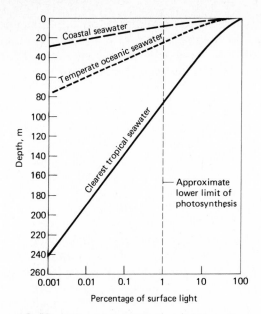

FIGURE 11.9   Penetration of sunlight in various types of seawater. (From Sumich, 1976, adapted from Jerlov, 1956.)

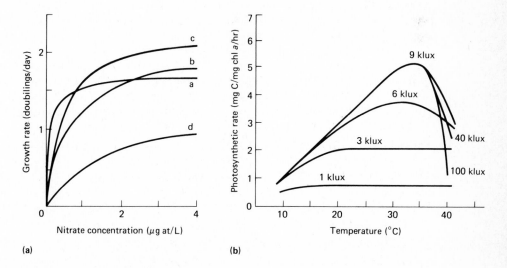

(a)

(b)

FIGURE 11.10   Laboratory evidence of the control of photosynthesis by light intensity, temperature, and nitrate concentration. (a) Growth rates of four algal species (a–d) over a range of different external nitrate concentrations. Internal concentrations of nitrogen also need to be considered. (b) Photosynthesis as a function of temperature (°C) and light intensity (klux) in cultured specimens of the alga *Scenedesmus*. Photosynthetic rate is expressed as mg of carbon fixed per mg of chlorophyll *a* per hour. (From Parsons and Takahashi, 1973, redrawn from Eppley et al., 1969 (a) and Aruga, 1965 (b).)

Over much of the ocean temperatures often lie below the temperature optimum for many species; hence the seasonal increases in temperature cause higher rates of photosynthesis. One study in the Pacific, for example, showed that certain algae had optimum rates at about 20°C, whereas temperatures in the study area ranged from −0.9°C to 18°C.

Nutrients   We must now examine the limitations imposed by material balance, that is, by the rate of supply of materials needed for photosynthesis. The fundamental requirements are water and carbon dioxide, the latter being distributed in various chemical species. Water, of course, is never in short supply in the ocean. Typical surface water in the open ocean contains the equivalent of about 90 mg of carbon dioxide per liter, mostly as bicarbonate. Could the photosynthetic rate be increased by supplying more total carbon dioxide? Experiments in which carbon dioxide was added to seawater did not result in an increase in the rate of photosynthesis. Moreover, in areas where the phytoplankton reproduction is unusually high (during an algal "bloom"), the total carbon dioxide concentration is below normal values. Thus we draw the important conclusion that total carbon dioxide is never in short supply in the ocean and that (for a given temperature and light intensity) the availability of other substances must regulate the photosynthetic rate.

As discussed in Chapter 9, these substances are the *nutrient elements*, that is, materials required by algae for growth, in addition to carbon dioxide and water. We may group them into two classes. The macronutrients are nitrogen, silicon, phosphorus, magnesium, calcium, and potassium; these are needed in relatively large amounts. Other elements needed in very small amounts are the micronutrients: iron, manganese, copper, zinc, boron, sodium, molybdenum, chlorine, vanadium, and copper.

Again, we may ask, which of these nutrients are ever present in such small amounts in relation to the amounts needed that the photosynthetic rate is reduced? In general, only three: nitrogen, phosphorus, and silica. Nitrogen and phosphorus are of major importance in the metabolism of plants, because both elements are needed to build cell tissue. In addition, phosphorus is continually turned over in the energetic processes of metabolism. More nitrogen is needed than phosphorus, because the nitrogen: phosphorus ratio in plants is about 16:1. In the ocean generally, the ratio of these elements is about the same. However, in shallow waters, the processes by which nitrogen is regenerated—that is, recycled into the water in a form suitable for plant uptake—are slower than those by which phosphorus is regenerated. Hence, in many near-shore areas, nitrogen is depleted with respect to phosphorus and is the rate-limiting nutrient. This nitrogen deficiency is seen most commonly in summer in shallow waters everywhere, and the year round in shallow subtropical and tropical waters. Experimental evidence on the role of nitrogen in algal growth is shown in Figure 11.10a.

Silicon is used in great abundance by two types of phytoplankton which build skeletons of silica: diatoms and silicoflagellates (see Chapter 9). Often, these plants extract the silicon so rapidly from surface waters that the resulting scarcity is itself the limiting factor in their growth. For this reason, a bloom of siliceous plankton is often followed seasonally by a bloom of nonsiliceous plankton.

Grazing Grazing by zooplankton has an immediate impact on the primary production, because each plant eaten no longer grows or reproduces. Thus one factor that influences the rate of primary production is the rate of grazing. In an equilibrium situation, the grazing rate would be just sufficient to maintain the phytoplankton population at a constant level. If the primary production were to increase, an increase in the number or efficiency of grazing animals would bring the system in balance. However, zooplankton reproduction takes time; hence, even if other environmental factors were constant, a steady state would never be achieved, and the numbers of grazers and phytoplankton would fluctuate about an equilibrium value (Figure 11.11).

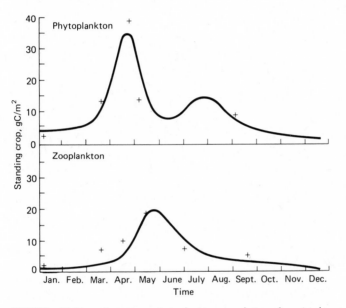

FIGURE 11.11 Impact on phytoplankton population of grazing by zooplankton. Observations on standing crops at Georges Bank (*crosses*), and (*smooth curves*) numerical model predicting these values. (From Riley, 1946 and 1947.)

GEOGRAPHIC VARIATIONS IN PRODUCTION

Very marked geographic variations occur in organic productivity, as is indicated in Figure 11.12. Areas of high productivity are (1) continental shelves and (2) areas in the open ocean where upwelling enriches the surface waters in nutrients. The shelves are productive for several reasons. Being relatively shallow, shelf waters are relatively warm and sunlit. Being near shore, these areas receive nutrient-rich river discharge. Moreover, regeneration from organic material on the sea floor can replenish the water column directly. In fact, in shallow estuaries there is evidence that production is largely controlled by the rates at which nutrients are released to the water column from the underlying sediment. In the open ocean, areas charac-

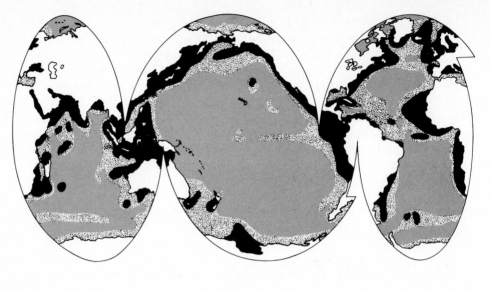

Regions of high productivity (greater than 100 gC/m²/year)   Regions of moderate productivity (50–100 gC/m²/year)   Regions of low productivity (Less than 50 gC/m²/year)

FIGURE 11.12   Geographic distribution of marine plant production. (From Sumich, 1976, adapted from FAO, 1972. Used with permission.)

terized by high productivity are quite restricted, because the anticyclonic gyre areas that occupy such a large fraction of the sea surface are areas of downwelling.

The highest production in the open ocean occurs in high latitudes, generally poleward of 50° latitude. Here, several effects combine to make this so. The most important is the breakdown of the seasonal thermocline in winter, with associated deep convective stirring. In addition, these areas represent portions of the ocean in which upward motions occur as part of the general circulation above the main thermocline, to balance the downward motions in the central waters (see Chapter 6). As still higher latitudes are reached, productivity is lowered as a consequence of cold temperatures, sun angles, and ice cover.

Areas of strong coastal upwelling have very high production. Typical of these are the regions of eastern boundary currents along the coasts of Peru, Oregon, Senegal, and southwest Africa. Where the flow of the western boundary currents is rapid, as in the core of the Gulf Stream and Kuroshio currents, turbulent effects increase surface productivity. Equatorial regions show relatively high levels, owing to turbulence associated with currents there.

SEASONAL VARIATIONS

At every locality in the ocean, the march of the seasons brings with it changes in primary production. The biological responses reflect in complex ways the seasonal changes in the physical environment, particularly changes in light intensity, wind strengths, and water temperatures. In temperate seas, these seasonal contrasts are the sharpest (see Figures 5.6a and b). A typical response pattern is shown in Figure 11.13. The thermal inertia of the ocean causes the curve of sea-surface temperatures to lag behind the atmosphere, so that its Northern Hemisphere maximum is attained sometime in August, and its minimum temperature in February. By the end of winter, low temperatures, together with low values of solar radiation penetrating the sea, result in greatly diminished stocks of diatoms and dinoflagellates. Meanwhile, the cool temperatures and winter storms have combined to cause a deep convective mixing of the upper layers of the ocean. The upwelling of deeper, nutrient-rich waters causes an increase in the level of nutrients in surface waters. Conditions are ripe, therefore, for the oceanic spring. As the waters warm and sunlight penetration increases, diatoms have optimum conditions for growth, and a phytoplankton bloom results.

By early summer, although temperature and sunlight conditions are optimal for diatom growth, several factors have combined to decrease the standing stock of these siliceous plants. First, grazing by the zooplankton, which has now reached a maximum, lowers crop levels. Second, the warm surface temperatures have stabilized the water column, thus inhibiting vertical movement of waters and preventing a replenishment at the surface of nutrient-rich waters from greater depths. The result is fewer diatoms in summer. Naked dinoflagellates (and other phytoplankton not requiring silica for skeletal growth) have their optimum following that of the diatoms. In autumn, when illumination is still strong enough for photosynthesis, the thermocline is disrupted by cooling surface waters, and convective mixing occurs.

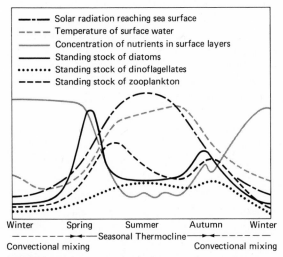

FIGURE 11.13 Seasonal changes of temperature, nutrients, phytoplankton, and zooplankton in surface waters of temperate seas. (From Tait and De Santo, 1972.)

Surface production increases, especially for diatoms, as a result of replacement of nutrients from deeper layers. As temperature and light intensity further decline, the stocks of all phytoplankton are reduced to low winter levels. Many species have special wintering-over stages which permit them to exist in the water column as "seeds" for the coming spring bloom.

In very high latitudes, where sea ice forms in winter, the cycle just described is reinforced by the depressing effect of the sea ice on plant growth. In low latitudes, conditions resembling the summer season at high latitudes occur all year round. Changes in production that do occur are relatively minor, and mainly reflect changes in vertical circulation. Surface waters are always warm; hence the thermocline is a strongly marked and permanent feature. As a result, surface nutrients are not renewed through vertical mixing of nutrient-rich waters below the thermocline. Away from upwelling areas, therefore, tropical seas are characterized by low productivity—in spite of the otherwise favorable conditions.

MARINE COMMUNITIES

Earlier parts of this chapter identified mechanisms and pathways by which materials and energy are transferred between all components of the marine ecosystem. In this trophic analysis it was emphasized that every individual organism in the sea is linked to all others, no matter how remote. Here we turn our attention to groups of individuals living together in one particular part of the ocean. Such assemblages are defined as *communities*.

Benthic communities, in particular, illustrate two kinds of ecological interactions which have not previously been emphasized: (1) the ecological impact many groups of animals and plants have on each other, quite apart from their trophic relationships, and (2) the shaping of the physical environment through interactions between biological and physical processes. These points will now be illustrated by examining Pacific coral-reef communities, a Bahamian rock-pavement community, and a mud-bottom community from the North Sea.

CORAL REEFS

Reefs are topographic structures on the sea floor shallow enough to be hazards to navigation. Coral reefs are built by a marine community in which corals and red algae play a dominant role. They are rigid, wave-resistant structures whose form at any moment represents an equilibrium between the destructive forces of the surf and the growth of the reef itself. The essential feature of a reef community, therefore, is its ability to build a rigid framework rapidly. The reef shown in Figure 11.14 is the surface expression of such a framework.

The activities of two groups of organisms are responsible for the frame-building potential of the reef community. The first group consists of certain corals capa-

FIGURE 11.14   A yellow-tailed snapper swims over a coral reef 10 m below the surface along the north coast of Jamaica. (Courtesy of W. Sacco.)

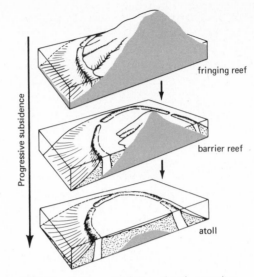

FIGURE 11.15 Evolution of reefs according to Darwin's theory of progressive subsidence. The volcanic island in the center is progressively submerged, thus allowing upward growth of the coral reef. (From Purdy, 1974, after Darwin, 1839.)

FIGURE 11.16 Islands of the Tonga Group, West Pacific. Eua Iki Island in the foreground is a fringing reef around a volcanic center. In the background is an atoll. (Courtesy of N.D. Newell.)

ble of rapidly forming stony skeletons of calcium carbonate. This ability, in turn, is due to a beneficial and obligatory association with brown algae. The second group consists of forms of red algae whose skeletons form hard crusts of calcium carbonate. These encrustations unite what would otherwise be a loose rubble of broken coral skeletons and individual corals. Once built, the reef plays host to a large number of other animals and plants that can take advantage of the desirable conditions in these warm, shallow seas, if some degree of shelter from the pounding surf is provided.

As noted by Charles Darwin, Pacific reefs assume three quite different forms: fringing reefs, barrier reefs, and atolls (Figure 11.15). Darwin's theory for explaining them was a progressive upward growth by the reef community on a progressively subsiding sea floor. Although other factors are also involved in shaping reefs, modern studies (including drilling of atolls) has confirmed Darwin's basic idea. The islands in the Tonga Group (Figure 11.16) show a typical example of a fringing reef and an atoll. Note that the reef-flat community has surrounded and partly cut off quiet-water lagoons; and that reef rubble, broken and tossed by storms, forms the nucleus for low-lying islands.

ROCK-PAVEMENT COMMUNITIES

Figure 11.17 shows an example of a community living near the Bahamian island of Bimini in relatively quiet, sunlit waters about 4 m deep. Two or 3 cm of sedi-

FIGURE 11.17   Rock-pavement community at a depth of 4 m near Bimini Island in the Bahamas is dominated by sponges and green algae. (Imbrie photo.)

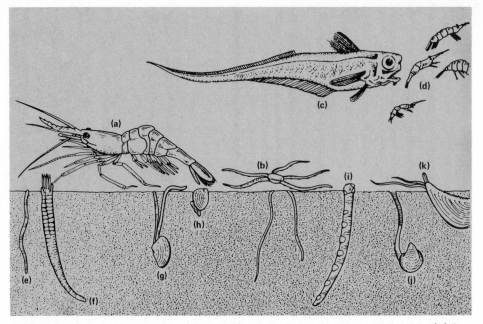

FIGURE 11.18 Mud-bottom community at 200 m in the Skagerrak of the North Sea. (a) red shrimp; (b) brittle star; (c) grenadier fish, with two sea-pen colonies behind; (d) hunting shrimps; (e) and (f) bristle worms; (g–k) burrowing clams. (From Thorson, 1971, after Thorson, 1968.)

ment cover a rock pavement. Burrowing organisms are relatively rare, and the easily visible portion of the community is dominated by sponges, green algae, and mollusks. Mud-sized spicules of calcium carbonate are embedded in the tissues of the green algae. On death, the algae release a rain of these particles on the sea floor. Also on death, the mollusk skeletons are broken up into a sand-sized rubble. But nearly as fast as sediment is produced by these organisms, the waves and currents transport it into deeper waters. Thus the dominant aspect of the community structure here is maintained by a balance between sediment production and sediment removal. Sponges and other organisms which require a hard substrate for attachment flourish, but burrowing organisms are rare because the sediment cover above the bedrock is too thin.

MUD-BOTTOM COMMUNITIES

Figure 11.18 shows a diagrammatic cross section through a mud-bottom community at a depth of 200 m in the Skagerrak of the North Sea. In contrast to the Bahamian community just discussed, here the sediment accumulates faster than currents and waves can remove it. As a result, burrowing filter feeders and deposit feeders, together with the carnivores that prey on them, are a conspicuous element in this and other mud-bottom communities.

SUMMARY

1. By transfers of energy and materials, marine organisms are linked to each other and to the physical and chemical properties of the sea.

2. Biological processes in the marine ecosystem include photosynthesis, feeding (grazing, predation, scavenging, deposit feeding, and decomposition), death, nutrient regeneration, and respiration.

3. Solar energy is absorbed by photosynthetic plants and transmitted in chemical form to various trophic levels on the main food chain. Only a small fraction of the energy content of one trophic level is transmitted to the next.

4. The downward settling of dead organic tissue drains the surface layers of the ocean of plant nutrients, which tend to accumulate in deep waters.

5. The amount of plant tissue formed over a specified time is the primary organic production. This production is controlled chiefly by the availability of light and plant nutrients. Secondary factors include the grazing rate and water temperature.

6. Productive regions of the ocean are concentrated mainly along the continental shelves and in those areas of the open ocean where upwelling enriches the surface waters in nutrients. Seasonal variations in production occur, notably in high latitudes.

7. Individuals living together in one part of the ocean form communities. In many areas the nature of the sea floor represents an equilibrium between physical processes at work there and the biological processes of the benthic community.

Fishing along the coast of Curaçao (left) and lobstering along the coast of Maine (right). (Courtesy of W. Sacco.)

Food From
the Sea

12

Some 15 percent of the four billion people alive today have a diet below that required for good health. With the likelihood that global populations will double in 40 years, food supply may soon become the central problem facing mankind. In this chapter we examine the relatively minor role the ocean now plays in human nutrition, discover why the harvest of food from the sea is so much smaller than that from the land, and ask if

CHAPTER

the marine harvest can be significantly increased.

MAJOR GLOBAL FOOD-PRODUCTION SYSTEMS

Because the oceans cover 71 percent of the world's surface and intercept more sunlight than the land, it is reasonable to expect that the contribution of seafood to our total nutrition would be substantial. But this expectation ·is not borne out by the facts. During 1970, for example, about 2892 million tons of food were produced on land, compared to a total marine harvest of about 60 million tons (Table 12.1). Yet these figures do not tell the whole story, because only about 36 million tons of seafood were consumed by man; the remainder of the marine production (about 26 million tons) was made into fish meal and fed to livestock. As explained in Chapter 11, energy is lost between trophic levels. Thus, the 26 million tons of fish meal are converted to about 5 million tons of livestock. The total marine contribution to human nutrition is therefore only 41 million tons: the 36 million tons used directly, in addition to the 5 million tons of livestock raised on fish meal. This total represents only a little more than 1 percent of man's total food consumption. However, seafood does make a substantial contribution to the *animal protein* consumed by man, because about 5 percent of this portion of our diet comes directly to the table from the ocean and another 5 percent arrives indirectly from livestock fed on fish meal.

Acre for acre, many areas in the sea compare favorably with land areas in terms of total annual primary productivity (Table 12.2). Why, then, is the harvest

TABLE   12.1   Human Food from Land and Ocean Production Systems, 1970.

Food Types	Categories of Production	Food Production, in 10^6 tons Land	Ocean
Plants {	Gathering	100	0.7
	Farming	2247	0.2
Animals {	Hunting	27	60.7
	Herding	518	0.6
Total		2892	61.3
	Less that used for fish meal		−26.5
	For human consumption		35.8

Adapted from Emery and Iselin, 1967; FAO Agriculture and Fisheries Reports, 1970; and Sumich, 1976.

TABLE   12.2   Primary Productivity of Various Ecological Systems

Ecosystem	Production in Organic Carbon per Year lb. per acre	metric tons per km²
Terrestrial		
European forests	2,004	225
Steppe (dry grassland)	428	48
Desert	53	6
Cornfield (Ohio)	7,640	862
Apple orchard, trees only (N.Y.)	4,696	526
Freshwater		
Hard water lake (Lake Mendota, Wisconsin)	4,275	480
Soft water lake (Lake Weber, Wisconsin)	2,138	240
Bog Lake (Cedar Bog Lake, Minnesota)	989	111
Marine		
Western Atlantic (23–41 N)	2,850–4,721	320–530
Western Atlantic (Georges Bank)	4,988	560
Long Island Sound	5,341–8,907	600–1,000

From Odum, 1953.

from the world oceans as a whole so much smaller? To answer this question we must find out what kinds of marine life are used as human food, consider the technical problems of harvesting the sea, examine the geography of fish production, and place fisheries in the context of the marine ecosystem.

HARVESTING THE SEA

VARIETIES OF SEAFOOD

The harvest from the sea includes many different species. Quantitatively, by far the most important category is *finfish* (Figure 12.1). This represents four distinct ecolo-

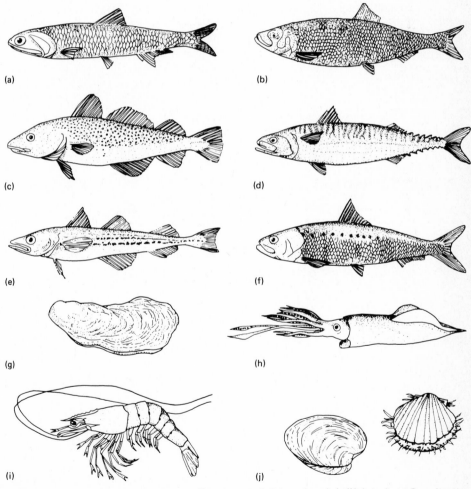

(a)

(b)

(c)

(d)

(e)

(f)

(g)

(h)

(i)

(j)

FIGURE 12.1   Examples of commercially important finfish (a–f) and shellfish (g–j). (a) Peruvian anchoveta; (b) Atlantic herring; (c) Atlantic cod; (d) mackerel; (e) Alaska walleye pollack; (f) South African pilchard; (g) oyster; (h) squid; (i) shrimp; (j) clams and scallops. (From "The Food Resources of the Ocean" by S.J. Holt. Copyright © 1969 by Scientific American, Inc. All rights reserved.)

gical types (Table 12.3). The *pelagic planktivores* are typically small fish, less than 25 cm long as adults, which form great schools in shallow water and feed, by means of gill rakers, on small planktonic organisms near the base of the food chain. Examples include the Peruvian anchoveta *(Engraulis ringens),* which for the most part feed directly on phytoplankton, and adult herring, which feed at the next higher trophic level on herbivorous zooplankton.

The commercially important *pelagic carnivores* include many larger fish which feed at various trophic levels. Mackerel of the genus *Scomber* and jack mackerel *(Trachurus)* are caught in great abundance. Both feed on pelagic planktivores and

TABLE 12.3 Ecological Classification of Seafood Organisms

Organism	Dominant Ecological Type
WHALES	Carnivorous mammals
FINFISH	
*Anchovies and *anchovetas*	
*Herring	Pelagic plankivores
Sardines and pilchards	
Menhaden	
*Mackerel	Pelagic carnivores
Tunas	
*Cod	
*Alaska pollack	
Flounder	
Haddock	
Sole	Demersal carnivores
Halibut	
Hakes	
Seaperches	
Salmon	
Capelin	Diadromous fish
Kilka	
SHELLFISH	
Clams	
Mussels	Benthic mollusks
Oysters	
Scallops	
Squid	Nectobenthic mollusks
Octopus	
Shrimps	Pelagic crustaceans
Prawns	
Lobsters	Nectobenthic crustaceans
Crabs	
PLANTS	
Brown algae	Benthic photosynthesizers
Red algae	

* Indicates the five species ranked highest in terms of world fish catch. Species in italics are used only for fish meal. In addition, much of the production of herring, anchovies, and sardines is used in this way.

invertebrates. Tunas are the largest commercial species and feed near the top of the food chain. *Demersal carnivores* live on or near the bottom, feeding on nektobenthic and benthic invertebrates as well as on smaller fish. Commercially important species are caught in shallow waters of the continental shelf. Flatfish, such as flounder, sole, and halibut, as well as cod have for years been taken from North Atlantic fishing grounds, including Georges Bank and the waters of the North Sea. In more recent times, the Alaskan pollack has been fished intensively from shallow waters around the Gulf of Alaska and the Bering Sea by Japanese, Russian, and U.S. trawler fleets. Finally, we should consider those fish which migrate between fresh and salt water. These *diadromous* species exhibit a fascinating diversity of life styles. Some, like the salmon, are marine fish that spawn in fresh water. Others, like the *kilka*, feed mainly as pelagic planktivores and live in brackish inland seas such as the Black and Caspian. And the freshwater eels, which are a significant element in many European diets, spawn in the ocean.

TABLE 12.4 Marine and Freshwater World Catch (Statistics for 1970 arranged by dominant ecological type.)*

Ecological Type	Catch Results	
	10^6 tons	% total marine
Pelagic planktivores		
Anchovetas and anchovies	14.419	
Herring	2.784	
Others	3.947	
Subtotal	21.150	34.0
Demersal carnivores		
Cod	3.076	
Pollack	3.057	
Others	9.067	
Subtotal	15.200	24.4
Diadromous fishes	2.940	4.7
Pelagic carnivores		
Mackerel	2.887	
Tunas	1.228	
Others	3.437	
Subtotal	7.552	12.1
Unsorted	9.578	15.4
Total marine finfish catch	56.420	90.7
Marine shellfish		
Mollusks	3.300	
Crustaceans	1.620	
Subtotal	4.920	7.9
Marine Plants	0.870	1.4
Total marine catch	62.210	100.0
Total freshwater catch	7.067	
Total world catch	69.300	

* Estimates made from publications of United Nations Food and Agriculture Organization.

Having considered the main ecological groupings of finfish, turn to Table 12.4 and consider the relative importance of these and other seafood types in human diet. Pelagic planktivores are by far the most important group, yielding about one-third the world catch. All the species used as fish meal (anchovetas, anchovies, sardines, menhaden, and herring) are in this group. The reason for this dominance is clear: Living close to the bottom of the food chain, pelagic planktivores must be more abundant than food fish living at higher trophic levels.

The second most important group is the demersal carnivores. These fish contribute about one-fourth the harvest from the sea. Considered as a whole, their diet includes many invertebrates which feed directly on small plankton. Pelagic carnivores rank third. These include lower level carnivores, such as the mackerel, and top-level carnivores, such as the tuna.

Several other important facts can be obtained from Table 12.4. First, marine shellfish form only a small part of the total world catch. Second, seaweeds and other benthic plants are insignificant. Phytoplankton, the numerically dominant form of plant life in the sea, plays essentially no direct role in human nutrition. Finally, we may note that commercial finfish are inhabitants of relatively shallow waters. Carnivores living at mid-depths in the ocean are not harvested. One important reason why the ocean plays such a small role in feeding man is already clear: the species eaten are not plants but animals removed at least one and usually several trophic levels from the primary producers.

FISHING TECHNIQUES

As any angler knows, fishing methods must be related to the habits of fish. Demersal fishing is accomplished mainly by bottom *trawling,* that is, by dragging a tapered bag of netting over the seabed (Figure 12.2a). *Long-line fishing,* which consists of putting out long lines with baited hooks attached, is much less efficient, but must be used where topography is too rough to permit bottom trawling (Figure 12.2b). It has also been effective in catching large, nonschooling fish above the bottom. Pelagic fishing is accomplished mostly by *gill nets, ring nets,* and *purse seines.* Gill nets are rectangular nets suspended near the surface by floats. The dimensions of the net are such that a fish, attempting to swim through it, is caught by the gills (Figure 12.2c). The net is played out and the fishing vessel allowed to drift through the water. Ring nets (Figure 12.2d) and purse seines (Figure 12.3) are fine-mesh nets suspended near the surface by floats. Fish are trapped rather than entangled. When a school of fish is detected, often by an echo-location device, the net is drawn up to encircle the fish. As the net is hauled in, messenger ropes close the bottom of the net and prevent fish from swimming away underneath. In this way, the high-speed purse seines used to catch tuna often trap and kill porpoises inadvertently.

A newer development in pelagic fishing is the use of *mid-water trawls,* which are drawn through the water at a particular depth below the surface. A knowledge of fish habits, supplemented by sonic location techniques, is required if mid-water trawls are to be used effectively.

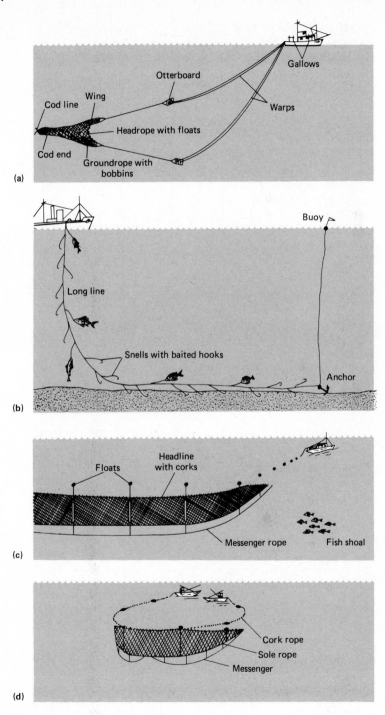

FIGURE 12.2   Common fishing techniques. (a) Bottom trawling for demersal species. (b) Long-line fishing for pelagic and demersal species. (c) Gill nets used to catch schools of pelagic fish. (d) Herring ring net. (From Tait and De Santo, 1972.)

FIGURE 12.3   A tuna seiner with a school of tuna trapped within its purse net. (Courtesy of U.S. Marine Fisheries Society.)

GEOGRAPHY OF MARINE FISHERIES

The fundamental control on the distribution of seafood resources is the distribution of primary organic productivity. In Table 12.5 the ocean is divided into three provinces: open ocean, the coastal zone, and areas of intense upwelling. On the aver-

TABLE   12.5   Division of the Ocean into Provinces According to Their Level of Primary Organic Production

Province	Percentage of Ocean	Area (km²)	Mean Productivity per Unit Area (grams of carbon/m²/yr)	Total Productivity (10^9 tons of carbon/yr)
Open ocean	90	326×10^6	50	16.3
Coastal zone*	9.9	36×10^6	100	3.6
Upwelling areas	0.1	3.6×10^5	300	0.1
Total				20.0

* Includes offshore areas of high productivity. (From Ryther, 1969).

age, upwelling areas are six times as productive as the open ocean; the coastal zone is intermediate. When allowance is made for the area of each productivity province, the open ocean yields 81 percent, the coastal zone 18 percent, and the upwelling areas only 0.5 percent of the total annual production.

We have thus far considered only the net organic production. Since man eats fish and lobsters rather than dinoflagellates and diatoms, we must now examine the processes by which this net primary production is transformed into fish and shellfish. Two aspects are critical: the number of trophic levels linking algae with the edible product and the ecological efficiency of energy transfers which occur between each trophic level. Table 12.6 shows an analysis of the average trophic structure of representative regions of the ocean. Typical food chains leading to commercial fish in the open ocean have an average of five trophic steps above the primary level. Because the average ecological efficiency is here estimated to be about 10 percent, 90 percent of the energy will be lost at every level. As a result, the productivity of the level cropped will be a tiny fraction of the net primary productivity—only one part in ten thousand. By contrast, a typical continental shelf community, such as the Grand Banks of Newfoundland, has an average of three trophic levels above the first. Two food chains are of interest. The pelagic chain leads to carnivores by way of pelagic plankton feeders such as the herring. The benthic chain leads to demersal carnivores by way of planktivorous invertebrates. From this analysis, we see immediately one of the reasons why herring and cod are more plentiful than tuna: They lie closer to the primary trophic level. In contrast, communities characteristic of upwelling areas have only one or two trophic levels intervening between the edible and primary levels. The shortest food chain of commercial importance is that occurring in the area of intense seasonal upwelling off Peru, where anchoveta feed directly on relatively large phytoplankton species.

TABLE 12.6   Estimated Fish Production in the Three Ocean Provinces Defined in Table 12.5

Province	Primary Production (tons of organic carbon)	No. of Trophic Levels Above First	Efficiency (%)	Fish Production (tons live weight)
Open ocean	16.3×10^9	5	10	0.16×10^7
Coastal zone	3.6×10^9	3	15	12×10^7
Upwelling areas	0.1×10^9	1½	20	12×10^7
Total	20.0×10^9			24×10^7

From Ryther, 1969.

These observations have important consequences in the distribution of oceanic food resources, as shown in Table 12.6. Here the average number of trophic levels and the average ecological efficiencies are shown, along with the calculated fish production per year. The important conclusion to draw from this calculation is that less than 1 percent of the total fish production *available* for cropping occurs in the open ocean. Virtually all of it occurs in the coastal and upwelling provinces. Surprisingly, it is about equally divided between them, in spite of the great difference in area (Figure 12.4). The economic advantages in fishing in such concentrated areas of production are obvious.

Thus the geography of oceanic food resources can be explained in terms of primary productivity and trophic structure: Fishing is best where nutrients are replenished rapidly and the number of trophic levels is small. These principles are well illustrated by the Peruvian anchoveta fisheries. In normal years, Peru now leads the world in the size of its annual catch, some 10 million tons—one-sixth the world's catch and three times that of the United States.

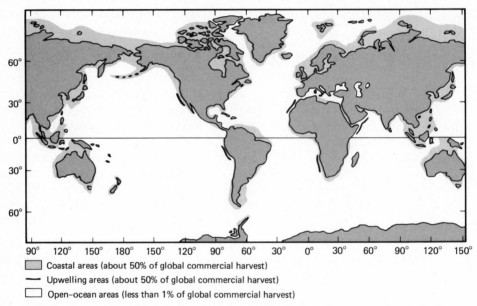

Coastal areas (about 50% of global commercial harvest)

Upwelling areas (about 50% of global commercial harvest)

Open-ocean areas (less than 1% of global commercial harvest)

FIGURE 12.4 Geography of main fishing areas. (From *Patterns and Perspectives in Environmental Science*. Report Prepared for the National Science Board, National Science Foundation, 1972.)

OPTIMUM FISHING RATE

Theoretical models For centuries man exploited the food resources of the sea, under the assumption that the ocean was so large that it made no difference how fast and intensively he fished. But the advent of modern fishing techniques and the needs of an expanding population have in many areas led to overfishing.

How do we recognize when an area is overfished? One answer lies in keeping track of the amount of fish caught per unit fishing effort and of the quality of the catch. If an area is overfished, so high a proportion of the breeding stock is removed each year that the population cannot replenish itself and stocks diminish from year to year. This condition can be recognized in two ways: The catch per unit fishing effort declines from year to year, and the proportion of small fish increases. These principles are clearly shown in the North Sea haddock fisheries statistics (Figure 12.5). Immediately following World War I, during which fishing virtually ceased, the catch per 100 hours effort was over 10,000 kg. This figure declined rapidly during the next few years and then continued a slow decline until fishing again ceased in 1940. At the same time the percentage of small fish caught rose from about 10 to nearly 80 percent. Clearly, a continuation of this level of effort would have resulted in a declining total catch per unit effort. The years of World War II, however, again saw an interruption of fishing. By 1945 the stock had replenished itself nearly to the level of 1920; after that, the declining trend recurred. We may conclude that this area is somewhat overfished.

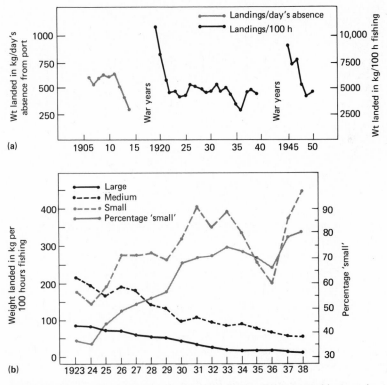

FIGURE 12.5   Catch statistics for the North Sea haddock. (a) Haddock catch per unit of fishing effort by Scottish trawlers. (b) Catch per unit fishing effort and percentage of small-size class caught. (From Graham, 1956.)

The aim of fisheries management is to secure the maximum catch which can be sustained for an indefinite period. From the biological point of view, the *optimum fishing rate* is that which results in the *maximum sustainable yield*. As we have seen, overfishing reduces the sustainable yield. But an area can also be underfished, for reasons that are in part determined by the process of growth in fish populations. As illustrated by the growth rates reported in Figure 12.6, herring increase in size more rapidly during their early years than they do later in life. This growth pattern is typical, and we may conclude that older fish use relatively more of their food for maintaining themselves, and less for growth, than do younger fish. Therefore, because the older fish are competing with younger fish for food, greater total growth in the population will occur if a portion of the older fish are cropped.

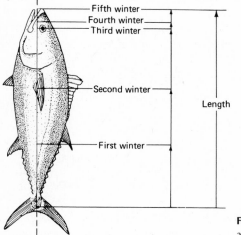

FIGURE 12.6   Growth of herring at various ages. (After Tait and De Santo, 1972.)

Figure 12.7 summarizes the theory of optimum fishing rate. If fishing effort is too low, too many older fish are left in the population and the yield is below the optimum. If the effort is too high, too much of the stock needed to recruit the next generation is removed and the yield is again below optimum. There are two practical problems in applying the theory. The first is to calculate the optimum rate; the second is to control fishing so that a yield near the calculated optimum is actually maintained. The problem of control has proved to be very difficult, especially where the ships of several nations are competing for the same resource. As explained in Chapter 20, the fundamental difficulty is the long-established principle of open access to marine resources. Because no one clearly owns the fish in many areas of the sea, there is great competition to exploit a resource before it is rendered less profitable (or eliminated) by the fishing activities of other operators or nations. Historically, fishing rates have usually been optimized not in biological, but in economic, terms; that is, fishing rates are maintained so that the cost of fishing is somewhat less than the financial return. This *economic equilibrium* fishing rate is usually much greater than the biologically optimum rate; hence the actual yields are often much less than could be obtained by proper regulation.

Let us now examine two case histories illustrating the consequences of overfishing: the history of Antarctic whaling and the exploitation of the Peruvian anchoveta.

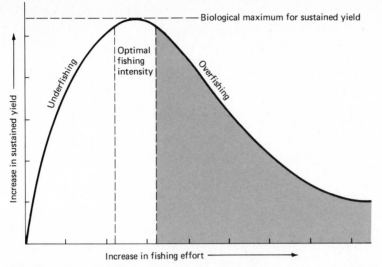

FIGURE 12.7   Sustained yield as a function of fishing effort. (© W. D. Russell-Hunter, 1970; *Aquatic Productivity;* Macmillan.)

Antarctic whaling   Rarely are the consequences of overfishing more clearly or more tragically illustrated than by the history of Antarctic whaling. Having seriously depleted the stocks of blue and fin whales from Atlantic and Pacific waters by the use of a harpoon with an explosive head fired from a cannon, the whaling industry, during the 1920s and 1930s, turned to a systematic exploitation of whales from their rich Antarctic feeding grounds. By 1930 the catch of blue whales was nearly 30,000 per year. During the succeeding decade, hunting continued unabated, in spite of clear evidence of overfishing and declining returns (Figure 12.8). Unlike the

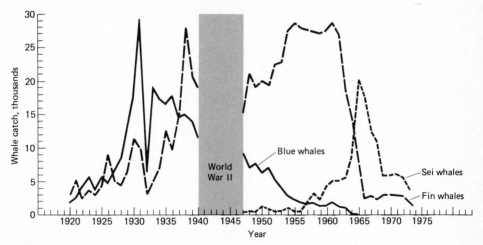

FIGURE 12.8   Catch of blue, fin, and sei whales in the Antarctic, from FAO catch and landing statistics. (From Sumich, J., *An Introduction to the Biology of Marine Life.* Dubuque, Iowa: Wm. C. Brown Co. Publishers. Used with permission.)

case of the North Sea haddock (Figure 12.5), the respite during World War II did not see a recovery of the stock. Presumably, the slow reproduction rates of these huge mammals (a female blue whale can produce only one young every four years) makes such a recovery a very slow process. In 1948, 20 whaling nations established an International Whaling Commission. Lacking powers of enforcement and setting quotas much higher than those recommended by biologists, this effort at control availed nothing. By the late 1960s the species was on the verge of extinction. As the stocks of blue whales became depleted, the fishing effort turned first to the smaller fin and then sei whales—with results that almost duplicated the exploitation history of the blue whale. Today, the Japanese and Soviet fleets are turning their attention to yet smaller whales.

Peruvian anchoveta. As discussed previously in this chapter, an area of intense upwelling occurs off the coast of Peru. Since the late 1950s, when the Peruvian anchoveta began to be fished intensively, this area has been the site of one of the world's great fishing industries. Peruvian fishermen compete with other predators—great numbers of boobies, pelicans, cormorants, and other seabirds which here are a conspicuous part of the ecosystem. Bird droppings on offshore islands leave a valuable deposit of fertilizer (guano), which is an important economic resource. Intense fishing during the 1960s yielded larger and larger catches until the peak year 1970, when some 13 million tons were harvested (Figure 12.9). The bird popula-

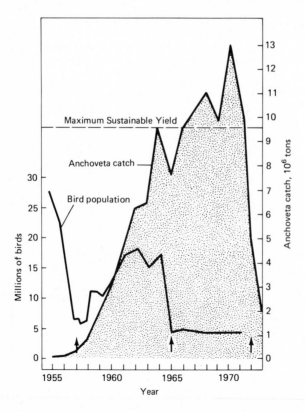

FIGURE 12.9 Changes in the guano bird population and the anchoveta catch along the coast of Peru. Arrows indicate El Niño years, when upwelling diminishes. (Adapted by Sumich, 1976, from Shaefer, 1970.)

tion, losing in the competition with man, meanwhile underwent a generally declining trend. Fearing the consequences of possible over-exploitation, a panel of experts recommended a maximum sustainable yield of about 9.5 million tons. The catch from 1966–1971 nevertheless exceeded this recommended limit. During 1972 and 1973 a catastrophic decline in the anchoveta harvest occurred, with economic consequences that reached halfway around the world. In the United States, the cost of food was increased because the lack of fish meal forced livestock growers to buy other, more expensive, feeds.

One explanation for the 1972–1973 anchoveta decline is that it occurred as a direct result of overfishing. Another explanation is that it occurred as a consequence of a decrease in the supply of nutrients to the phytoplankton, on which anchoveta feed. As will be explained in Chapter 19, such decreases in nutrients are known to occur in certain years when the rate of upwelling diminishes during the summer—a phenomenon known as El Niño. The El Niños of 1957 and 1965 were both accompanied by severe drops in seabird populations. Presumably, the impact on anchoveta was completely overcome in 1957 by the intense and accelerating fishing effort, and largely overcome in 1965 for the same reason. According to this explanation, the anchoveta disaster of 1972 occurred because the El Niño of that year affected a stock which had been overfished for six years.

IMPROVING THE HARVEST

During the interval 1950–1973 the increase in world fish catch was greater than the increase in global population (Figure 12.10). These facts encouraged belief that the problem of feeding future generations might be solved, or at least significantly alleviated, by turning to the sea. We are now in a position to return to this question and ask how much food the sea can produce.

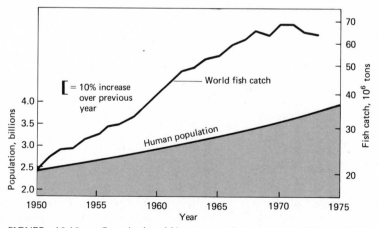

FIGURE 12.10   Growth of world human population and global fish catch, 1950 –1973, from FAO catch and landing statistics. (From Sumich, J., *An Introduction to the Biology of Marine Life*. Dubuque, Iowa: Wm. C. Brown Co. Publishers. Used with permission.)

TECHNIQUES

Four basic ways of improving the harvest can be recognized. The first, and simplest, is to increase the fishing intensity on underfished stocks. This approach, implemented by modern methods and equipment, led to the spectacular increases in fish catch shown in Figure 12.10. There is much evidence to indicate that further gains can be expected along this line—especially by catching species not now used as human food—provided fishing intensities can be held to optimal rates. How great an increase can be expected we will examine in the following section.

The second, and most basic, approach to improving the marine harvest is to increase primary productivity by adding nutrients to the water. In Scotland, experiments carried on in estuaries by adding chemical fertilizers to the water have had mixed results. In some cases, production increased among algae rather than fish. In others, flounders flourished but other species died. In one loch, where circulation to the sea was not inhibited, growth rates of flatfish were significantly increased, but adults escaped into the open ocean. Bolder schemes call for some form of pumping of deep, nutrient-rich ocean waters to the surface. At one site in the Caribbean such an experiment was tried (Chapter 21), but unfortunately the benefits achieved were small compared to the cost. It has also been suggested that a nuclear reactor, placed on the seabed, would heat the water, promote an artificial upwelling, and so increase surface productivity.

The third way of increasing the harvest is to utilize species lower in the food chain. Ideally, we should eat diatoms rather than tuna. Pilot projects carried out in British waters have actually attempted to harvest plankton, but costs far exceeded benefits. More ambitious projects, undertaken by Japan, the U.S.S.R., and other countries, are now harvesting large forms of zooplankton *(krill)*, which are extremely abundant in Antarctic waters and are the main food of many whales. It remains to be seen whether the practical difficulties of harvesting, processing, and transporting can be overcome.

Finally, *mariculture* might be practiced; that is, a complete system of producing marine food might be organized. Such a system would include the selective breeding of species, techniques for herding mobile organisms, and procedures for introducing nutrients. Practical difficulties, however, will restrict such activities to protected, near-shore waters or to the hatching and culturing of salmon and other fish in coastal streams. Limited but successful forms of mariculture are practiced today, notably with shellfish.

LIMITS

If a maximum effort is made to fish more intensively in underfished areas, how much food could be obtained from the sea? One way of answering this question is to extend our experience in the North Sea fisheries area to the world ocean. As we have seen in Chapter 11, intensive pelagic and demersal fishing there yields a maximum sustainable yield on the order of 0.066 percent of the gross primary production. The estimate from Table 12.5 of the gross production of the world ocean of 20×10^9 metric tons C/yr, the equivalent of 272×10^9 metric tons (wet weight) of fish per year. Assuming that the harvesting effort in the North Sea area could be extended globally, we would estimate a total yield of 0.066 percent of this figure, or about 180×10^6 metric tons of fish per year. The annual catch of fish is now about 60 million metric tons, or about one-third our estimate. Practical considerations, however, would suggest that we could not achieve this intensity on a global basis. A doubling of the present yield seems a reasonable possibility. To allow for uncertainties in this estimate and for the possibility that organized mariculture in nearshore and protected waters might someday yield one-half the present total world fish catch, let us assume that the present harvest could be tripled. The 180 million metric tons/yr produced would provide only 3 percent of the 6000 million tons required to maintain 8 billion people at current nutritional levels. It is projected that such a doubling of world population will occur in 40 years.

All ecosystems, including one as huge as the world ocean, have limits. Although increased fishing can be expected to allieviate food shortages, it is unrealistic to expect to solve the problem of human food supply by increasing the harvest from the sea.

SUMMARY

1. The 60 million tons of seafood havested each year constitute only about 1 percent of human food.

2. Most seafood is fish belonging to three ecological groups: pelagic planktivores, demersal carnivores, and pelagic carnivores.

3. Fishing is best where nutrients are replenished rapidly and the stocks fished are low in the food chain.

4. The open ocean contributes little to the harvest from the sea, because upwelling is slow and the species of fish taken there generally feed at high trophic levels. Most of the catch is taken about equally from two areas of very unequal size: regions of intense upwelling near the coast and coastal areas.

5. Optimum fishing rates, calculated to give a maximum sustained yield, are often exceeded owing to economic competition which prevails under an open-access policy.

6. Calculations indicate that if underfished areas were fished more intensively, the current harvest might be doubled. Although organized systems of mariculture, practiced in protected waters near shore, would further increase seafood production, it is unrealistic to expect to solve the problem of human food supply by increasing the harvest from the sea.

Kilauea Crater in the Hawaiian Islands. (Photo courtesy of the Hawaii Visitors Bureau.)

Sediments: Where They Come From and What They Are

13

Our experience with sediments may range from the sight of sand or accumulations of seashells on the seashore to the feel of soft mud at the bottom of a lake. The experience of our toes and our eyes is that sediments may range from soft oozes to rough gravels. If we carry our observational acumen from sea level to the hills, in many places we will notice sedimentary rocks resembling, in virtually all features except hardness, the sediments we sensed at the seashore, the river,

CHAPTER

or the lake. Sedimentary rocks record the fact that sediments have been produced throughout geologic time by processes we can observe today. These sediments have been hardened into rocks by nature's unrelenting processes of squeezing, heating and cementing.

In this chapter we discuss the sources of sediments and their constitution.

SOURCES

There are three ultimate sources of the material found on the ocean floor: the continents, which supply detrital and dissolved materials; indigenous sources in the ocean basin associated with vulcanism; and material derived from extraterrestrial sources.

The most important of these volumetrically is the continents and the least important is the meteoritic debris which falls into the sea. We will thus concentrate on the first two sources only.

THE CONTINENTS AS A SOURCE OF SEDIMENTS

Weathering and erosion   The most obvious source of sedimentary material is the continents, where weathering and erosion clearly attest to a denudation of the land and transport to the sea. Weathering is the action of plants and bacteria in chemically altering and physically disaggregating rocks. The major chemical expression of this attack is the reaction of soil acids, mainly carbonic acid, with silicates to form clay minerals and dissolved chemical species. An idealized reaction for weathering can thus be written:

$$\text{Na feldspar} + H_2CO_3 = Na^+ \text{ (aqueous)} + HCO_3^- \text{ (aqueous)} +$$
$$SiO_2 \text{ (aqueous)} + \text{kaolinite (a clay mineral)}$$

This reaction indicates that every time one "molecule" of a sodium feldspar— a common component of most rocks—is transformed into one "molecule" of a clay mineral—in this case indicated as kaolinite—by reaction with one "molecule" of carbonic acid (H_2CO_3), a sodium ion (Na^+), a bicarbonate ion (HCO_3^-) and soluble silicon oxide (SiO_2), will be transported by the stream flowing over the terrain undergoing weathering. Eventually the clay mineral will also reach the sea, but its path may be more circuitous.

Sediments deposited at the continental margins or directly on the continents at times of submergence by the sea are subject to uplift and exposure above sea level as the result of mountain building or broad regional warping of the continent. During the period of deposition and burial and uplift, the sediment may be altered by pressure, temperature, and reactions with subsurface waters. When sediments are so processed, they are called *sedimentary rocks*. A sedimentary rock made of calcium carbonate is called *limestone*, one made up of clay minerals is called *shale*, and one made of disaggregated mineral grains from the original rock (other than the clay minerals) is called *sandstone*. Most sedimentary rocks actually are mixtures of these three.

Sedimentary rocks, like the rocks from which they were ultimately derived, are also subject to the weathering forces at the Earth's surface. In addition to the action of carbonic acid on the silicate-bearing sedimentary rocks, the limestone component will dissolve according to the following reaction:

$$CaCO_3 + H_2CO_3 = Ca^{++} \text{ (aqueous)} + 2\ HCO_3^- \text{ (aqueous)}$$
$$\text{limestone} \quad \text{carbonic}$$
$$\text{acid}$$

This is the process by which caves are formed and limestone generally is weathered. The released calcium transported by streams to the oceans is the source for much of the calcium carbonate deposited by marine organisms.

Shales and sandstones that are relatively less consolidated than igneous rocks such as granites and basalts are also subject to erosion by the action of water. They are transported as particles or *detritus*. The major amount of detrital material brought to the oceans is derived from the erosion by streams of ancient sedimentary deposits now uplifted above sea level.

Denudation of the continents   The rate of weathering and erosion of the continents provides an estimate of the supply rate of sedimentary material to the ocean. The most obvious way of obtaining the supply rate of dissolved chemicals and detritus carried by streams to the oceans is to set up a station at the mouth of each river to record the volume of water passing per unit time, its chemical composition, and suspended detrital concentration, and to sum up all these figures over long periods to determine the total water transport by streams, the total supply of dissolved species, and the total supply of detritus by the flowing streams for any time period.

Table 13.1 shows the estimate of the volume of water transported per unit time by the rivers of the world together with the weighted average concentration of dissolved chemicals and suspended detrital material. Table 13.2 gives the average concentrations of the dissolved material in streams. If we assume that the chloride ion concentration is all due to atmospherically transported sea spray, we correct for the amounts of the other recycled elements from the ocean by using the elemental ratios to chlorine in seawater as shown in Table 4.1. We thus arrive at the concentration of "noncycled" elements to the ocean from the continent.

TABLE 13.1   Riverine Fluxes of Water, Dissolved Solids, and Suspended Material from Each of the Continents

Continent	Water (10^{15} l/yr)	Dissolved Solids (10^{15} g/yr)	Suspended Solids (10^{15} g/yr)
North America	6.0	0.8	2.0
South America	8.0	0.3	1.2
Europe	3.0	0.5	0.3
Asia	12.0	1.7	15.9
Africa	6.0	0.7	0.5
Australia	0.5	0.03	0.2
Total (rounded)	36	4.0	20

Atmospheric transport of material to the ocean   Anyone who has been in a dust storm or a sandstorm cannot doubt that the wind has the capacity to erode and transport material. Some of this material can be wafted high enough into the atmos-

TABLE 13.2   Average Composition of Streams

Component	Measured Concentration		Amount Due to Marine Aerosols	Derived by Weathering	
	Milligrams per liter	Millimoles per liter	Millimoles per liter	Millimoles per liter	Milligrams per liter
HCO_3^-	58.4	0.957	—	0.957	58.4
$SO_4^=$	11.2	0.116	0.011	0.105	10.1
Cl^-	7.8	0.219	0.219	0	0
NO_3^-	1.0	0.016	—	0.016	1.0
Ca^{++}	15.0	0.375	0.004	0.371	14.8
Mg^{++}	4.1	0.171	0.022	0.149	3.6
Na^+	6.3	0.274	0.188	0.086	2.0
K^+	2.3	0.059	0.004	0.055	2.1
SiO_2	13.1	0.211	—	0.211	13.1

phere to be transported far out to sea (Figure 13.1). The farther from the source of windblown material, the lower the accumulation rate of deep-sea sediments transported by this mode. If other modes of sediment supply to the deep ocean are overwhelming, the windblown component may not be noticeable. The effectiveness of wind in transporting detritus to the ocean basins is clearly seen on islands where soils contain minerals that are not derivable from the weathering of native rock—the red soils of Bermuda established on pure marine limestone and the presence of quartz in Hawaiian soils which cannot be derived from the weathering of quartz-free basaltic rocks typical of the Hawaiian islands.

Windblown quartz in deep-sea sediments can be unequivocally identified by the microscopic surface features of the quartz grains found in the sediments (Figure 13.2a).

Glaciation as a source of sediment   Our planet has permanent ice caps in Antarctica and Greenland. During the glacial times, the last of which ended 11,000 years ago, major parts of North America and Europe were also covered with ice sheets. The erosive powers of glaciers and large glacial ice sheets is seen by the vast amount of rock debris found at the bottom of glacial valleys and even across the continents in the case of large scale glaciation by ice sheets. Much of this material, called *till*, can be remobilized fairly easily by streams and the erosion products reaching the sea can retain the imprint of a glacial origin. Pebbles, sharply faceted quartz grains, and the presence of minerals which under slower weathering and disaggregation would have disappeared, all attest to the massive physical degradation and grinding action of glaciers (Figure 13.2b).

Where glaciers with their erosion products embedded in their bases reach the ocean, a process of iceberg formation called *calving* occurs. The sediment-laden

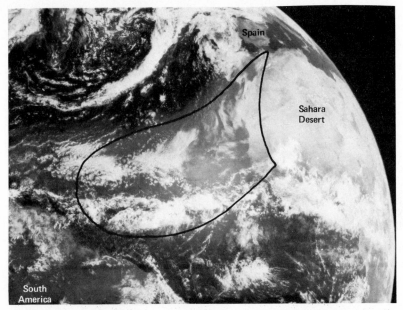

FIGURE 13.1 Satellite photograph showing dust from the Sahara Desert moving westward (SMS Visible Satellite Image, June 30, 1974). The effect on local and worldwide climate by such dust clouds is one of the studies included in the Global Atmospheric Research Project, Atlantic Tropical Experiment. (NASA photograph.)

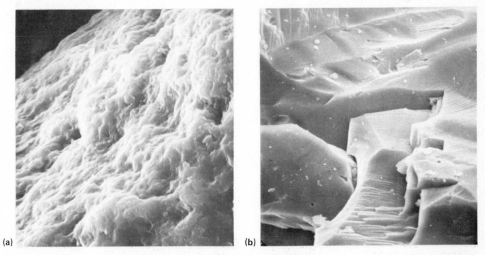

FIGURE 13.2 Quartz grains in deep-sea deposits reflect in their surface features the way in which they were processed during erosion and deposition. These scanning electron micrographs show two diagnostic surface features. (a) "Hot" desert environment: The upturned plates subdued by rounding are characteristic of this type of action. (b) Glacial erosion: The very high relief conchoidal breakage pattern and other features mark this as being formed by the crushing and grinding action of glaciers. (Enlargement in both photographs is about 3000 times.) (From Krinsley, Biscaye, and Turekian, 1973.)

icebergs can float away from the point of calving quite a distance before melting completely, so that glacially derived material in deep-sea sediments can be expected as far as icebergs are found in the oceans.

VOLCANOES WITHIN THE OCEAN BASINS

Throughout all the oceans and along the margins of the Pacific, there is continuing volcanic activity (Figure 13.3). Volcanic islands are found near or on the large oceanic ridge systems such as Iceland and Galapagos. They also occur as island chains within the Pacific Ocean Basin—the Hawaiian Islands and the islands of the South Pacific are examples. Volcanic island arcs rimming the Pacific Ocean are so striking a feature that it has been called "the ring of fire."

In addition to these volcanic islands there is strong evidence of submarine vol-

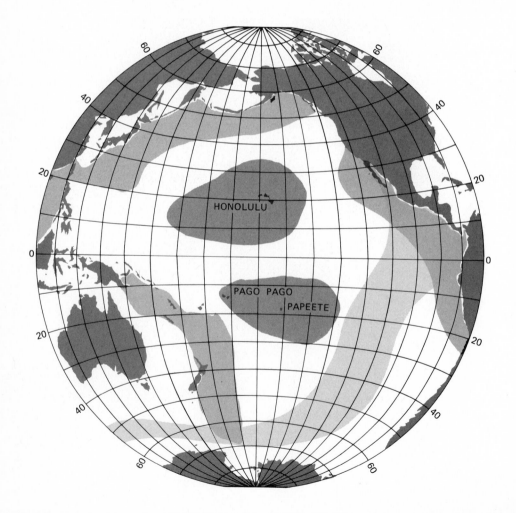

canism. The rugged features of the major oceanic ridge systems, for example, are partly due to volcanic activity and partly to structural movements of the basaltic blocks. Underwater photographs taken by a remote camera or a submersible clearly attest to considerable underwater outpouring of lava from fissures.

The basaltic rock reacts with seawater to produce characteristic alteration products as well as rock and glass fragments. Some of these resemble the clay minerals (see below) found in weathering profiles. Others, such as the zeolites (Figure 13.4), however, are uniquely diagnostic of altered basaltic material. (Zeolites are characterized as having large structural holes in their lattices, thus providing easy passages for ions and molecules to enter. They are of commercial value now because they are useful in many research and production applications as a kind of "molecular sieve" to capture specific classes of organic molecules while letting pass other gaseous substances.)

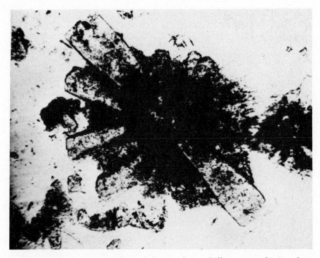

FIGURE 13.4   Basaltic glass with needles of the zeolite, phillipsite, radiating from it. It is believed that phillipsite and some montmorillonite are formed by the devitrification of hydrated basaltic glass. Magnification 190 times. (Photo courtesy E. Bonatti.)

FIGURE 13.3   Three types of volcanic sources supply materials to the Pacific Ocean bottom. (1) *Dark shading:* Volcanism creating the islands in the center of the Pacific also deposits debris on the ocean bottom, much of which is altered to form phillipsite and montmorillonite, as shown in Figure 13.4. (2) *Medium shading:* Ejecta from explosive volcanism along the continental margins of the Pacific is transported with the prevailing winds at each latitude eastward between 20° and 50° and westward within 20° on either side of the equator. (3) *Light shading:* Submarine volcanism along the East Pacific Rise results in basaltic flows and from the disintegration of the volcanic outpouring into seawater glass is distributed widely in the sediments.

MINERALOGY OF SEDIMENTS

The sedimentary components in the ocean are derived directly as detritus, without significant subsequent alteration, either from the weathering and erosion of the continents; from the alteration of volcanic rocks in the ocean basins; or from precipitation from seawater. Because the first two are mainly silicates and the last is dominated by biologically derived material, a convenient division for our discussion of the mineralogy of sediments can be made on this basis.

SILICATE MINERALS

The sediments in the ocean are made up of different grain-size components (Table 13.3). Although a certain amount of sorting occurs during transport in a fluid, this is never complete. It is useful, however, to talk about the major mineralogy typical of the different grain-size sediments.

TABLE 13.3 Sizes of Sedimentary Components*

Name	Particle Diameter (mm)
Boulders	Greater than 256
Cobbles	64 to 256
Pebbles	4 to 64
Granules	2 to 4
Sand	0.062 to 2
Silt	0.004 to 0.062
Clay	Less than 0.004

* Commonly called the "Wentworth scale."

The smallest size of sediment particle is called clay size and is dominated by the clay minerals. These are the common degradation products of the action of acids on the more "primitive" rock-forming minerals such as feldspar, pyroxene, amphibole, and mica. The acids may be soil acids such as carbonic acid or the acid solutions associated with volcanic activity.

The clay minerals have in common the atomic structure typical of ordinary muscovite mica (Figure 13.5). The dominant feature, obvious to anyone who has looked inside a toaster or peeled a "book" of mica, is the sheet-by-sheet structure. This is a reflection of the atomic stacking. In the case of mica, the structure is highly ordered and the "cement" between the sheets is potassium ions firmly held. The clay-size equivalent of muscovite mica is called *illite*.

Variations in this basic mica structure result in the clay minerals shown in Figure 13.5. *Chlorite* is a common mineral in continental and some submarine rocks but can also be the product of weathering or hot acid water activity on pyroxene, amphibole, and biotite mica. The more disordered clay mineral called *montmorillonite* can be formed in weathering profiles and by hot acid water activity on vir-

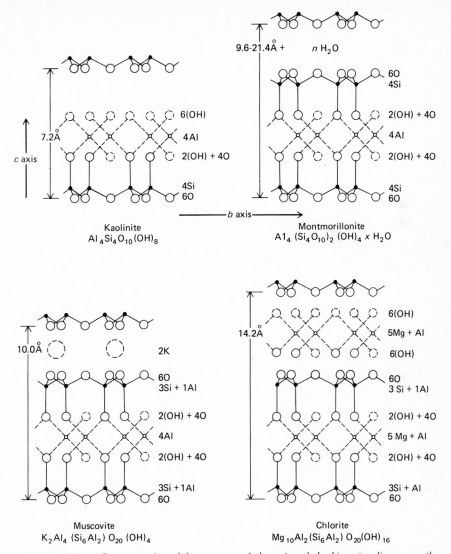

FIGURE 13.5   Representation of the structure of clay minerals looking at a slice across the sheets. The imaginary line perpendicular to the sheets is the "C axis" and the length between each repeating sheeting along this axis is given in Angstroms (= 10^{-8} cm). These characteristic layer repetition distances provide reflection planes for x-rays. Thus clay minerals can be identified by recording the multiple reflection planes.

tually any common primary mineral found in continental rocks or submarine volcanic rocks. It is characterized by having exchangeable ions between the sheets. The easy mobility of the ions also permits the interjection of water or organic molecules between the sheets, and for this reason it is called "expandable" clay.

Kaolinite is the final product of clay formation from which all the metals except aluminum and silicon have been stripped. The sheets of silicon oxide and aluminum oxide are held together through hydrogen bonds in the OH group in the sheets. Kaolinite is formed massively in tropical and semitropical environments because of the intense weathering there.

When complete silicon oxide leaching occurs, the residual mineral is a virtually pure oxide of aluminum called *gibbsite,* with the chemical formula $Al(OH)_3$. Gibbsite commonly occurs with kaolinite in tropical or subtropical soil profiles (or fossil soil profiles which were once formed under mild climatic regimes). This is the common aluminum ore which was called bauxite after the French town Les Baux, where it was first designated. Iron oxide as goethite, $FeOOH$, is commonly associated with the gibbsite and kaolinite, and the general term for such a strongly leached soil mineral assemblage is *laterite*.

Although many sedimentologists subscribe to the size scale classification of Table 13.3, others make the formal distinction between size types at the 2, 20, and 500 micrometer diameters. The grain-size range between a diameter of 2 micrometers and 20 micrometers (or "microns") is commonly called *silt* and between that of 20 micrometers and 500 micrometers is called *sand*.

A common mineral of both the silt and sand sizes is quartz, which is a ubiquitous mineral found in continental rocks and is relatively stable to weathering. In rapidly disaggregated rocks the chemically less resistant minerals also contribute to the silt and sand fractions.

Gravel (greater than 500 micrometers) is typical primarily of coastal sediments, because there is no way such a large-size grain can be transported to the deep ocean under normal conditions. If we do see layers of sand or gravel-size sediments in deep-sea sediments, we can be sure that strong bottom transport processes have been operating or that they were deposited from melting icebergs.

MINERALOGY OF BIOGENIC COMPONENTS OF MARINE DEPOSITS

Calcium carbonate, silicon oxide, and to a lesser extent calcium phosphate are the major chemical compounds deposited by marine organisms in their hard parts (Figure 9.4).

Calcium carbonate as found in the skeletal parts of most marine organisms occurs in one of two crystallographic forms; that is, for the same approximate chemical composition, two different atomic arrays of the calcium and carbonate ions can occur. These are called *polymorphs,* and are given different mineral names. The names of the calcium carbonate minerals are *calcite* and *aragonite*.

At Earth's surface or ocean bottom pressures and temperatures, calcite is theoretically the stable form, just as graphite is the stable polymorph of carbon in relation to diamond. But whereas diamond is known to form as a stable polymorph in the mantle of the earth and is found at the surface only because of deep-seated forces pushing it to the surface, aragonite in marine shells exists primarily because of the peculiarities of the calcium-carbonate-depositing process of living organisms.

Table 13.4 shows the polymorphs of calcium carbonate found in each of the major marine organisms. In addition to calcite and aragonite there is also a distinct difference between low magnesium calcite and high magnesium calcite. The former is typical of the *tests* (or shells) of pelagic organisms found in deep-sea sediments. High magnesium calcite is found in echinoderms (for example, starfish).

TABLE 13.4 Mineralogical Forms of Calcium Carbonate Deposited by Marine Organisms

	Aragonite	Calcite	High-Magnesium Calcite
Corals (scleractinian)	X		
Snails	X	X	
Clams	X	X	
Echinoderms (e.g., starfish)			X
Foraminifera (pelagic)		X	
Coccoliths		X	

Silicon oxide in a form resembling opal is found in the tests of diatoms, radiolaria, and silicoflagellates. Sediments made up primarily of these siliceous tests are converted over time to dense crystalline material which is really fine grained quartz —it is called *chert*. Contemporary sediments contain siliceous tests, but the older the sediment the greater the likelihood of the occurrence of chert as a result of this *recrystallization* process.

SUMMARY

1. Weathering, as the result of the action of life, and erosion by streams, wind, and glaciers act to denude the continents and to supply materials to the ocean basins.

2. Volcanism, around the Pacific Ocean in particular and within all the ocean basins at the major ocean ridges and elsewhere, contributes material to the oceans.

3. Marine sediments are composed of detrital and indigenously formed silicate minerals and biogenic materials, including calcium carbonate and silicon oxide.

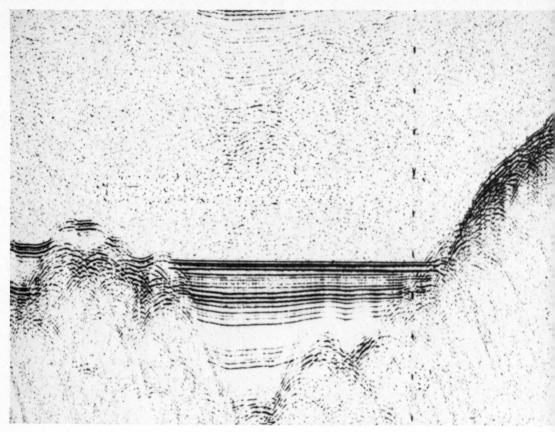

Reflection profile of the flank of the Mid-Atlantic Ridge. (Courtesy of the Teledyne Exploration Company.)

Deep-Sea
Deposits

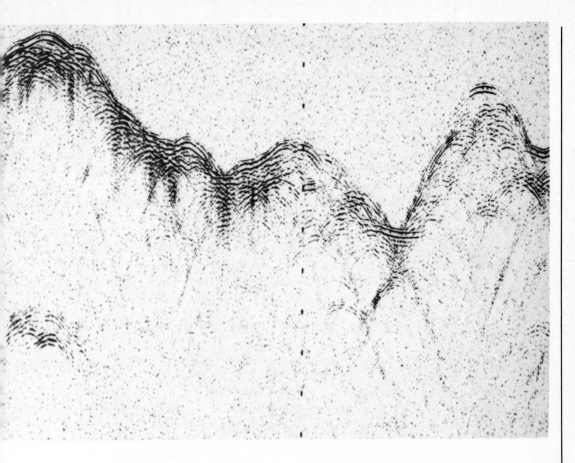

14

If we were to scrape the top of the sediment pile at different places in the ocean, as was first done systematically on the *Challenger* expedition, we would find differences in composition. The major types of sediments from the deep sea are made of different proportions of three major components: (1) calcium carbonate of biological origin (that is, foraminiferal tests and coccoliths); (2) opal-like silicon oxide, also of biological origin (that is, radiolaria, diatoms, and silicoflagel-

CHAPTER

lates); and (3) silicate minerals of nonbiological origin. In addition, our scoop might occasionally bring up black-brown nodules made largely of complex manganese and iron oxide minerals.

By what means did the deep ocean become the repository of such ensembles, and what are the patterns of distribution there? These are the questions for which we seek answers in this chapter.

ASSESSING SEDIMENT REPOSITORIES IN THE OCEAN

As we have seen, sedimentary materials may be carried to the oceans by various means: through the atmosphere, in solution or suspension in waters entering the seas; or they may be produced in the oceans by physical, chemical, or biological processes. The distribution of sediments in the oceans is uneven and depends upon the methods by which the materials are produced and how they are transported and deposited. The uneven distribution results in variations in sediment thickness from zero to more than 14 km; hence the sediments have a strong influence on the topography of the ocean basins.

The principle of echo sounding can be employed to investigate the sediments on the ocean floor by increasing the amount of energy in the outgoing signal and by recording at frequencies lower than are used for measuring depth only (see Chapter 3). This allows us to see deep into the sedimentary column and identify reflecting horizons within it (Figure 14.1). We measure the time required for an acoustic signal to travel to each reflecting horizon and to return again to the surface. If we know the velocities of sound in the water and in the sediments, we can convert these times to depths by extending the relationship used in depth recording

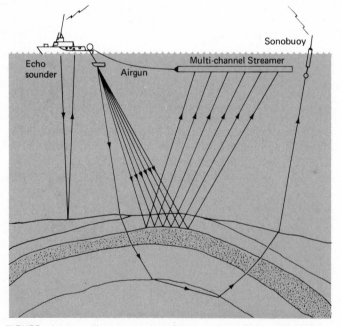

FIGURE 14.1   The topography of the sea floor is determined from echo soundings. An acoustic signal sent from a ship reflects off the sea floor and returns to the surface. If the velocity of sound in water is known and the signal is accurately timed, the depth of water under the ship can be determined.

Higher powered sound sources and multiple receivers are used to find the configuration of the layers beneath the sea floor. Seismic reflection techniques have been used to determine structures as deep as 10 km. For deeper determinations of sub-bottom structure, seismic refraction measurements are used (see Chapter 16).

(see Chapter 3). Thus we obtain the depth to any reflector in the sediment pile by the following equation:

$$\left(\begin{array}{c} \text{Depth to} \\ \text{reflector} \end{array}\right) = \frac{\text{reflection time to ocean bottom}}{2} \times \text{sound velocity in sea water}$$
$$+ \frac{\text{reflection time, bottom to reflector}}{2} \times \text{sound velocity in sediment} \tag{14.1}$$

If there is more than one reflecting horizon, the depth to each can be determined by adding additional terms to this equation.

The velocity of sound in the sediments, which can be determined by field or laboratory measurements, is dependent principally upon the porosity and water content. Variations in these properties with thickness are reasonably predictable in the deep-ocean basins, but where thick piles of sediments are found, as on the continental margins, the properties are more variable and velocities must be measured in place. With these velocities and the mean velocity of sound in the water column, the recorded times can be converted to true depths.

The first deep-sea seismic reflection work was done in the course of antisubmarine warfare studies during the 1940s using small explosive charges as the sound source. The inherent difficulties of handling and transporting large quantities of explosives and the need for a more rapid rate of generating signals led to the development of new sound sources ranging from explosive mixtures of gases, to high-voltage electrical sparks, and to various mechanical or electromechanical devices for rapid release of a bubble of highly compressed air. With these devices it is now possible to make subbottom seismic reflection profiles with a data density approaching that of the echo sounder used in charting the ocean bottom topography.

OCEANIC PROCESSES AFFECTING DISTRIBUTION OF SEDIMENTS

The processes of sediment transport and deposition where streams meet the ocean are dependent on the local geography and offshore topography as well as the sediment supply rate and the strength of offshore currents. These are of importance in understanding the vicissitudes of coastal events and are discussed in a later chapter.

Some of this sediment, however, makes its way to the deep ocean. The processes effecting this transfer, as well as the modes of supply of sediment to the deep ocean and their subsequent fate, are the subject matter of this section. During glacial time when sea level was lower, the continental shelves were exposed above sea level, the stream gradients were steeper, and the sediment load of rivers reaching the deep ocean must have been greater. This effect can be seen in the record in the deep-sea sediment cores. Even today, only a few streams flow into the deep ocean directly in areas where no continental shelf of any consequence exists. These rivers, the Congo or the Magdalena (in Colombia), for example, provide some insight into the behavior of stream-borne sediments supplied to the ocean at times of generally low sea level.

Sediments falling from the open-ocean surface, whether of biogenic or non-biogenic origin, are called *pelagic* sediments. These sediments may accumulate in this particle-by-particle fashion, or they may be mobilized with other sediments at the bottom of the ocean.

SUBMARINE CANYONS AND TURBIDITY CURRENTS

In the eastern United States, the edge of the shelf is cut by submarine canyons (Figure 14.2), some of which, such as the Hudson Canyon, appear to connect to present onshore drainage patterns, whereas others have no obvious relation. The cause of these canyons has been the subject of considerable controversy over the years. One general point of agreement is that they owe their major development to lowering of sea level. During the Ice Age, when about 2.5 percent of the water on the Earth's surface was removed from the oceans and frozen into thick continental glaciers, sea level was 100 m or so lower and much of the shelf was exposed. This is verified by peat deposits and oyster shells found on the present shelves and the numerous mastodon teeth recovered from the bottom by the trawls of fishermen. At the time of maximum lowering it would have been relatively easy for the shelf to be notched by rivers.

The submarine canyons are deeply incised into the continental slope, in places rivaling the Grand Canyon in size. Some continue beyond the continental rise to the abyssal plain. Although the slope looks quite steep on cross sections, because of the vertical exaggeration, the angle is only 4°–6° from horizontal.

One imaginative geologist, R. A. Daly, suggested as early as 1936 that the canyons might have been cut by currents of sediment-laden water. The normal sediment-laden currents are totally incapable of eroding canyons, but laboratory experiments by the Dutch geologist P. H. Kuenen showed that a higher density mixture of water and sediment yielded high enough velocity currents to erode canyons and to transport sediments great distances. These are called *turbidity currents*.

In order to test these laboratory observations with field data, the results of an earthquake that occurred at the edge of the Grand Banks of Newfoundland in 1929 were examined using data on cable breaks thought to have occurred in association with it. Most of the submarine telegraph cables from North America to Europe run through the area southeast of the Grand Banks. When the earthquake occurred, a number of cables within 50 miles of the epicenter broke immediately and, 59 minutes later, cables started to break in sequence downslope from the epicenter over a period of 13 hours and for distances up to some 400 miles. If the times of breaks

FIGURE 14.2   Physiographic diagram of the seafloor off the eastern United States. (From Heezen and Tharp, 1968.)

are plotted against distance from the epicenter (Figure 14.3a), the velocity of propagation of the disturbance can be determined. The velocities ranged from 55 to 12 knots (100 to 22 km /hr), with the slowest velocity occurring on a slope of less than 1 to 500. Maurice Ewing and Bruce C. Heezen suggested that this was a full-scale analogue of Kuenen's tank experiments.

When cable ships went to repair the cables, they did not find simple breaks. In some cases long sections of cable were completely missing; in others the cables were buried in the sediments so firmly that they could not be retrieved and had to be cut. The uppermost sediment layer in cores of the sediments from the abyssal

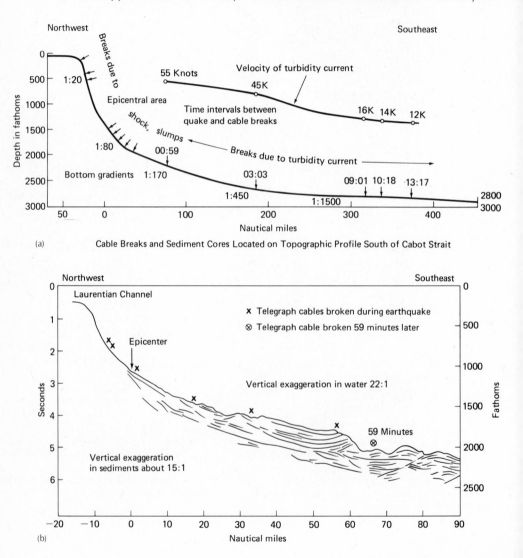

(a) Cable Breaks and Sediment Cores Located on Topographic Profile South of Cabot Strait

(b)

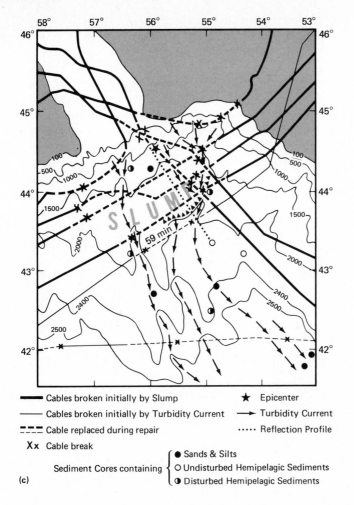

Cables broken initially by Slump — ★ Epicenter

Cables broken initially by Turbidity Current — → Turbidity Current

Cable replaced during repair — ····· Reflection Profile

Xx Cable break

Sediment Cores containing { ● Sands & Silts / O Undisturbed Hemipelagic Sediments / ◑ Disturbed Hemipelagic Sediments

(c)

FIGURE 14.3 (a) Submarine cable breaks following the Grand Banks earthquake of 1929. Topographic profile has a vertical exaggeration of sixty to one. Velocities calculated from the time of break after the earthquakes and distance from the epicenter are given on the upper curve. (From Heezen and Drake, 1963, after Heezen and Ewing, 1952.) (b) Tracing of seismic reflection profile on the continental slope and rise near the epicenter of the 1929 Grand Banks earthquake. The record clearly shows a slump 30–50 miles in length and several thousand feet thick. Cables on the slump mass broke immediately; one just seaward of the toe of the slump broke 59 minutes later as a result of the ensuing turbidity current. (After Heezen and Drake, 1963.) (c) Map of the Grand Banks slump area. Long sections of cable were missing or buried in the slump area. The subsequent turbidity current tended to be confined to channels on the continental rise, depositing sand and silt, while sediments on neighboring elevated regions were undisturbed. (After Heezen and Drake, 1963.)

plain below the epicenter contained silt, sand, and even gravel. This layer was graded, the sediment being coarser at the bottom than at the top, a characteristic of sediments transported by a current and later deposited.

On the continental rise (Figure 14.3b) it was found that the sediments that had moved downslope were channeled. They did not move down in a broad sheet, but were primarily confined to submarine canyons. Cores taken in the canyons were sand and silt; those on the elevated areas between them were disturbed or undisturbed clays. Since the turbidity currents travel in the canyons and since they are of sufficient strength to break and carry away submarine cables, it is not unreasonable to conclude that they have the ability to erode canyons.

A seismic reflection profile in the area allows us to reconstruct the events (Figure 14.3c). When the earthquake came, slumping occurred and a huge mass of sediments, 350 m in thickness and 100 km in length, moved downslope, breaking the cables that lay on it. This slump undercut the sediments at the shelf edge; they were thrown into suspension, became a turbidity current, and moved rapidly downslope, carrying away cables as it encountered them.

The characteristic sediments of abyssal plains are graded silts and sands intermittently produced by turbidity currents interbedded with fine-grained pelagic sediments continuously raining down on the sea floor. Reflection records (Figure 14.4)

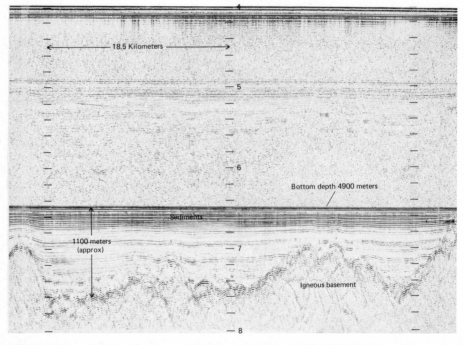

FIGURE 14.4   Seismic reflection record taken over an abyssal plain. Turbidite layers show as closely spaced horizontal reflections. The bedrock surface is typically quite rough. (Courtesy of Teledyne Exploration Company.)

from the abyssal plains characteristically show an abundance of closely spaced horizontal reflectors within the sedimentary column in contrast with records from areas covered by pelagic sediments, which typically show no internal reflectors. These horizons within the sediments of the abyssal plain indicate the presence of multiple turbidity layers. The rock surface beneath the sediments shows considerable topography, and the flatness of the surface of the abyssal plains can be attributed to the action of the turbidity currents sweeping across them.

In 1855 Matthew Maury wrote that a deep cable "would lie in cold obstruction without anything to fret, chafe or wear, save alone the tooth of time." The existence of turbidity currents shows that this prediction was too optimistic.

SEDIMENT TRANSPORT BY DEEP CURRENTS

The sediments composing the continental rise off the eastern United States are predominantly detrital and have accumulated to thicknesses that exceed 10 km. Underwater photographs of the surface of the rise reveal the presence of current-produced features—ripple marks, scouring, linear features, bare rock, bending over of stemmed organisms (Figure 14.5), all suggesting the presence of southward-moving bottom currents, flowing parallel to the bottom contours of sufficient strength

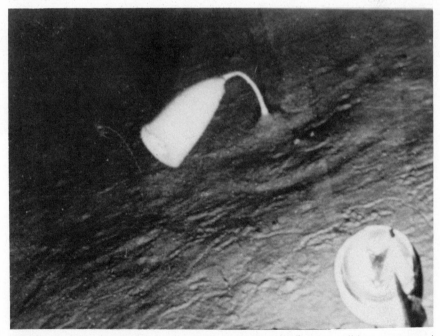

FIGURE 14.5   The bottom current near the base of the continental rise off the eastern United States is sufficiently strong to cause the bell-shaped glass sponge and the smaller organisms in this photograph to bend over. Grooves in the soft sediment indicate that at times they have been deflected into the sediment. (Photo courtesy of Lamont-Doherty Geological Observatory.)

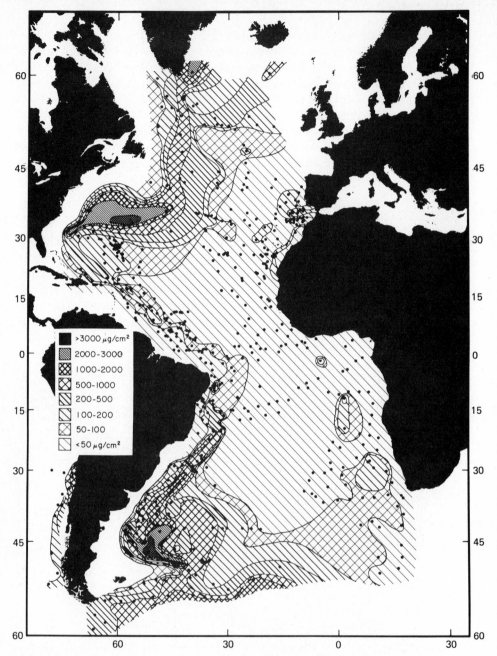

FIGURE 14.6 Distribution of the total amount of particulate matter in the deep water of the Atlantic Ocean shows that in the regions of the western boundary currents the highest resuspension and transport of sediments occur. The net standing crop of particulate matter is defined as the amount of material below the minimum in particulate concentration in the water column on descending from the particulate (mainly organic) rich surface waters to depth. (From Biscaye and Ettriem, 1977.)

to transport sediment. Such currents had been predicted by the physical oceanographers (see Chapter 6), who contended that because of the rotation of the earth, currents flowing from the polar regions toward the equator would tend to flow along the bottom on the western flanks of the ocean basins. As a result they are called *western-boundary currents*. Photographic and light-scattering data indicate that the bottom water of the lower continental rise in the region affected by the western-boundary current is clouded by suspended sedimentary material, whereas on the upper rise and much of the rest of the ocean basin the water is quite clear (Figure 14.6, facing page).

DISTRIBUTION OF SEDIMENTS

If for the time being we ignore the marginal waters around the continents, we can get an overview of the distribution of the major sedimentary components in the world oceans, as shown in Figure 14.7. It is obvious that there are definite patterns of distribution of calcium carbonate, siliceous deposits, and silicate minerals. In this section we discuss the observed distribution of these three major types of deep-sea components in terms of sources and mechanisms of accumulation.

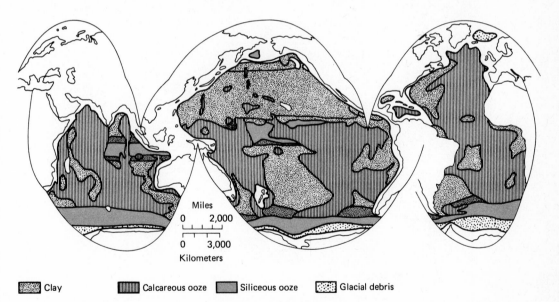

Miles
0   2,000

0   3,000
Kilometers

Clay   Calcareous ooze   Siliceous ooze   Glacial debris

FIGURE 14.7   Distribution of the major sedimentary types in the deep sea. Calcareous-ooze and siliceous-ooze regions are so designated if they each have at least 30 percent of calcium carbonate or siliceous fossils respectively. (Adapted from W. Berger, F. Shepherd, and other sources.)

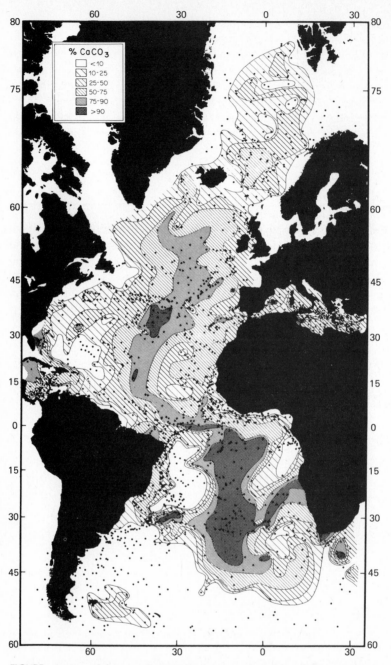

FIGURE 14.8   Calcium carbonate distribution in deep-sea sediments of the Atlantic Ocean. The highest concentrations are found along the ridges and underneath regions of upwelling and high biological productivity. (From Biscaye, Kolla, and Turekian, 1976.)

CALCIUM CARBONATE

The major contribution to the pelagic (deep ocean) sediments are the calcitic tests of foraminifera and the coccoliths. Sediments rich in these tests are sometimes called "oozes." In water depths of less than 3500 m, shells of snaillike aragonitic pteropods can be found. Because they are aragonite, they are more effectively dissolved during their descent through the ocean than the calcitic tests.

Figure 14.8 shows the observed distribution of calcium carbonate in the sediments of the Atlantic Ocean. An examination of this map and similar ones for the Pacific and Indian Oceans, leads to the following generalizations: (1) calcium carbonate concentration is highest along the major ridge systems of the ocean and (2) calcium carbonate concentration is high in areas of very high biological productivity virtually independent of depth of water.

The first observation has been generalized in Figure 14.9 which shows that there is a level in each of the oceans where there is a sharp transition from shal-

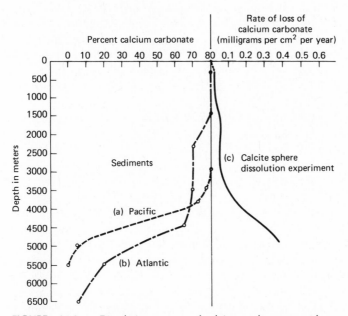

FIGURE 14.9   Dissolution patterns of calcium carbonate as a function of depth is recorded in the calcium carbonate concentration of the sediments (a) Atlantic; (b) Pacific. (After Bramlette, 1961; Turekian, 1965.) A comparison of the Pacific sediment composition curve with the results from an experiment in which calcite spheres were suspended for several months at various depths in the Central Pacific (c) shows a striking similarity. (After Peterson, 1966.) The results imply a depth at which dissolution of calcite begins to proceed rapidly. This is thought to correspond to the lysocline. Here the calcium carbonate compensation depth is deeper than the lysocline, as inferred from the sediment data.

lower high calcium-carbonate sediments to deeper sediments virtually devoid of calcium carbonate. The sharp-transition region has been called the "compensation depth." Above the compensation depth the dissolution rate of calcium carbonate is sufficiently slow to allow its accumulation in the sediment, whereas below the compensation depth the dissolution rate is sufficiently high that virtually no calcium carbonate is preserved in the sediments.

The meaning of the compensation depth has troubled marine scientists ever since it was first defined by Murray and Thomson from the samples obtained on the *Challenger* Expedition. There are three models to explain the compensation depth that can be subjected to testing. The first is that the compensation depth is determined strictly by the level at which seawater is undersaturated with respect to calcium carbonate. The second is that seawater below the top several hundred meters is undersaturated with respect to calcium carbonate but that the kinetics of dissolution result in effective dissolution only at the greatest depths. Finally we can impute to the bottom some property that would accelerate or retard dissolution.

As the result of laboratory work, theoretical calculations, and shipboard measurements it appears that large parts of the oceans are undersaturated with respect to calcium carbonate below several hundred meters depth. Thus the saturation model, although perhaps applicable in some parts of the ocean, is not generally thought to be the answer for all parts of the ocean. The third model is ad hoc and requires special properties for different parts of the ocean, and for this reason is not generally useful, although there is utility to it, especially in explaining the composition of coastal deposits.

We are left then with a kinetic model to explain the distribution of calcium carbonate in the oceans as the most general one. Aside from the chemical data supporting the idea that seawater is undersaturated at depths below several hundred meters, certain other observations confirm the kinetic model.

One of these experiments involved the measurement of extent of dissolution of calcite spheres and foraminiferal tests suspended at various depths along a wire in the deep Pacific (Figure 14.9). The observations indicated that dissolution occurred at all depths along the wire below several hundred meters but that there was the most loss from dissolution at depths greater than about 4000 m.

In another critical experiment, observations were made on the presence or absence in deep-sea deposits, of easily dissolved foraminiferal tests in relation to more *robust* ones. The depth of water at which there was a marked decrease in the fragile tests in relation to the robust ones was designated the *lysocline* and was interpreted as a chemical boundary in the water column influencing the intensity of dissolution of calcium-carbonate tests.

On the basis of laboratory experiments on the rate of dissolution of calcium carbonate in undersaturated solutions it has been shown that there is a critical degree of undersaturation at which point the rate of dissolution of calcium carbonate increases markedly. This has been designated the *chemical lysocline* to distinguish it from the lysocline derived from field observations and the examination of the state of preservation of foraminiferal test assemblages.

We can now return to the compensation depth and relate it to the lysocline. It is evident that if the oceans are undersaturated with respect to calcium carbonate at most depths and the supply rate of calcium carbonate from the productive surface layers to depth is very small, the amount of calcareous material preserved on even a relatively shallow sea floor will be very small and the "compensation depth" will be very shallow. If, on the other hand, the productivity is very high, the rate of supply of calcium carbonate to depths below the lysocline might be faster than the dissolution rate under which conditions the compensation depth is very deep. For the normal oceanic regions the compensation depth will be at about the level of the lysocline or below it, depending on the flux of calcium carbonate from above.

SILICEOUS DEPOSITS

Diatoms and radiolaria are the primary siliceous tests found in deep-sea deposits. Of lesser importance are the silicoflagellates. Unlike calcium carbonate there is no compensation depth for these siliceous tests. Rather their abundance in sediments is controlled by the level of biological productivity in the surface waters and the dissolved silicon concentration of the deep water through which the test falls. The solubility of "amorphous" silicon oxide of the type making up the tests of diatoms and radiolaria is very high even in cold seawater, so only the flux of tests downward and the degree of departure from saturation of the deep sea water determines the preservation potential of a siliceous test.

It is not surprising then to find that siliceous deposits in the deep sea are virtually restricted to the upwelling areas around Antarctica and in the high northern latitudes, the eastern parts of basins, and the eastern equatorial regions of the oceans (Figure 14.7).

SILICATE MINERALS

Detrital materials from the continents arrive at the ocean boundary by streams, wind, or glaciers. Whatever the mode of injection, the greatest opportunity for transport in the surface currents of the ocean is for the finest grained fraction. The slow accumulation of these clay-rich materials in association with oxidized iron gives it a reddish appearance in many places, and for this reason this sediment has sometimes been called "red clay."

As we have seen in the previous chapters, sediments, once deposited on the ocean floor, can be remobilized by submarine processes such as turbidity currents, downslope movements, and boundary current transport. All these processes affect the distribution of continentally derived materials, and the distribution of the minerals in the clay and silt sizes are charts of the point and mode of injection of sediment into the ocean and its ultimate fate under the influence of these ocean-bottom forces.

The overwhelmingly abundant clay mineral in the North Atlantic and the North Pacific is illite, derived mainly from the erosion of ancient shale deposits on the continents (Figure 14.10). Second in importance is montmorillonite, whose abundance is due to the supply of weathered materials from the continents and the supply from submarine volcanic activity. This is most striking in the South Pacific. Montmorillonite in the South Atlantic is closely accompanied by microscopic volcanic glass shards, and in the South Pacific with a zeolite, phillipsite, an alteration product of basalts (Figure 13.4). Since most of the continents drain north of about 20°S, this is the area dominated by illite, whereas the sediments south of this have as much or more montmorillonite in many parts of the ocean basins indicating the importance of volcanogenic materials in relation to continental materials.

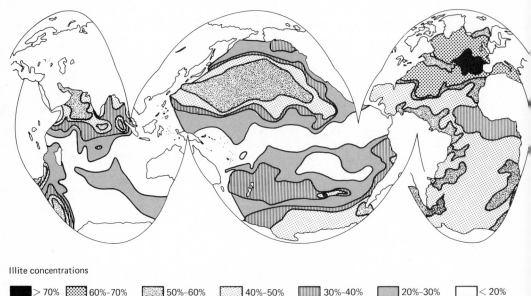

Illite concentrations

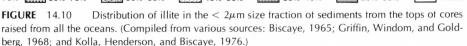

■ > 70%   ▒ 60%–70%   ░ 50%–60%   ⬚ 40%–50%   ▥ 30%–40%   ▤ 20%–30%   ☐ < 20%

FIGURE 14.10   Distribution of illite in the < 2μm size fraction of sediments from the tops of cores raised from all the oceans. (Compiled from various sources: Biscaye, 1965; Griffin, Windom, and Goldberg, 1968; and Kolla, Henderson, and Biscaye, 1976.)

Imposed on the main clay mineral assemblage made up of these two minerals are the other clay minerals. Kaolinite and chlorite, although generally representing smaller fractions than illite, are the indicators of climatic regime differences as a function of latitude. This is most strikingly seen in the Atlantic Ocean (Figure 14.11). Chlorite is preserved or made in weathering profiles at high latitudes, whereas kaolinite is the dominant clay mineral of tropical latitudes; then the kaolinite/chlorite ratio is the strongest indicator of climatic regime and in particular weathering intensity.

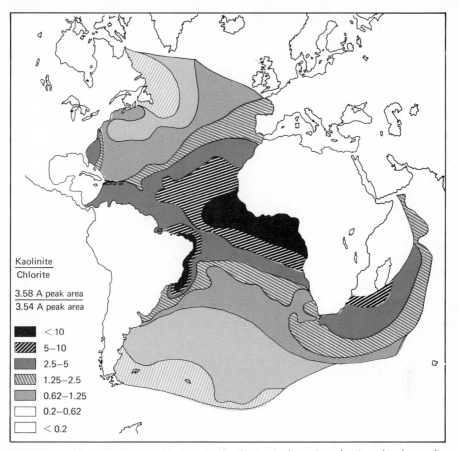

Kaolinite
Chlorite

3.58 A peak area
3.54 A peak area

- ▨ < 10
- ▨ 5–10
- ▨ 2.5–5
- ▨ 1.25–2.5
- ▨ 0.62–1.25
- □ 0.2–0.62
- □ < 0.2

FIGURE 14.11   Kaolinite-to-chlorite ratio distribution in the < 2μm fraction of surface sediments of the Atlantic Ocean. High kaolinite-to-chlorite values are found in the tropics and reflect the intense weathering on the adjacent continents. (After Biscaye, 1965.)

FERROMANGANESE NODULES

In addition to the sediments on the deep-sea floor there are found over large areas what have been called manganese nodules (Figure 14.12). These are spherical or quasi-spherical lumps of iron and manganese oxide and hydroxide minerals showing a virtually concentric growth pattern and are more properly called ferromanganese nodules. The common size range of these small nodules is between 1 and 20 cm in diameter. There are other types of ferromanganese deposits in the deep oceans, ranging from microscopic coatings on calcareous tests in the sediment to large-scale encrustations on volcanic rock outcrops. Whatever the form of these ferromanganese deposits, they indicate a chemical deposition from the aqueous

FIGURE 14.12   Ferromanganese   nodules
from a depth of 5676 m in the western North Atlantic. The organism is an actinarian. (Photo courtesy
T. Amos, University of Texas Marine Sciences
Institute.)

medium surrounding them. Many nodules have high concentrations of nickel, copper, and cobalt, which make them economically interesting (see Chapter 21).

The source of the iron and manganese may be different for different parts of the oceans. There is an indication that along the mid-oceanic ridges where volcanic activity is the most intense, the iron and manganese have an association with the emplacement of the volcanic rocks. The ferromanganese deposits on the vast deep abysses of the ocean floor, however, may have a different source.

Other than the localized volcanic-associated ferromanganese deposits the source of the iron and manganese appears to be the same as the source of the major part of the sediments—the weathering of the continents.

Ferromanganese nodules grow at the rate of about several millimeters per million years, so people have worried about how these nodules can keep from being buried by the surrounding more rapidly accumulating sediment. The answer may lie in the fact that benthic organisms are always burrowing in the sediments. This process, in addition to the action of bottom currents, may turn over the nodules frequently enough to keep them "afloat" in the sediment pile. Nodules, however, are also commonly found at depth in cores, so eventually burial can occur when the sediment regime changes.

SUMMARY

1. By the use of principles of echo sounding, but with lower frequency sound waves, the structure and thickness of the marine sedimentary pile can be evaluated.

2. Sediments are distributed in the ocean basins as the result of currents. These range from turbidity currents originating at the continental margins to the western boundary currents associated with the general circulation of the deep ocean.

3. Deep-sea biogenic deposits are distributed according to a number of controls. Calcium carbonate is restricted to ridges and areas of high surface productivity. Siliceous deposits are found mainly at the high latitudes but also in areas of high surface productivity.

4. The clay minerals reflect sources of sediment. Montmorillonite is highest in areas of volcanic activity, and illite is the overwhelming clay mineral derived from continental erosion.

5. Ferromanganese deposits occur throughout the oceans, but large concentrations of "nodules" are restricted to areas of low sediment accumulation.

The Grand Canyon of Arizona, showing a "layer-cake" stratigraphy of ancient marine deposits. (Courtesy of the Union Pacific Railroad.)

Stratigraphy of Marine Deposits

15

In Chapter 14 we examined the sediment on the ocean floor and found that all of its characteristics vary from place to place. These variations form geographic patterns which can be explained in terms of chemical, physical, and biological processes known to be operating in the ocean today. This fact suggests that it might be possible to

CHAPTER

study the sediments lying below the sea floor and use them as clues to reconstruct past oceans. In fact, just such a detective technique is extensively applied to sediment samples raised in deep-sea sediment cores, and forms one important branch of *historical oceanography*. In this chapter we take our first look at this subject. First, we will see how various coring techniques are used to sample the pile of sediment at the bottom of the sea. Then we will see how stratigraphic methods, combined with techniques of radioactive geochronometry, make it possible to transform these tubes of mud into an organized record of the history of the world ocean.

SAMPLING THE SEDIMENT
COLUMN IN THE OCEANS

SEDIMENT CORING

In the earliest investigations of the deep-sea floor, samples of the sediments were recovered by dredging or trawling. These gave indications of the surface distribution of sediment types, but not of variations with depth. As a result, sediment sampling tubes were devised as early as the latter part of the nineteenth century in order to obtain vertical sections of the sediments.

Most of the devices consisted of a hollow pipe with a heavy weight on top and some sort of core-retaining device at the bottom. These were lowered to the sea floor and penetrated up to 6 ft (2 m) into the sediments under the force of gravity. These gravity corers were used quite successfully to recover clay samples from the sea bed, but had little ability to penetrate sandy layers.

A major breakthrough was made by Börje Kullenberg, who designed the piston-coring apparatus for the Swedish *Albatross* expedition in 1948. In this apparatus (Figure 15.1), a piston inside the core pipe is attached to the wire that lowers

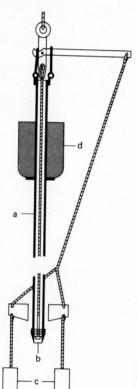

FIGURE 15.1   Diagram of a piston corer. (a) is the coring tube. (b) is a snugly fitting piston that is connected to the cable by means of a strong wire. When the counterweights (c), touch the bottom, a release mechanism, from which the coring tube is suspended, is activated and the tube drops rapidly in free fall. The weight (d), gives the falling tube enough energy to be driven into the sediment while the piston is stationary. The sediments are sucked into the tube by the piston as the tube penetrates the sediment. Cores as long as 20 m can be obtained in virtually all depths of water. (After Dietrich, 1963.)

the corer to the bottom. The corer itself is held in place temporarily by a release mechanism activated by a trigger weight hanging about 6 ft below the bottom of the core pipe. During lowering, the piston is at the bottom of the pipe. When the trigger weight reaches the bottom, the coring apparatus free-falls while the piston remains stationary at the sediment-water interface. Thus the piston breaks the friction and provides a pumping action that allows recovery of long cores. Modified versions of this device are the standard corers in use today. Cores of deep-sea deposits up to 70 feet (20 m) in length and 2 to 3 in (5-8 cm) in diameter have been recovered from the sea floor.

DEEP-SEA DRILLING

In order to obtain still longer cores of the sediments, the *Joides* Deep-Sea Drilling Project (DSDP) was begun in 1964. DSDP employs the dynamically positioned drilling vessel, *Glomar Challenger*, which maintains her position in relation to a sonic beacon on the sea floor by means of large thrusters that can move it fore and aft or sideways (Figure 15.2). Drilling is accomplished by lowering the drill string through the center of the vessel to the bottom, and then penetrating the sediment by rotating the drill string and bit. Seawater is used as the drilling fluid.

Glomar Challenger can handle up to 25,000 ft (7600 m) of drill pipe. Deep-sea drilling operations have penetrated almost 2000 ft (600 m) into the sea floor.

FIGURE 15.2 *Glomar Challenger,* the ship used in the Deep Sea Drilling Project. (Managed by the Scripps Institution of Oceanography under a contract with the National Science Foundation.) (Photograph courtesy of Scripps Institution of Oceanography.)

Most of the holes are drilled with a single bit. When the bit wears out, it is necessary to stop drilling. However, a reentry device has been successfully tested that will allow the drill string to be recovered, a new bit to be mounted, the hole reentered, and drilling continued. Thus the potential limit of penetration using this method is the amount of drill pipe that can be handled by the ship and the drilling rig.

THE FUNDAMENTAL
STRATIGRAPHIC PROBLEM

Whenever we penetrate the pile of sediments lying on the sea floor, we find that the biological, detrital, and chemical properties of the pile change from level to level. The changes may be subtle or sharp, and are sometimes seen in one property and not in another. In effect, these changes are part of a natural recording system sensitive to environmental changes at and above the sea floor. The historical oceanographer uses physical evidence from the sediments to find out what happened, and his knowledge of the contemporary ocean to find out why.

If marine sediment accumulated everywhere at a known and constant rate, it would be relatively easy to reconstruct the history of the ocean. For under such conditions any measurable sediment property could be mapped for any time in the past by plotting all core data from a common depth in the sediment. However, this ideal is never—not even closely—approached in the sediment layers of the real ocean. Hence the fundamental problem of historical oceanography is the basic *stratigraphic* problem: to determine how sediment age varies with depth below the sea floor (Figure 15.3). Once this age-depth function is known for any core, a chronological sequence of events can be reconstructed for that site; and if it is known for a set of cores, a map of an ancient seabed can be reconstructed.

The methods by which this problem is solved vary from one situation to another; in general, however, four different procedural steps are involved: (1) recognition of sedimentary layers, (2) correlation of the layers, (3) analysis of layer geometry, and (4) radiometric dating of key levels. How these procedures are applied to solve the stratigraphic problem can be appreciated by considering the structure of a block of wood (Figure 15.4). If we ignore the fact that the wood is built up by yearly growth layers (Figure 15.4a), we have no way of reconstructing its history. However, if we can identify the layering (Figure 15.4b) we can see how the block developed by tracing the layers. The first-formed layers are inside the tree, at the bottom of our block, with younger layers deposited on top in succession. In this case, since we know that trees lay down one layer a year, we can actually measure the rate of growth. Moreover, if we have just cut down the tree and can identify the outermost layer as the one just formed, we can count back and obtain an accurate age for any layer. The knots in the grain show where the growth of branches has distorted the layers and that some layers are discontinuous. Such gaps reflect accidents of abrasion or some local condition where the tree did not grow for a time.

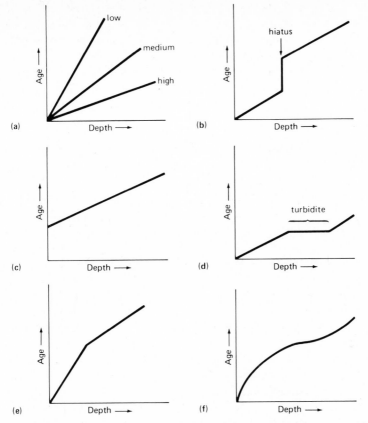

FIGURE 15.3   Diagram showing various ways in which age may vary with depth in deep-sea sediments. (a) Cores with different but constant accumulation rates. (b) Core with unconformity. (c) Core with top missing due to sea-floor erosion. (d) Core with turbidite layer. (e) Core with abrupt change in accumulation rate. (f) Core with gradually changing accumulation rate.

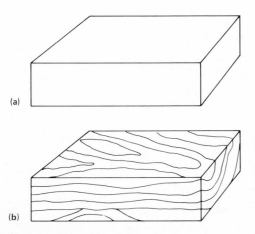

FIGURE 15.4   Stratigraphy in a block of wood. If the block is painted white, as in (a), the growth pattern of the tree cannot be observed. If the growth layers can be distinguished, as in (b), the geometry of the layers records the development of the tree. Some layers are discontinuous, recording abrasion prior to later growth.

All the situations illustrated in Figure 15.4 have their counterparts in the study of marine sediments (Figure 15.5). These sediments also form in layers, one on top of another, with the youngest layer on top. The sequence of layers is known as the *stratigraphic sequence* or the *geological column*. Gaps in the sequence are called *unconformities*. Such gaps are very common. The technical name for each layer is a *stratum* (plural, *strata*). In marine sediments, this term is applied to layers that have accumulated over a wide range of time intervals, from a few hours to many millions of years.

We are, of course, not privileged to observe *directly* a large cross-section of marine sediments on the ocean floor, as we did the block of wood. Instead, we must study and sample the pile at individual geographic points at which deep-sea cores have been taken. A major problem here is that we cannot, by direct observation of the layer geometry, establish that an unconformity is present in the core. To do this we must first *correlate* the layers recognized in different cores, that is, establish which layer in one core was formed at the same time as a given stratum was formed in another. On land, where ancient marine deposits have been uplifted and exposed by erosion, we can sometimes make correlations directly by observing the stratification. But, for deposits now on the sea floor, correlation must be accomplished by indirect procedures to be explained in the following section. When these correlations have been made, the *stratigraphy* (layer geometry) of the deposit is known, and it is possible to establish certain aspects of the history of the region.

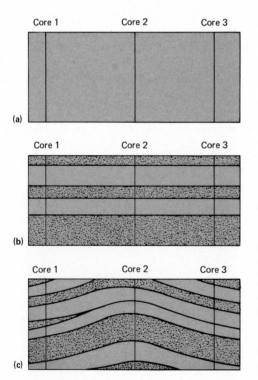

FIGURE 15.5   Stratigraphic problems in a pile of sediments penetrated by three cores. In (a), the sediment is assumed to be homogeneous; in this case, no stratigraphic analysis is possible. In (b), five horizontal strata of constant thickness can be correlated across the area, and there is no evidence of a hiatus in the sequence. In (c), the correlations drawn indicate a more complex history than in (b): an upward flexure (after deposition) of sediments in the vicinity of core 2 and a depositional hiatus in cores 2 and 3. This hiatus is indicated not by an inspection of cores 2 and 3 alone, but also in conjunction with core 1, where a layer not present in the other cores is observed.

We could, for example, identify unconformities in the record or, by correlating a 1-m thick stratum in one place with a 10-m thick stratum in another, we could prove that the accumulation rates during the formation of the layer varied across the study area by a factor of 10. But to fit the history of one portion of the ocean into the history of the whole, to relate that history to events in the rest of the universe, and to establish the rate at which ocean processes operated, the absolute ages of key strata in the sequence must be established.

STRATIGRAPHIC CORRELATION

All correlation techniques must be based on some process which leaves a permanent and distinctive record in the sediment. Results will be best if the entire sea floor is affected simultaneously.

For example, if a volcanic ash with a distinctive chemical composition were to be spread globally by a gigantic eruption, that stratum would provide a key correlation level wherever it could be identified. A hundred ash layers, each with a distinctive composition and spaced at appropriate intervals of Earth history, would provide an ideal stratigraphic framework. Not surprisingly, no such ideal has been found—although ash beds do occur in marine sediments, and, over limited distances, are extremely useful correlation tools. Similarly, correlation techniques based on other chemical or mineralogical changes in the sediment are important in solving local correlation problems.

Three geological correlation techniques—*biostratigraphy, magnetic stratigraphy*, and *oxygen-isotope stratigraphy*—have proved generally useful in worldwide correlation. Used alone, none are foolproof; but employed in combination, they provide a reliable stratigraphic framework for deciphering the history of the ocean. All involve laboratory observations on samples of sediment. The *geophysical correlation* techniques discussed in Chapter 16 are on a very different footing, because they are operated remotely and sense discontinuities in the physical properties of adjacent bodies of sediment. Where these discontinuities follow boundaries between strata, the geophysical methods provide a remarkable, large-scale view of the stratigraphy. The problem is to determine when the discontinuities sensed remotely are the boundaries between strata and when they originate differently. This problem can generally be solved by direct sampling of the sediments and applying geological correlation methods to key beds.

BIOSTRATIGRAPHY

The fact that animal and plant species have limited life spans, terminated either by extinction without issue or by evolution into different, descendant species provides the basis for *biostratigraphy*. Some species have longer life spans than others, and are therefore less useful than their more unfortunate relatives who became extinct sooner. The ideal qualities for a biostratigraphic species are two: a short life span and a wide geographic range during its life. The second quality is typical of many planktonic species (see Chapter 10), whose biostratigraphic record is therefore an important basis for correlating deep-sea deposits. Because most planktonic species live near the sea surface, their geographic distribution is relatively independent of conditions on the seabed, provided they are not dissolved after deposition.

Because any single species has a geographic range that is limited by its ecological tolerances, no species can be used to correlate sediments from all parts of the ocean. Global biostratigraphy is therefore based on many species with overlapping geographic ranges.

By combining information on the point of origin or the point of extinction of planktonic species representing different groups (chiefly radiolaria, coccoliths, foraminifera, and diatoms), marine deposits of the past 150 million years have been subdivided biostratigraphically into zones averaging about three million years long.

MAGNETIC STRATIGRAPHY

Because the Earth is like a gigantic magnet, free-swinging small magnets in compasses align themselves with the magnetic field at the Earth's surface. Thus the north-seeking pole of the compass magnet points North wherever you may be on Earth. But this was not always the case. For example, if you were transported by a time machine back 800,000 years into the past, and sought to navigate with the aid of a compass, you would find that the north-seeking pole of the compass was actually pointing South, because the Earth's magnetic polarity at that time was reversed (Figure 15.6).

Such magnetic reversals have occurred many times in Earth history. Fortunately for historical oceanography, a stratigraphic record of these ancient polarity

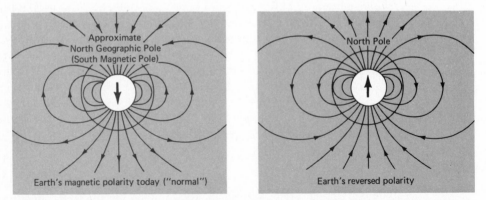

FIGURE 15.6 The Earth as a magnet. At present the north magnetic pole of the Earth is found in the vicinity of the south geographic pole. That is why the north poles of magnets point north, following the lines of force of the Earth magnet. The reversed polarity (in relation to the present) occurs when the north magnetic pole and the north geographic pole are coincident.

epochs has been formed in various kinds of sediments and in lava flows on land and on the sea floor. As these deposits form, minute magnetic grains act as miniature compasses and orient themselves according to the ambient magnetic field. After deposition of the sediment or freezing of the lava, the record of magnetic polarity is fossilized. The record in lava flows on land has been particularly useful, for here the potassium-argon dating technique (Table 15.1, below) often permits the age of the flow (and hence the age of one point in the magnetic polarity epoch) to be determined. When many such records of magnetic polarity and age are compiled, a consistent picture of the Earth's magnetic polarity history emerges (Figure 15.7).

Once the age of the magnetic reversals has been established, these events form the basis for a magnetic stratigraphy, because the changeover from one polarity regime to the other occurs rapidly (on the order of 1000 years) and affects all places on the Earth's surface simultaneously. Since the pattern of normal and reversed magnetic polarities is recorded in sediments, as well as lava flows, the magnetic record of deep-sea cores becomes the basis for correlating these strata against a standard compilation of magnetic polarity history. Thus the recognition of a particular magnetic polarity reversal in two deep-sea cores provides not only a correlation between the two cores but also an absolute date for that level in the cores determined by correlation with the magnetic polarity epochs dated on land (Figure 15.8).

One shortcoming of the method is that, given a short section of sediments, it may not be possible to distinguish one reversal from another, because the magnetic polarity record is a simple, two-state pattern. (In contrast, the processes of radioactivity or biological evolution are typified by time-dependent trends.) This shortcoming can be circumvented, however, by applying biostratigraphic and paleomagnetic techniques to the same sedimentary sequence. Together, the two methods provide an unambiguous method of stratigraphic correlation (Figure 15.9).

Back to about 5 million or 10 million years ago, the patterns of polarity reversals and their absolute ages have been worked out in detail by the methods discussed. As we shall see in Chapter 16, the longer time scale patterns are studied by examining magnetic patterns on the ocean floor.

Although many details of this standard history remain to be filled in, and the absolute age of many boundaries determined more accurately, the combined paleomagnetic and biostratigraphic method has already provided a simple and rapid method of working out the stratigraphy and approximate chronology of deep-sea deposits. The impact of this development is very large, because it is now possible to determine the rates of many processes (sea-floor spreading, climatic change, and organic evolution), which are too slow to be measured by the normal methods of laboratory science.

OXYGEN ISOTOPE STRATIGRAPHY

Atoms of a chemical element which differ only in mass are called *isotopes* of that element. Oxygen has three isotopes: ^{16}O, ^{17}O, ^{18}O. Of these, ^{16}O is by far the most

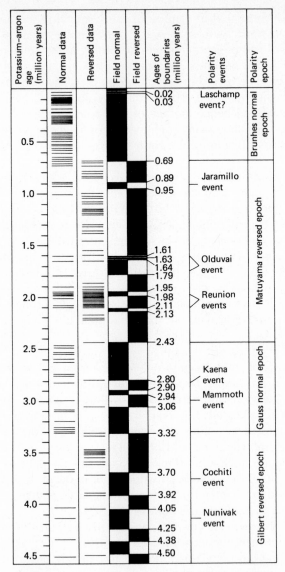

FIGURE 15.7   The reversal of the Earth's magnetic field with time. The diagram was made by determining the magnetic polarity of volcanic materials whose ages were determined by potassium-argon dating. (After Cox, 1969 with later additions.)

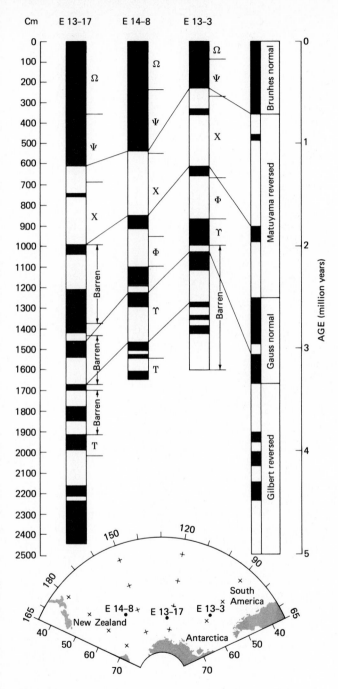

FIGURE 15.8 Magnetic stratigraphy and biostratigraphy in three antarctic cores. The location of the cores is shown on the inset map. Depth scale (in cm below the sea floor) is shown on left. Normally magnetized sediments are indicated in solid black; reversed polarity is shown in white. Correlation with the standard magnetic polarity sequence, with ages indicated, is shown on right. Greek letters denote biostratigraphic zones characterized by distinctive species of radiolaria. Note that any possible ambiguity in correlating the magnetic zones is eliminated by the biostratigraphic determinations. (From Hays and Opdyke, 1967.)

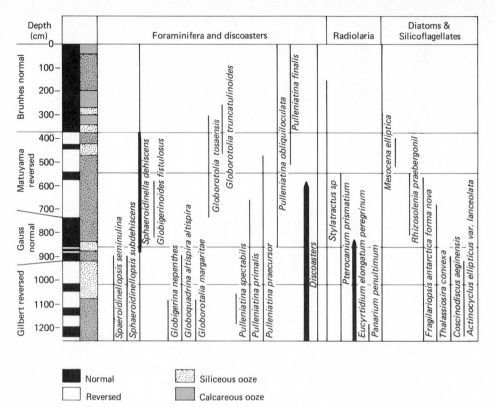

FIGURE 15.9   Magnetic, faunal, and floral stratigraphy of core V24–59, from the equatorial Pacific Ocean. Although the core is only 12 meters (1200 cm) long, observations on magnetic, lithologic, and biological properties make it possible to recognize numerous strata. The thin vertical lines indicate the range of various planktonic species at low abundances. The heavy lines indicate that a species is present in high abundance. (From Hays et al., 1969.)

abundant and ^{18}O the next most abundant. Natural variations in the ^{18}O to ^{16}O ratio occur. These have been thoroughly assessed, and the reasons for the variations are now fairly well understood. Marine calcium-carbonate shells (such as foraminiferan tests and coccoliths) deposit oxygen that is about 3 percent heavier than oxygen in the water molecules of seawater, that is, the $^{18}O/^{16}O$ ratio is greater in the shells. At any given temperature the difference in the oxygen isotope ratio in the shell and the water is fixed. Thus if the oxygen isotopes in the ocean water get heavier and the temperature remains constant, the oxygen isotopes in the shells deposited from that water will also get heavier. On the other hand, if we keep the oxygen isotope composition of the water constant, and drop the temperature of the ocean, the deposited calcium carbonate becomes enriched in ^{18}O in relation to the seawater.

Both the temperature effect and the water-composition effect just discussed work together to provide a technique of stratigraphic correlation applicable to deep-sea sediments deposited over the past 700,000 years. Over that interval of time (as we will see in Chapter 19) the Northern Hemisphere ice sheets exhibited large fluctuations in volume, giving rise to a sequence of ice-age climates separated by warmer periods. As the ice sheets expanded, the water was derived from the ocean by evaporation, a process which leaves the ocean reservoir isotopically heavier as a result of the preferential storage of the light isotope in the ice sheets. At the same time, the water during the ice age became cold, thus making the oxygen isotopic composition of marine carbonate shells still heavier. When the ice melted, the isotopically lighter water was returned to the reservoir.

Thus, by measuring the $^{18}O/^{16}O$ ratio in the calcium-carbonate tests of planktonic and benthic foraminifera, it is possible to obtain an isotopic record of this ice-age succession in deep-sea cores. The resulting isotopic curve (Figure 15.10), in which the isotopic ratio is plotted as a function of core depth, has been found to have essentially the same shape wherever the core is located—and regardless of whether benthic or planktonic fossils are used. This can only mean that the water-composition effect, reflecting changes in the volume of global ice, dominates the temperature effect. Thus, the isotope curve reflects mainly the changing volume of the continental ice caps.

In Chapter 19, we will return to this isotopic paleoclimate curve. Here we take note of an important and purely stratigraphic use of the method. The climatic history revealed by the isotope record can be subdivided into intervals with generally low oxygen isotopic ratios (interglacial stages) and high isotopic ratios (glacial stages). For convenience of reference, these isotopic stages were numbered from the top down, so that the odd-numbered stages represent ice-free intervals and the even-numbered intervals represent glacial intervals. Fortunately, the detailed shape of each isotopic stage is distinctive. Given a previously unstudied core, therefore, an investigator can compare the isotopic record in that core with a standard stage sequence developed in long and continuous cores used for reference. Thus, core-to-core correlations can be made, and unconformities identified. As shown in Figure 15.10 the Matuyama reversal (dated c. 700,000 years ago) occurs at the Stage 19-Stage 20 boundary. Over this time span, the isotopic stages give a twentyfold increase in precision over magnetic stratigraphy, and provide an important stratigraphic tool for working out the detailed history of the ice ages. On the average, each isotope stage represents about 35,000 years.

RADIOACTIVE
GEOCHRONOMETRY

Radioactivity is the successful attempt of an unstable nucleus to transform itself to a stable one. Two features of radioactivity are of importance in its use as a clock: (1) the conversion of the unstable atomic nucleus to a stable one proceeds according to a definite rule, and (2) each radioactive species "decaying" to a stable species has a characteristic constant which tells how fast the conversion is proceeding.

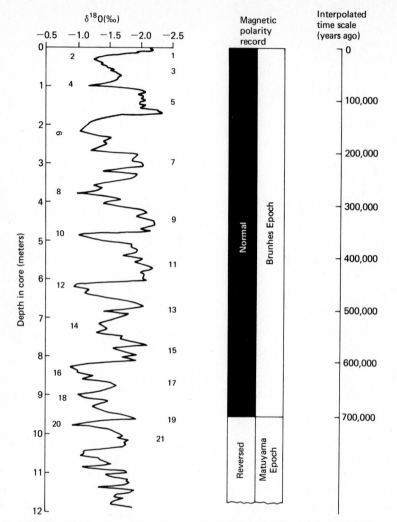

FIGURE 15.10   Isotopic and magnetic stratigraphy from Pacific core V28–238. Variations of the oxygen isotopic ratio δ^{18}O are plotted at the left. More negative values indicate times of low ice volume. Isotopic stages 1–21 are indicated by the small figures placed near the center of each stage. The magnetic polarity record is shown, and a time scale interpolated between the core top and a known 700,000-year date for the Brunhes-Matuyama boundary. (From Shackleton and Opdyke, 1973.)

NOTE: The use of δ^{18}O as a measure of ^{18}O/^{16}O is common practice because the actual variation in this ratio is really quite small. It is a quantitative expression of the difference of the sample from a standard. A common standard used is a specific fossil shell (a belemnite from the Pee Dee Formation designated PDB). The relationship is:

$$\delta^{18}O \text{ (in parts per mil)} = 1000 \left[\frac{(^{18}O/^{16}O) \text{ sample}}{(^{18}O/^{16}O) \text{ standard}} - 1 \right]$$

The statement describing this is: "The rate of decay of a radioactive species at any time is proportional to the number of atoms of that species present at that time. The proportionality constant is characteristic of the species." Mathematically this is written:

$$\text{rate of decay} = \frac{\Delta N}{\Delta t} = -\lambda N \qquad (15.1)$$

where N is the number of atoms at any designated instant of time and λ is the "decay constant" (the minus sign at the right-hand side of the equation indicates that the number of atoms is *decreasing* with time).

Since the number of radioactive atoms is decreasing with time, the above statement can be converted to another relatively simple mathematical expression which describes what happens to the quantity of a radioactive species over time:

$$N = N_0 e^{-\lambda t} \qquad (15.2)$$

Here N_0 is the number of radioactive atoms at the start of the isolation and initiation of decay of the species of interest, N is the number left after a length of time, t, has elapsed, and λ is the "decay constant"—that is, the proportionality constant described above. The symbol e is the so-called "natural base." Figure 15.11 shows how N varies with time. Clearly if we know N, N_0, and λ, t will be determined by the equation:

$$-\frac{2.303}{\lambda} \log \frac{N}{N_0} = t$$

$$(15.3)$$

(where 2.303 is the necessary conversion factor from natural base logarithms to the more commonly used base 10 logarithms).

Natural radioactive species are made by high energy interactions of neutrons with stable isotopes. The radioactive isotopes useful in dating geologic deposits were produced during the formation of the elements in stars by such reactions. These include the two uranium isotopes (U-235 and U-238) and a potassium isotope (K-40), all of which are radioactive clocks that have been used in geologic dating. These clocks are good for measuring time scales up to several billion years —the age of the Earth itself!

The uranium isotopes decay ultimately to stable lead isotopes, but they do so through a string of intermediate radioactive species that have their own characteristic decay constants and also their own characteristic chemical properties (Figure 15.12). The chemical separation, in nature, of these intermediate radioactive species provides a group of clocks capable of measuring events on the time scales of tens of years to several hundred thousand years.

Neutrons as well as radioactive isotopes are produced in the Earth's atmosphere when cosmic rays bombard the atoms composing the atmosphere high above the Earth's surface. The neutrons subsequently react to produce radioactive isotopes in the atmosphere, adding to those produced by the direct interaction with

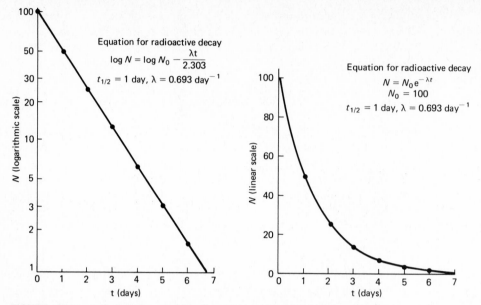

FIGURE 15.11 Radioactive nuclides decay to more stable nuclides according to a definite law. This is called exponential decay, and is shown in the right-hand plot. N_0 is the number of original atoms (assumed to be 100 in this example) and N is the number of atoms left after a length of time t has elapsed. In the example, the half-life, $t_{\frac{1}{2}}$, is chosen as one day; hence the decay constant, λ, is 0.693 day^{-1}. The left-hand curve is a plot of the logarithm of N against time.

cosmic rays. The best known and geologically most useful product of this type of reaction is carbon-14. Once formed, it makes its way into the carbon reservoir at the Earth's surface and can be used as a clock after it is firmly a part of a tree or a seashell. The limit of dating by this method is about 40,000 years.

Table 15.1 shows how various radioactive species have been used in determining time scales in oceanic deposits. In addition, K-Ar methods have been extensively used on land to date magnetic-reversal boundaries. These boundaries, correlated with reversals observed in deep-sea cores, then provide an indirect way of obtaining ages for marine sediments.

TABLE 15.1 Radiometric Techniques of Dating Marine Deposits

Method	Feasible Dating Range (years)	Types of Material Datable
^{14}C	0–40,000	Carbon-bearing materials—shells, organic matter
^{32}Si	0–2,000	Diatom- and radiolarian-rich deposits
^{231}Pa	0–120,000	Deep-sea sediments, ferromanganese nodules, corals
^{230}Th	0–400,000	Deep-sea sediments, ferromanganese nodules, corals
K-Ar	60,000 and older	Volcanic materials
^{10}Be	Up to 10 million	Ferromanganese nodules

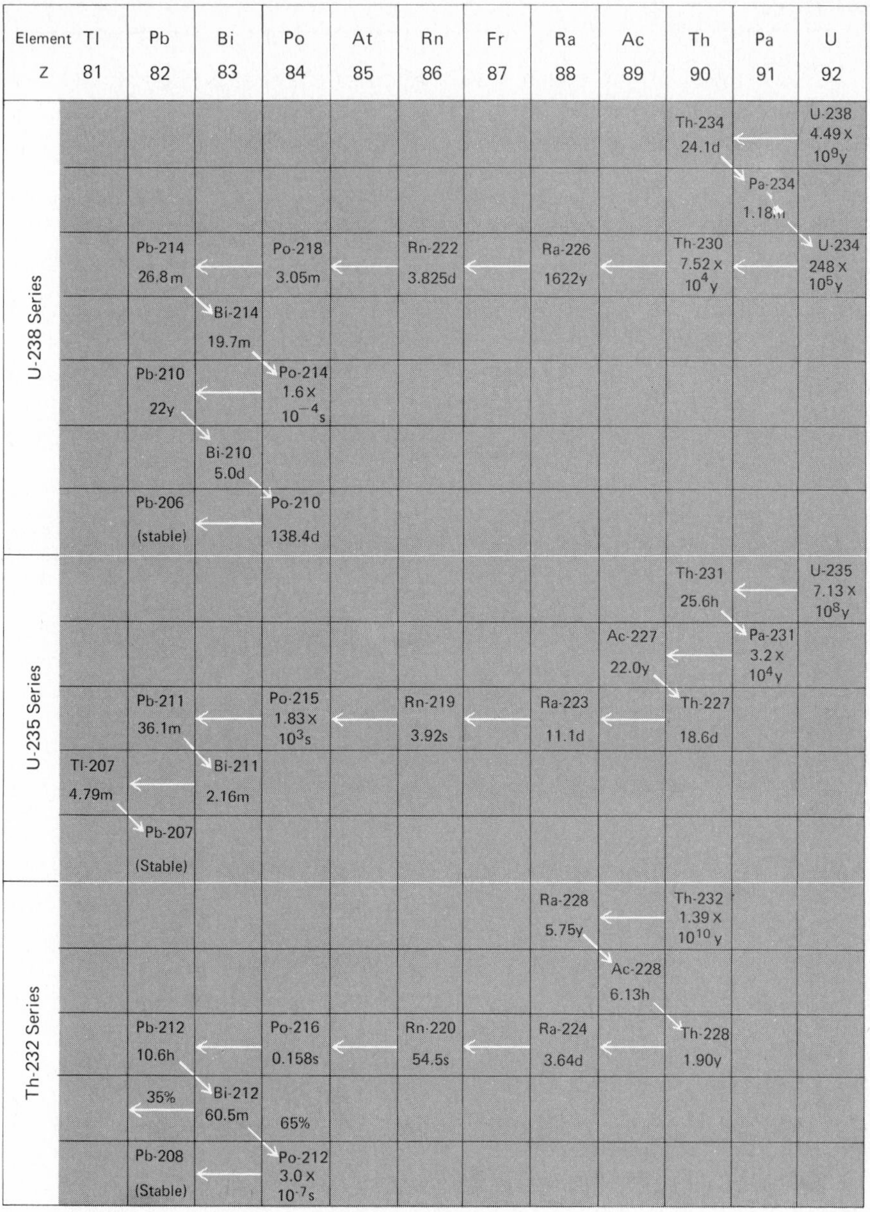

Element	Tl	Pb	Bi	Po	At	Rn	Fr	Ra	Ac	Th	Pa	U
Z	81	82	83	84	85	86	87	88	89	90	91	92

U-238 Series

Tl	Pb	Bi	Po	At	Rn	Fr	Ra	Ac	Th	Pa	U
									Th-234 24.1d		U-238 4.49×10^9 y
										Pa-234 1.18m	
	Pb-214 26.8m		Po-218 3.05m		Rn-222 3.825d		Ra-226 1622y		Th-230 7.52×10^4 y		U-234 248×10^5 y
		Bi-214 19.7m									
	Pb-210 22y		Po-214 1.6×10^{-4} s								
		Bi-210 5.0d									
	Pb-206 (stable)		Po-210 138.4d								

U-235 Series

Tl	Pb	Bi	Po	At	Rn	Fr	Ra	Ac	Th	Pa	U
									Th-231 25.6h		U-235 7.13×10^8 y
								Ac-227 22.0y		Pa-231 3.2×10^4 y	
	Pb-211 36.1m		Po-215 1.83×10^3 s		Rn-219 3.92s		Ra-223 11.1d		Th-227 18.6d		
Tl-207 4.79m		Bi-211 2.16m									
	Pb-207 (Stable)										

Th-232 Series

Tl	Pb	Bi	Po	At	Rn	Fr	Ra	Ac	Th	Pa	U
							Ra-228 5.75y		Th-232 1.39×10^{10} y		
								Ac-228 6.13h			
	Pb-212 10.6h		Po-216 0.158s		Rn-220 54.5s		Ra-224 3.64d		Th-228 1.90y		
	35%	Bi-212 60.5m	65%								
	Pb-208 (Stable)		Po-212 3.0×10^{-7} s								

FIGURE 15.12   The uranium and thorium decay series.

SUMMARY

1. Changes in the ocean are recorded as changes in the chemical, isotopic, and biologic properties of marine sediment. The history of the ocean can therefore be deciphered by examining piston and drill cores.

2. Because the rates of sediment accumulation vary from place to place in the ocean, and because layers formed may later be removed or disturbed, a historical study of the sediment pile at any place must be accompanied by an analysis of the geometry of the sediment layers (a stratigraphic analysis).

3. A stratigraphic analysis aims first at establishing the age of events recorded in one sedimentary section with respect to other sections. Such correlations are made by the techniques of magnetic, biostratigraphic, and isotopic stratigraphy.

4. Finally, the absolute age of events recorded in deep-sea cores is estimated by the techniques of radioactive geochronometry.

U.S.S. Bergall on gravity measuring cruise, Brisbane, Australia, 1943.

Exploring the Structure of Ocean Basins

16

The basic tools of geology, in its formative years, and indeed well into maturity, were a hammer, a notebook, a pair of stout boots, and keen powers of observation. Extensive use of these tools in the field provided the basic framework of geologic knowledge and a description of the character, age, and presumed history of the exposed rocks.

These tools sufficed for land geology,

CHAPTER

but in the oceans the rocks are not easily approached, because they are hidden beneath great depths of water. As a result, it was necessary to develop remote sensing techniques in order to study the rocks and sediments of the sea floor. These techniques are of two types: (1) those, described earlier, which sample directly the rocks and sediments of the sea floor and allow them to be returned to the surface for examination, and (2) those which measure some physical property of these materials or some property of the environment in which they are found. Taken independently these techniques can reveal a great deal; collectively they can provide major insights into the structure and the history of the ocean basins.

TECHNIQUES OF DETERMINING PHYSICAL PROPERTIES OF THE OCEAN FLOOR

Indirect, or remote sensing, methods of examining the sea floor and the rocks beneath it depend upon measurement of some physical parameter, such as magnetism, gravitational attraction, elastic, or thermal properties. Although a number of methods are useful, by far the most important are those that depend upon the elastic (or acoustic) properties of the water and the underlying rocks.

ACOUSTIC METHODS

Just as the atmosphere is transparent to light or microwaves—hence remote sensing devices using light or very high frequency electromagnetic waves are used to probe land areas, so are the ocean waters relatively transparent to sound—hence acoustic techniques are most useful in probing the ocean floor. Most of the techniques and devices had their origin in the problems of submarine detection. These techniques use sound waves at different frequencies to determine different properties at different depths beneath the ocean floor. Two of these techniques, echo sounding and seismic reflection profiling, have been described in Chapter 14. These methods, which employ the higher frequency portion of the usable sound spectrum, are most useful to determine the topography of the ocean basins and the distribution of sediments within them.

SEISMIC REFRACTION METHODS

To penetrate deeper into the crust beneath the ocean basins, we must use lower frequency waves and with them take advantage of the fact that sound travels at different velocities in different media. As a result, sound rays behave like optical rays and are bent, or refracted, when they encounter media in which the sound velocities vary. The useful range of frequencies in this case varies with the path of the sound rays and varies from 1 Hz to 1 kHz (1 to 1000 cycles per second).

The seismic refraction method utilizes this property of sound waves to measure the average properties of the sediments or the rocks beneath the sea floor along a profile some kilometers in length (Figure 16.1). It is based on the following assumptions:

1. That the sediments and rocks are layered.
2. That the boundaries between layers are planar over the length of a refraction profile.
3. That the properties of any layer are uniform and constant.
4. That the velocity of sound increases with layer depth.

These assumptions are not strictly true, because we know that the rock surface beneath the sediments is rough and velocity variations within a layer or velocity

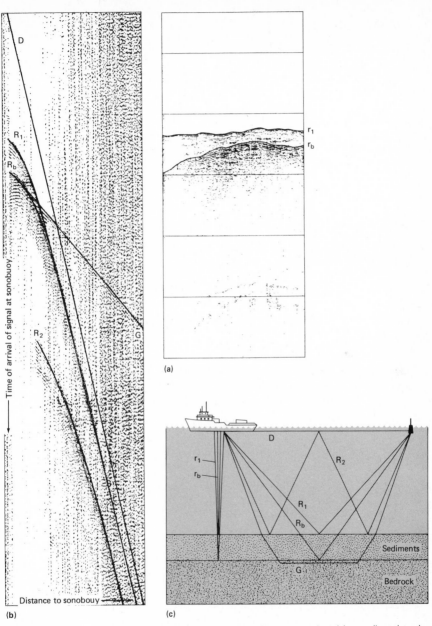

FIGURE 16.1 Sound waves are used not only to map the topography of the sea floor, but also to determine the thicknesses and sound velocities in the sediments and rocks beneath the sea floor. (a) A record of vertical reflections showing the ocean bottom (r_1) and the bedrock surface beneath the sediments (r_b). (b) A record made by steaming away from a sonobuoy which records the sound of an airgun fired at regular intervals by the ship and radios the signals back to the ship for recording. (c) Ray paths of the sound waves recorded in (a) and (b). Refracted arrivals (G) travelling through the layers beneath the bottom can be used to determine the thicknesses and the sound velocities in these layers. It is through measurements like these that we can determine the thickness of the crust beneath the oceans. (After Houtz, 1967, from Ewing and Worzel, 1967.)

inversions are not unknown. They are sufficiently true that serious errors in interpretation will not generally result from following them.

The seismic refraction method is a very powerful tool for examining the nature of the ocean crust since it will provide us with the depths to the different layers that make up the crust and upper mantle, will give us the seismic velocities in these layers, and will give us the average structure over the length of the profile.

EARTHQUAKE SEISMOLOGY

Nature generates useful (as well as destructive) acoustic energy in earthquakes that can be employed to determine additional facts about the structure of the ocean basins. This is very low frequency sound compared to artificial sources, falling in the range 1 Hz to 0.0003 Hz, or to periods of individual waves ranging from 1 second to 56 minutes. This sound can be used in several ways to tell us something about the nature of the sea floor.

First, the location of earthquakes tells us something (Figure 16.2). The earthquakes in the oceans are not randomly distributed, but are concentrated along the mid-ocean ridges and in the vicinity of island arcs and the continental margins particularly of the Pacific Ocean. Nor are they randomly distributed in depth. The

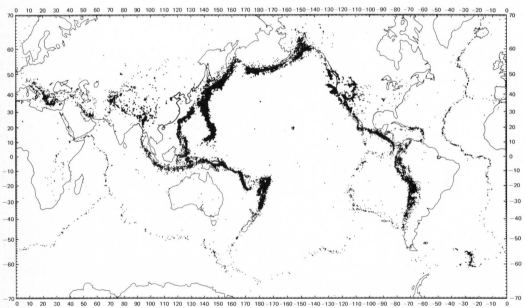

FIGURE 16.2   Distribution of earthquakes during the period 1961–1967. Earthquakes along the ocean ridges are shallow (0–70 km), while those along the island arcs and seismically active continental margins extend from near surface at the deep sea trenches to as much as 700 km beneath the islands and the continents. (After Barazangi and Dorman, 1969.)

earthquakes along the axes of the ridges are shallow focus (0-70 km in depth); in the island arcs and continental margins they fall in zones extending to depths exceeding 700 km. (The actual three-dimensional location of an earthquake is called the *focus*. The geographical location on the Earth's surface made by a radial line through the focus is called the *epicenter*.)

Second, the nature of the motion at the source of an earthquake gives us additional information. If the earthquake is recorded by seismographs located in many different directions from it, it is possible to determine whether the predominant motion at the source was vertical slippage, or normal faulting; horizontal slippage, or transverse faulting; underthrusting, or even pure compression (Figure 16.3).

When an earthquake occurs, it generates many different phases or groups of waves that travel to a seismograph station by different paths. Basically all these waves are of three types, compressional, or *P*, waves, shear, or *S*, waves, and surface waves. The velocity of *P* waves in a medium is directly related to the bulk modulus, a measure of compressibility, and the shear modulus, a measure of rigidity, and inversely related to the density. Thus the more rigid and less compressible a

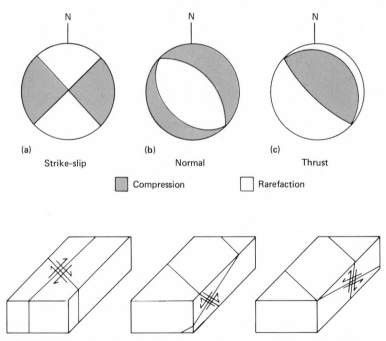

FIGURE 16.3   Focal mechanism diagrams for common fault systems. First, motion data from seismograph stations at varying distances and directions are plotted on a sterographic diagram. Then two mutually perpendicular planes are drawn which separate the data into quadrants of compressional or rarefactional first motions. One of these planes is a fault plane, the other an auxiliary plane. It is not possible to tell which is the fault plane without additional data. (a) Strike-slip motion, parallel to the ground surface. (b) Normal faulting on a plane dipping 60°NE or 30°SW. (c) Thrust faulting on a plane dipping 80°SW or 10°NE.

body is, the higher the *P* velocity. In principle, rocks of higher density should have lower seismic velocities, but in practice, the elastic constants usually increase more rapidly than the density. The *S* velocity is directly dependent on the shear modulus and inversely dependent on the density; hence if the shear modulus falls to zero, as in a liquid, no shear waves are propagated. If it is decreased—for example, because the rocks are near their melting point—the shear waves will be attenuated. Surface waves are elastic waves that travel along the surface of the Earth, like gravity waves in the ocean waters. Their amplitudes diminish exponentially with depth as a function of their wavelength.

The body waves from earthquakes are most useful in determining the structure of the Earth's interior (Figure 16.4). The rocks we see at the surface are part of the thin outer crust whose base is marked by a discontinuity in velocity, which has been named the Mohorovičič Discontinuity after its discoverer. Beneath this, the mantle, composed of silicate rocks, extends down to 2900 km. Here a sharp decrease in *P* velocity marks the outer boundary of the iron core, and since no *S* waves propagate through the outer core it is presumed to be liquid. Another lesser discontinuity at a depth of 5100 km marks the boundary of the inner core, also iron, but solid. The fact that the outer core is liquid iron, which is a good electrical conductor, has led to the conclusion that motions in the core produce the magnetic field of the Earth.

Surface waves are of special interest because they are dispersed. An earthquake occurs as an impulse or a step function made up of energy in many different

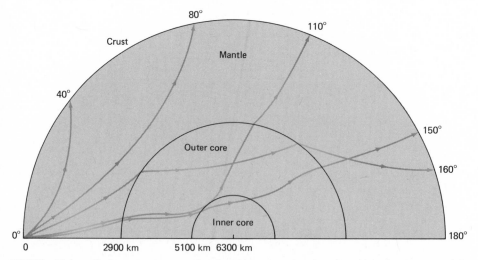

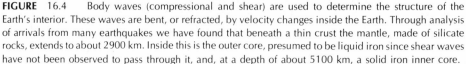

FIGURE 16.4   Body waves (compressional and shear) are used to determine the structure of the Earth's interior. These waves are bent, or refracted, by velocity changes inside the Earth. Through analysis of arrivals from many earthquakes we have found that beneath a thin crust the mantle, made of silicate rocks, extends to about 2900 km. Inside this is the outer core, presumed to be liquid iron since shear waves have not been observed to pass through it, and, at a depth of about 5100 km, a solid iron inner core.

wavelengths. The depth of penetration of the surface waves is a function of wave-length; the shorter wavelengths are influenced only by the properties of the near surface rocks, whereas the longer waves feel the presence of the deeper layers of the Earth. Since velocities in general increase with depth, the average sound velocity encountered by the long waves will be greater than for the short ones, and at great distances from the earthquake the initial impulse will be stretched into a long-wave train with the long waves arriving first, the shorter ones later. By use of the electronic computer, it is possible to construct detailed models to duplicate this stretching, or dispersion, and to determine the velocity structure of the outer part of the Earth (Figure 16.5). Surface-wave studies have been especially useful in defining the channel found at depths of 100 to 200 km within the mantle, in which the sound velocities are lower than in the layers above and below. It has been found necessary to include such a channel in most models in order to match the data, and, as we shall see in Chapter 17, this finding has great significance with regard to current ideas about Earth deformation and formation of the oceans.

GRAVITY METHODS

Although the acceleration of gravity had been measured on the continents since the eighteenth century, until 1923 the oceans were devoid of accurate gravity measurements. In that year, F. A. Vening Meinesz, who had developed a multiple-pendulum apparatus for measuring gravity on the unstable land of the Netherlands, borrowed a submarine from the Dutch Navy to try it at sea. By diving to depths where wave action was minimal, he found that his apparatus would work after some modifications (Figure 16.6). Subsequently gravimeters were developed that can measure gravity from surface vessels (Figure 16.7).

Part of the variation in gravity from place to place is due to the shape of the Earth. The mutual gravitational attraction between two bodies is given by

$$G = \gamma \frac{mM}{r^2}$$

(16.1)

where G = gravitational acceleration in cm/sec²
m,M = masses of the two bodies in grams
r = distance between the centers of mass of the two bodies in cms
γ = gravitational constant, 6.67×10^{-8} cm³/g²/sec²

The gravitational acceleration depends upon latitude, since the Earth is rotating and is not truly spherical. The equation for the variation of gravity with latitude takes these factors into consideration:

$$G = 978.049 \ (1 + 0.0052884 \ \sin^2 \phi - 0.0000059 \ \sin^2 2\phi) \ \text{cm/sec}^2$$

where ϕ = latitude.

(16.2)

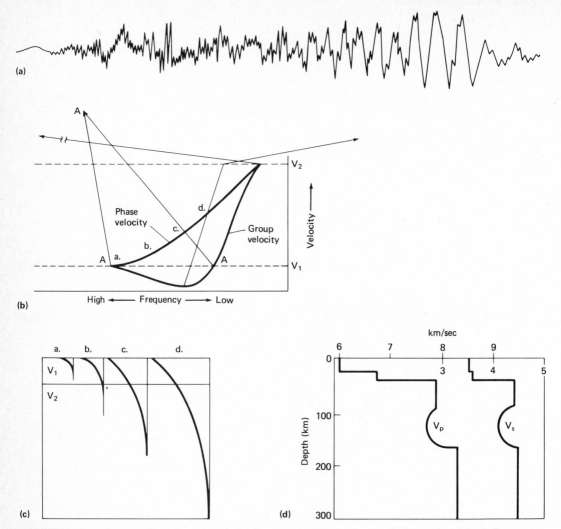

FIGURE 16.5   An example of dispersion of seismic waves, in this case the sound of an explosion in shallow water. Although the original sound was a sharp impulse, the waves arriving at the recording ship are spread out, or dispersed, because waves of different frequencies travel at different velocities. Models can be constructed to fit the observed data and to determine the velocity structure of the subbottom. (a) Portion of the record of a shot at a distance of 30 km. (b) A plot of velocity versus frequency showing a dispersion curve for an appropriate model. (c) The amplitude of surface waves decreases exponentially with depth. The high frequency waves, a, travel entirely in the upper layer, hence with the sound velocity in this layer. Very low frequency waves, d, travel mostly in the lower layer and approach the velocity of sound in it. Intermediate frequency waves travel at intermediate velocities. Individual waves travel with the wave or phase velocity. (d) Structure calculated from earthquake surface wave data from the Basin and Range province in the western United States. The model must include a low velocity layer between 50 and 150 km depth in order to fit the data. (After E. Herrin, in *Nature of the Solid Earth*, E.C. Robertson, ed. Copyright 1972. Used with permisssion of McGraw-Hill Book Company.)

FIGURE 16.6   The submarine USS S-48, used in an early gravity measuring cruise in the Caribbean by Hess, Ewing, and Vening Meinesz. Measurements were made using a submarine gravity measuring pendulum apparatus invented by Vening Meinesz. Measurements were made while submerged in order to provide a reasonably stable platform for the apparatus, and each measurement required about 30 minutes of observation. (From U.S. Hydrographic Office, 1933.)

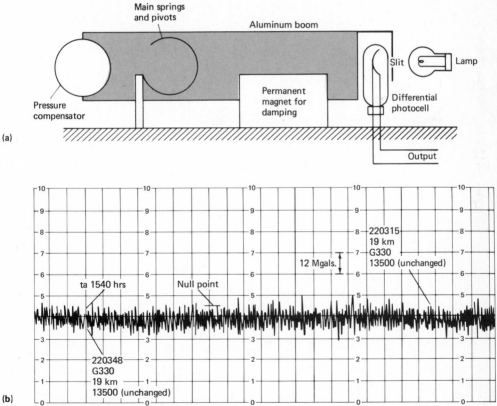

FIGURE 16.7   Surface ship gravity meter. (a) A boom supported by a curved spring responds to changes in gravitational acceleration by moving up or down. In order to minimize the effects of wave action, the boom is damped by a permanent magnet. Changes in the position of the boom are measured by a photocell. (b) A record of gravity measurements in the North Atlantic. Gravity change is very small on this record. The oscillations are due to wave motion not completely damped out. (From *IGY Bulletin*, **8**, Feb. 1958.)

The above equation defines the gravitational acceleration at sea level as a function of latitude in units of cm/sec², or gals, named after Galileo. To make geologically significant observations, we must measure to the order of 0.001 gal, or 1 milligal, which is one millionth of the Earth's total field. Because we are interested in the effects of the rock structures and not in the shape of the Earth, we take the difference between the measured value and that given by the equation for the same latitude. This difference, or gravity anomaly, is due to density inhomogeneities in the rocks of the Earth's crust and upper mantle, and by mapping these differences we can determine fundamental facts about the structure of the Earth.

MAGNETIC METHODS

A major problem during World War II was the detection of submerged submarines. With the development of the snorkel submarine and its ability to steam long distances at relatively high speeds with only a small breathing tube reaching the surface, a new detection device was needed. This need was answered by the magnetometer, a sensitive instrument that could be towed at low altitudes behind an aircraft and which would detect the magnetic field of the submarine even though it was submerged (Figure 16.8).

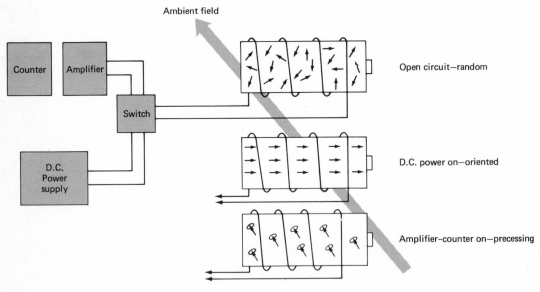

FIGURE 16.8   Most magnetometers now used at sea take advantage of the fact that hydrogen nuclei have gyromagnetic properties. The detector consists of a coil of wire wrapped around a bottle of water. When a strong, direct electrical current is applied to the coil, a magnetic field is created and the nuclei align themselves along this field. When the current is switched off, the nuclei try to orient themselves in the direction of the Earth's magnetic field, but because they are spinning, they behave like tiny gyroscopes and precess rather than flopping immediately over. The combined effect of these nuclei precessing produces a weak alternating current, the frequency of which is directly related to the absolute strength of the Earth's magnetic field.

At sea the magnetometer can be towed behind aircraft, which has the advantage of rapid coverage, but the disadvantage that the instrument is farther away from the sources of magnetic effects and the relationship of the magnetic effects to bottom topography is not completely clear. It can be towed behind ships, on which the topography is measured simultaneously, or, by use of deep towed vehicles, it can be towed close to the sea floor and to the sources of the magnetic effects, but only at very slow speeds.

Magnetic measurements are the sum of the total field of the Earth plus the magnetic effect of the crustal rocks. Since we are interested in the effect of the rocks, we subtract a standard main field from the measurements, and the remainder is the magnetic anomaly.

We are fortunate in that the rocks of the ocean crust are predominantly basaltic, with a reasonably high content of magnetic minerals, and thus the magnetometer is a very useful tool for exploring the oceans. The magnetic effect of these rocks may be induced—that is, produced—by the present Earth's field, and hence in the direction of the present field; or it may be remanent, frozen in the rocks at an earlier time and thus representative of the direction and the magnitude of the field at that time. The Earth's field is not fixed in time or in space, but shows secular variations in amplitude and in position. It is too strong and too variable to be produced by permanent magnetism of the crustal rocks, and current thought places its origin in the outer liquid core of the Earth. Motions in the core, with a small exciting field and given orientation by the rotation of the Earth, can produce a field of the observed magnitude. One would expect this field to be roughly oriented along the rotational axis of the Earth, but the polarity of the field must be related to the excitation and could be as it is at present or the reverse (see Chapter 15).

There is a temperature, called the Curie point, above which materials capable of retaining permanent magnetization become demagnetized. Similarly, if these materials are cooled through this temperature in the presence of an external magnetic field, they will become magnetized in the direction of that field. The basaltic rocks of the sea floor were originally molten, so when they cooled through the Curie point, they became magnetized in the direction, and reflected the magnitude, of the Earth's main field at that time. Thus measurements of the magnetic effects of these rocks tell us something not only of their nature and their structure but also of their relation to the history of variation of the total Earth's field. This residual magnetism, known as *paleomagnetism*, has been studied extensively both in the oceans and on the continents and has led to startling new concepts of the history of the ocean basins.

THERMAL METHODS

Temperature increases with depth within the Earth, and heat is constantly escaping, or flowing, from the interior through the surface. This flow is very small, of the order of millionths of a calorie/cm²/sec, and is not everywhere the same. Thus measurements of its variations with area and geologic structure will provide us with additional facts about the nature of the sea floor.

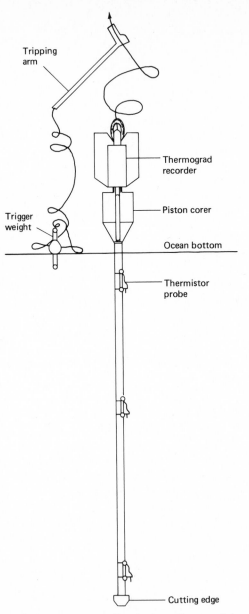

FIGURE 16.9   To measure the temperature gradient in the sediments of the sea floor, thermistors (temperature-sensitive resistors) are strapped to the pipe of a piston corer. These data, combined with measurements of thermal conductivity of the cores which are made aboard ship, determine the amount of heat flowing out from the sea floor. The amounts are very low, of the order of millionths of a calorie per square centimeter per second.

Since the temperature of the bottom waters of the deep ocean remains constant for long periods, measurements of the temperature in the sediments to depths of a few tens of feet, coupled with measurements of the thermal conductivity of the sediments, are sufficient to provide accurate determinations of the heat flow from the interior. These data may be obtained by mounting thermistor (temperature-sensitive resistor) probes for temperature measurement at intervals along the piston core pipe (Figure 16.9), and both a sediment sample and the temperature gradient can be taken on a single lowering. Temperature measurements of the water column can also be made as the apparatus is lowered.

OCEAN BASIN STRUCTURE

THE OCEAN BASIN PROPER

Each technique described above tells us something of the nature of the ocean basins. An accurate model of the structure of the ocean basins must be compatible with the data from all these techniques. Thus the thickness and the conformation of the sedimentary and crustal layers must correspond to those indicated by seismic refraction and reflection; the rocks must be of the correct density distribution to agree with gravity data and must have magnetic properties appropriate to the magnetic results. The seismic velocities in these rock types must be appropriate and the surface rocks and sediments must be as sampled. The model must agree with observed thermal anomalies and must account for observed earthquake effects. Collective application of all the geological and geophysical techniques narrows the possibilities and produces a model that must be very close to the real structure.

The first geophysical technique applied at sea, other than the echo sounder, was the measurement of gravity. Now, if there are major differences in the densities of crustal materials, one might expect to find large gravity anomalies. If we compare seawater, with a density (ρ_w) of 1.03 g/cm³ and continental crustal rocks, with a mean density (ρ_c) of about 2.67 g/cm³, we find a density difference of 1.64 g/cm³. The gravitational attraction of an infinite horizontal slab with thickness h is given by

$$G = 2\pi\gamma\rho h \qquad (16.3)$$

where γ = gravitational constant, 6.67×10^{-8} cgs
ρ = the density in g/cm³ (ρ_c = crustal density; ρ_w = water density)
h = thickness in cm
G = gravitational acceleration in gals

Let us suppose that the *only* difference between continent and ocean basin is in the presence of the water and this is 5 km thick in the ocean basins. Below this depth everything is uniform (Figure 16.10a). Then the *difference* in gravity between land at sea level and the ocean basin at some distance from the shoreline is:

$$\Delta G = 2\pi (\rho_c - \rho_w)\gamma h = 2 \times 3.14 \times (2.67 - 1.03) \times 6.67 \times 10^{-8}$$
$$\times 5 \times 10^5 = .344 \text{ gals} = 344 \text{ milligals}$$

$$(16.4)$$

So if the above supposition is correct, we would expect to find a difference in gravitational acceleration between continent and ocean of some 344 milligals.

We do not find this difference. On the contrary, the values on continents and in oceans are approximately equal if we are some distance away from the continental margin. Thus everything below 5 km cannot be the same.

Now, suppose the crust of the Earth is thinner under the oceans than under the continents. This would mean that the mantle rocks beneath the oceans, with a density in the upper part of 3.27 g/cm³, would be contrasted with crustal rocks of density 2.67 g/cm³ for a density difference of 0.6 g/cm³ (Figure 16.10b). If the difference in crustal thickness were 13.6 km, then:

$$G = 2 \times 3.14 \times 6.67 \times 10^{-8} \times 0.6 \times 13.6 \times 10^5 = .342 \text{ gal}$$
$$= 342 \text{ milligals} \qquad\qquad (16.5)$$

or very nearly the same value as the difference caused by the water-crust contrast (Figure 16.10c). Thus, if we take the sum of these two effects, the two regions are essentially in balance except near the continental margin, where the difference in depth of the two influencing masses results in an edge effect. Gravity measurements indicate that most of the Earth's surface is close to being in what is called "isostatic balance," mountains being underlain by low-density material and the oceans by high densities. This balance may be produced by variations in thicknesses or densities (Figure 16.11), but we cannot determine which is the cause from gravity measurements alone.

Empirical relationships have been established, however, between seismic compressional velocity and density (Figure 16.12), so if we make seismic refraction measurements we can determine the approximate densities for the rock layers. Moreover, from these measurements we can determine the thicknesses of the various layers in the crust and the upper mantle. Many such measurements have been made both on the continents and in the oceans, and from these we can determine mean structures and compare them (Figure 16.13).

At sea level the continental crust is about 30 km thick. Seismic velocities in the upper half are similar to those found in granitic rocks; those in the lower part are appropriate for basaltic rocks. The high velocity below the Mohorovičič Discontinuity in the mantle is appropriate for rocks such as dunite, peridotite, and eclogite, rocks considerably denser than those making up the crust. In the oceans, beneath about 5 km of water are an average of 0.5 km of sediments, 0.5 km of volcanic "basement," 4 km of oceanic crust, and at a depth of about 10 km we encounter the mantle. If we compare the masses of 1 cm² sections from continents and oceans to a depth of 30 km (below which differences are small), we find that these masses are very nearly the same. Thus it is not surprising that gravity measurements give similar results.

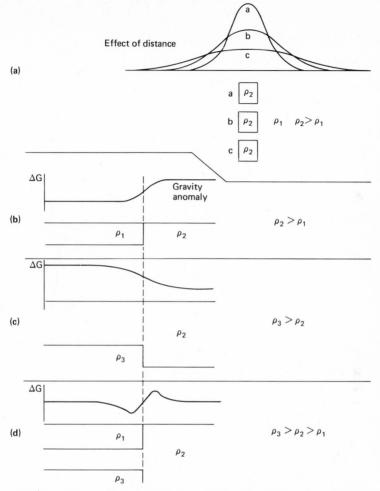

FIGURE 16.10   The gravitational attraction of a body depends on its mass, and is inversely proportional to the square of the distance between its center of gravity and that of the gravity measuring device. Thus, the further away the body is, the lower in amplitude and the broader in width its gravitational effect will be, as shown in (a). If the only differences between continent and ocean were the level of the rock surface and the presence of the seawater, we would expect a gravitational effect as shown in (b) where ρ_1 represents the density of seawater and ρ_2 the density of crustal rock. However, we do not observe this; the gravitational effect on land at sea level and in the ocean (away from the margin between the two) is approximately the same. If the thickness of the crust is greater under the continent than under the ocean, we would expect a gravitational effect as in (c), where ρ_2 is the density of the crust and ρ_3 the higher density of the underlying mantle rocks. The combined effect of crustal thickness and water depth (d) gives equal gravity values on the two sides of the continental margin, with an anomaly due to the edge effect at the boundary.

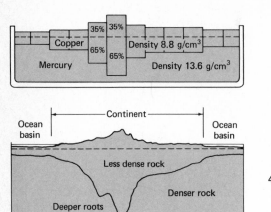

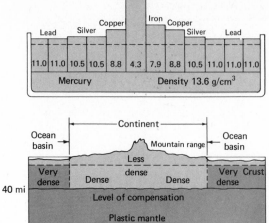

(a)

(b)

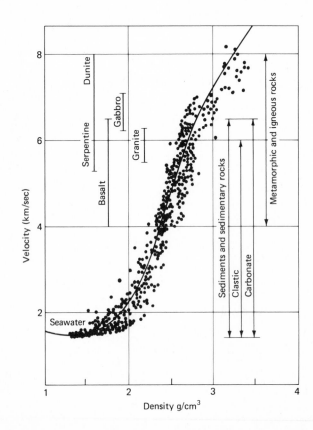

FIGURE 16.11   Two hypotheses have been offered to explain the fact that most of the Earth is very close to being in isostatic balance. (a) Airy suggested that great elevations are supported by masses of rock of the same density at depth, just as an iceberg is supported by ice beneath the sea surface, and that crustal density is the same everywhere. (b) Pratt, on the other hand, suggested that there was a level of compensation above which everything floated and that elevation differences were due to differences in mean crustal density. It is not possible to discriminate between these two models from gravity data alone.

FIGURE 16.12   Empirical data indicate that an increase in seismic compressional velocity is accompanied by an increase in density. This curve can be used to estimate the densities of subsurface rocks from seismic velocities. Because rocks of many types may have similar velocities, it is not usually possible to determine the kind of rock from the seismic velocity. (After, Ludwig, Nafe, and Drake, 1970.)

FIGURE 16.13   Generalized crustal structure of continents, ocean basins and oceanic ridges. These features are essentially in isostatic balance and it can be seen that the masses of columns one centimeter square are approximately equal at a depth of about 58 km.

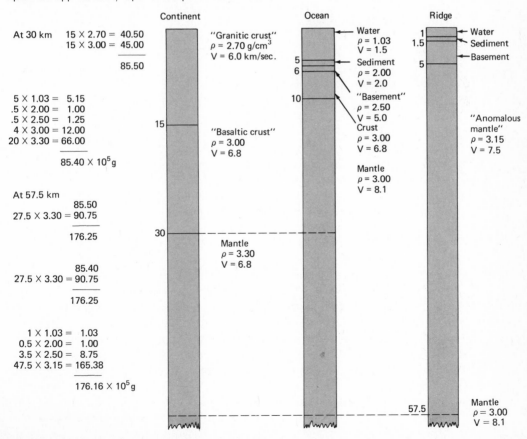

At 30 km
15 × 2.70 = 40.50
15 × 3.00 = 45.00
―――――
85.50

5 × 1.03 = 5.15
.5 × 2.00 = 1.00
.5 × 2.50 = 1.25
4 × 3.00 = 12.00
20 × 3.30 = 66.00
―――――
85.40 × 10^5 g

At 57.5 km
85.50
27.5 × 3.30 = 90.75
―――――
176.25

85.40
27.5 × 3.30 = 90.75
―――――
176.25

1 × 1.03 = 1.03
0.5 × 2.00 = 1.00
3.5 × 2.50 = 8.75
47.5 × 3.15 = 165.38
―――――
176.16 × 10^5 g

Continent

"Granitic crust"
ρ = 2.70 g/cm^3
V = 6.0 km/sec.

"Basaltic crust"
ρ = 3.00
V = 6.8

Mantle
ρ = 3.30
V = 6.8

Ocean

Water
ρ = 1.03
V = 1.5

Sediment
ρ = 2.00
V = 2.0

"Basement"
ρ = 2.50
V = 5.0

Crust
ρ = 3.00
V = 6.8

Mantle
ρ = 3.00
V = 8.1

Ridge

Water
Sediment
Basement

"Anomalous mantle"
ρ = 3.15
V = 7.5

Mantle
ρ = 3.00
V = 8.1

OCEAN RIDGES

The ocean ridges present a different problem. The ocean ridge system extends across the Arctic Ocean, through the middle of the Atlantic and Indian Oceans, between Australia and Antarctica, and across the Pacific to Baja, California. Seismic refraction measurements show a crustal structure for the ridges similar to that in the ocean basins, but velocities in the "mantle" rocks at depths of about 5 km are anomalously low—7.5 km/sec rather than 8.1—and densities in these rocks will be correspondingly low. Seismic refraction measurements do not reach deep enough to give us the depth of the bottom of this low-velocity mantle, but we can make some estimates from gravity measurements. Gravity measurements across the ridges indicate that they are essentially in balance with the adjacent basins. If the ridge were merely an elevated piece of ocean floor, we would expect to find large positive gravity anomalies. Since we do not, it is not unreasonable to assume that the elevation is compensated for by some mass deficiency at depth, and a logical source is the low-density mantle which must extend to some depth.

From these measurements, then, we can conclude that the ocean crust is qualitatively different from that of the continents. It is about one-third as thick and appears to contain no granitic rocks. From its measured physical properties and from sampling by drilling and dredging, we know that the upper part of the oceanic crust consists of basaltic lavas. While the crustal rocks of the ridges are similar to those of the ocean basins, the mantle beneath the ridge appears to be chemically or thermally altered to some depth. The similarities in character and gradational nature of the transition from ridge to ocean basin suggest a common or related history; the completely different character and the abrupt transition between continent and ocean suggest an entirely different history.

Additional information comes from seismicity—the location of earthquakes in the oceans or on their margins (Figure 16.2). The continuity of the ridge system was first inferred from this distribution of earthquakes. Where the ridge comes ashore, as in East Africa, or protrudes above the surface, as in Iceland, the dominant features are normal faulting, or rifting, and abundant volcanism. In Iceland, which lies astride the axis of the ridge, the youngest currently active volcanism occurs in rifts in the middle of the island, whereas the volcanics on the eastern and western flanks are tens of million years old. From these examples and from the topography of the ridges, which frequently includes a central valley or rift, we can conclude that they were formed by a tensional mechanism and were built up by volcanic processes and that the present activity is located along their centers. Since the anomalously low mantle velocities are often found beneath volcanic seamounts as well as beneath Iceland and the East African rifts, and since the heat flow near the centers of the ridges is as much as five times the norm for ocean basins, we can conclude that the anomalously low mantle velocities are created by thermal processes.

Because the ridges are tensional in character, with cross-cutting fracture zones and often a central rift valley, they provide us with an opportunity to study the crustal rocks of the oceans that are in most areas covered by a blanket of sediments. These rocks have been examined by dredging for some years, have more recently been drilled by the Deep Sea Drilling Project, and have been examined in place during Project FAMOUS, a joint United States-French expedition employing three research submersibles, *Alvin, Cyana,* and *Archimede* (Figure 2.17).

The results of these studies have verified the conclusions of the geophysical measurements. The upper part of the ocean crust is made up of basaltic lavas in the pillowlike form typical of underwater extrusions (Figure 16.14). A deep-sea drilling project borehole about 30 km from the axis of the ridge in the North Atlantic drilled through 600 m of alternating pillow lavas and sediments with the amount of sediment decreasing toward the bottom of the hole until only the lavas remained.

The nature of the deeper layers is less certain, since a variety of basaltic and "ultrabasic" rocks (dunites, peridotites, eclogites—see Glossary) have been dredged or sampled, some by *Alvin* in the Cayman Trough in the Caribbean Sea. It is probable that the deeper portions of the crust are made up of more coarsely crystalline basaltic rocks of the same composition as the fine-grained volcanics, whereas the mantle is made of ultrabasic rocks.

CONTINENTAL MARGINS AND ISLAND ARCS

Some geologists like to divide continental margins into two types, Atlantic and Pacific. Pacific margins are marked by instability—by earthquakes and volcanoes, by deep-sea trenches, and by geological structures that usually parallel the coasts. Frequently island arcs, with the above features, separate the main basin of the Pacific from marginal seas off the shores of the continents. Atlantic margins, on the other

(a)   (b)

FIGURE 16.14   (a) Pillow lavas from the mid-Atlantic ridge (Lamont-Doherty Geological Observatory). (b) Pillow lavas now above sea level from Iceland (Drake photo). The characteristic pillow-like structure results when the outer surface of an underwater lava flow is quickly chilled.

hand, are quiescent—few earthquakes, no volcanoes or deep-sea trenches, and geological structures frequently, though not always, truncated at the continental margin. Along these margins and arcs are located the major discontinuities in crustal thickness and, often, the greatest accumulations of sediments to be found in the ocean basins. These sedimentary accumulations are of great interest, since the major mountain systems of the world are made up of great thicknesses of sedimentary rocks. James Hall first called attention to this when he was mapping the state of New York, and the regions where these thick accumulations occur became known as *geosynclines*. There are many different types of geosynclines, but those of particular interest to us are the ones called *orthogeosynclines* (or "true" geosynclines): long, linear sediment-filled depressions. Hans Stille identified two types of orthogeosynclines in the geologic record, often occuring as a couple—one type, the *eugeosyncline,* was marked by volcanic activity, the other, the *miogeosyncline,* by the absence of volcanic activity. There were also differences in the nature of the sediments and in the way they subsequently became deformed, but the original definition was based on the presence or absence of volcanism.

A question that bothered geologists over the years was "Where are the modern geosynclines"? These are filled with marine sediments, so the sea must have had access to them. Some of these sediments are of shallow-water character, but others are more similar to those found on continental rises or abyssal plains, so the continental margins would appear to be good candidates as modern geosynclines if their other characteristics are appropriate and a suitable mechanism for deforming them can be found. The stable Atlantic margins lend themselves better to studies of properties; the unstable Pacific margins are more appropriate for studies of process. Let us look at examples of each type.

Atlantic margin of North America   Seismic refraction and reflection profiles (Figure 16.15) indicate that beneath the continental shelf of eastern North America there is a thick wedge of sediments and another beneath the continental rise. There are major differences between the two sections of Figure 16.15 due primarily to the fact that clastic sediments—clays, muds, and sands—predominate in the north and carbonate sediments—limestones—in the south.

If we compare the sections with a reconstruction of the Appalachian geosyncline from New York through Maine prior to its major deformation (Figure 16.16), we find some remarkable similarities. The trough under the present continental shelf has been shown from studies of outcrops and drill cuttings to be filled with sedimentary rocks with an abundant bottom fauna deposited in shallow water. They compare very well with the sedimentary rocks of the folded Appalachians, which are also of shallow-water character and appear to have been deposited on a continental shelf. The sediments of the present continental rise are of quite different character and contain very few remains of bottom fauna, most of which have probably been swept off the continental shelf. This is also true of the metamorphic Appalachians to the east, where few fossils are found, and the character of the sediments here, when it can be determined in the strongly metamorphosed rocks, is

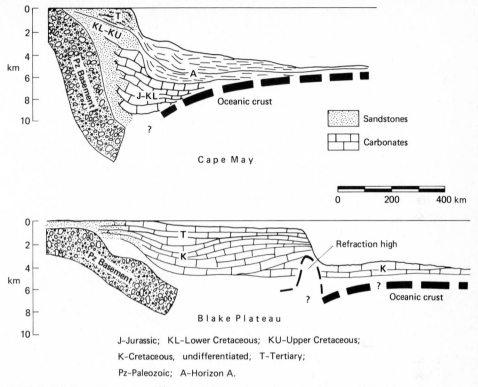

J–Jurassic; KL–Lower Cretaceous; KU–Upper Cretaceous;

K–Cretaceous, undifferentiated; T–Tertiary;

Pz–Paleozoic; A–Horizon A.

FIGURE 16.15   Structure profiles across the continental margin of eastern United States from geological and geophysical measurements. (After Bally, Shell Oil Co., 1975.)

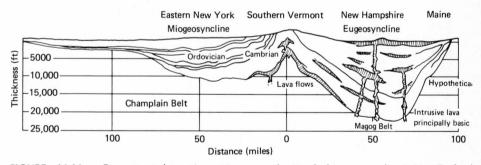

FIGURE 16.16   Reconstructed structure section across the Appalachian geosyncline in New England prior to its deformation. (After Kay, 1951.)

similar to that of the sediments of the modern continental rise and abyssal plain. There is little evidence of volcanics under the present margin, although we may presume that basaltic rocks lie beneath the sediments of the continental rise and abyssal plain, but it is not unreasonable to attribute this apparent absence to the fact that the area has not yet been subjected to deforming forces.

Pacific margin of Central America. To study the deforming forces, we must move to the Pacific, where activity is currently occurring. The geological evidence of the deformation is not everywhere the same but the principles are well-illustrated in Central America, which is one of the most volcanically active regions in the world.

The Pacific margin of Central America is marked by a deep sea trench (Figure 16.17). These trenches are characteristic of island arcs and are the deepest parts of the oceans. Typically they are marked by large negative gravity anomalies, too

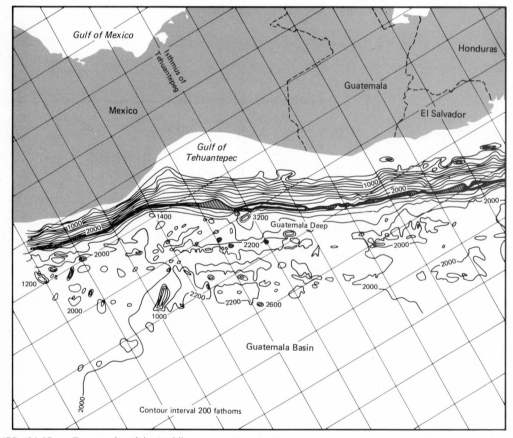

FIGURE 16.17 Topography of the Middle America Trench off Mexico and Central America. (After Heacock and Worzel, 1955.)

large to be accounted for by water depth alone. In some cases the trench has become filled with sediment and it still has a negative gravity anomaly. Since the trenches have been in existence for some time, this indicates that active forces are keeping them depressed.

This activity is reflected by numerous earthquakes that fall into a narrow zone extending from near the surface at the trench to depths exceeding 300 km beneath Central America and in other island arcs to depths in excess of 700 km. Studies of the mechanisms of these earthquakes indicate normal faulting for the shallow ones near the trench—which might be expected if the crust were bent downward in this area—but underthrusting for the deeper earthquakes under Central America.

The results of this underthrusting have been demonstrated by deep reflection studies made in connection with petroleum prospecting (Figure 16.18). Off Guatemala the result has been the formation of imbricate or piled-up thrust sheets accompanied by crustal thickening, subsidence of the crust in response to the increased load, and steepening of the thrust planes as one proceeds shoreward. The structural style varies from place to place along the Pacific margin, but everywhere gives evidence of compressional forces at work.

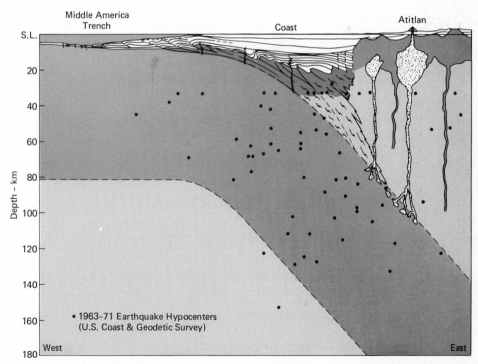

FIGURE 16.18   Interpretation of structure across the Middle America Trench off Guatemala. The data suggest that the sediments and the oceanic crust are thrust-faulted and stacked in imbricated layers by the convergence process. (After Seely, Vail, and Walton, 1974.)

The volcanoes of Central America tend to be located near the region where earthquakes occur at depths of about 100 km and a plot of earthquakes for 1961–1972 shows a high density of earthquakes in the vicinity of volcanoes that erupted during the same period, thus indicating a genetic relationship.

The deforming forces on these active continental margins are of compressional character, produce folded and thrust-faulted rocks, and are accompanied by thermal events of which the most obvious manifestation is volcanism and the most permeating is large-scale thermal metamorphism of the rocks.

Sometimes the thrusting along the active continental margins brings up rocks from the depths and exposes them at the surface. Among these are rock sequences that consist of pillow lavas, coarsely crystalline basaltic rocks, and ultrabasic rocks ordered so that the ultrabasic rocks are always the lowest member. Often these are associated with siliceous rocks containing the tests of radiolarians, pelagic organisms whose presence suggests that these rocks had their origin in the open ocean.

The similarity of these rock units in physical properties and character to those measured or sampled on the deep-sea floor imply that they represent ocean crust and uppermost mantle that has been thrust onto the continents. If so, they provide us with a unique opportunity for learning about the nature of the ocean crust.

SUMMARY

1. A number of geophysical techniques have been developed to determine the properties and the structure of the sediments and the rocks beneath the sea floor. These include explosion and earthquake seismology and measurements of the gravity, magnetic, and thermal fields.

2. The ocean crust is considerably thinner than that of the continents and is predominantly basaltic, while the continental crust is largely granitic.

3. The structure of the ocean ridges is similar to that of the ocean basins, and their elevation seems to be due to thermal effects.

4. The margins of the continents are often the location of thick accumulations of sediments. These sediments are similar to those of ancient geosynclines that have become mountain systems.

5. The processes taking place in seismically active continental margins appear to be those which have turned ancient geosynclines into mountain systems.

Rift in western Iceland, linear eruptions, steam (right); closeup of rift (left).

Plate Tectonics
and History of
the Ocean Basins

17

In 1782, Benjamin Franklin wrote in a letter to Abbé Soulavie,

Such changes in the superficial parts of the globe seemed to me unlikely to happen if the earth were solid to the centre. I therefore imagined that the internal parts might be a fluid more dense and of greater specific gravity than any of the solids we are acquainted with; which therefore might swim in or upon that fluid. Thus the surface of the globe would be a shell, capable of being broken and disordered by the violent movements of the fluid on which it rested. . . . If (these thoughts) occasion any new inquiries and produce a better hypothesis, they will not be quite useless. You see I have given a loose to imagination; but I approve much more your method of philosophizing, which proceeds upon actual observation, makes a collection of facts, and concludes no farther than those facts will warrant.

Franklin's comments were directed toward the long-recognized observation that the rocks at the Earth's surface were not all in the same position or of the same charac-

CHAPTER

ter as when they were first formed. Some showed evidence of remarkable changes in elevation, others had been badly distorted or fractured, still others had been drastically changed in character through subjection to heat or pressure or a combination of both. With continued study, patterns began to emerge and it was found that the activities reflected by these changes occurred in rather specific and limited areas over the course of geologic time.

EARLY FRAMEWORK

CONTRACTING EARTH

As the fundamental geology of the various parts of the world became known, it was apparent that at various times in the geologic past, strong tectonic activity occurred in relatively narrow mobile belts (Figure 17.1). There appeared to be a synchronism of movement among many of these tectonic pulses, and they were described as orogenies or revolutions, such as the Caledonian, Hercynian, Alpine, and others. It became accepted generally that accumulation of sediments and epeirogenic movements, or general changes of relative sea level (Figure 20.7), occurred over a very long period and that the orogenic activity had a much shorter time constant.

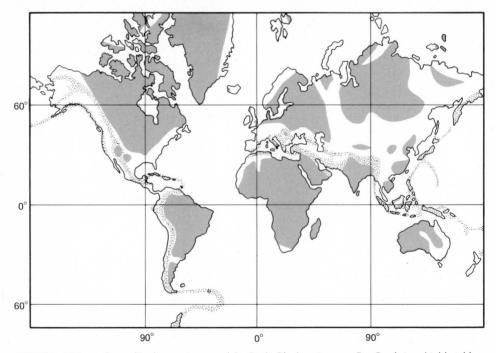

FIGURE 17.1   Generalized tectonic map of the Earth. Black regions are Pre-Cambrian shields (older than 600 million years); white regions are those deformed during Paleozoic time (600 m.y.–230 m.y.) and dotted areas are those deformed during Mesozoic and Cenozoic time (230 m.y. to the present). (After Drake, 1970.)

The features preserved in these deformed orogenic belts were predominantly of a compressional nature—folds and thrust faults—and these were accompanied by thermal events that produced volcanism, intrusion of igneous rocks, and metamorphism. In 1840 in the Alps of Switzerland the overthrusting of Jurassic rocks over Eocene was first recognized and the concept of nappes, huge sheets of rock overthrust for tens of kilometers, was born. Detailed mapping in this and other deformed areas permitted unraveling of the structures and the construction of predeformation restored sections. Inevitably, the restored sections occupied much more horizontal space than the existing deformed remnant. This led to the concept of crustal shortening, and because the scale of the deformed belts of a given age was of global proportions, a global cause of the deformation was sought. This in turn led to the idea of a contracting Earth in which its outer shell, which was already cool, and the innermost part, which was still hot, were maintaining their dimensions, while the part in between was cooling and shrinking. The outer shell was then too large and wrinkled, like the skin of a dehydrating apple.

EXPANDING EARTH

The idea of a contracting Earth caught the fancy of many geologists and geophysicists and was in good repute until more began to be known about the ocean floor. In the 1950s the continuity of the ridge system throughout the world ocean was established on the basis of seismicity. The earthquakes in the oceans were concentrated about the axis of the ridge, and this axis was also marked by high values of heat flow and frequently a central valley or rift. The ridge-rift system appears to come ashore in East Africa and the Middle East, where rift valleys extend as a continuous feature from Rhodesia through the Red Sea and on to the borders of Turkey. In East Africa, normal faulting predominates and there is little evidence of horizontal movement. In the Red Sea, horizontal movement is evidenced by the absence of continental crustal rocks in the wide central trough. The separation is accentuated in the Gulf of Aden, and it is not difficult to picture a progression of structures from East Africa to the ocean basins in which horizontal motions become greater and greater. This apparent genetic relationship and the global scale of the ridge-rift system led to revival of the concept of an expanding Earth.

CONFRONTATION

At this time a confrontation developed. There are two active systems, active in the sense of volcanism and seismicity. One corresponds to the young mountain systems with associated deep-sea trenches and island arcs that surround the Pacific and extend through the Himalayas into the Alps and the Mediterranean; the other runs through the middle of the ocean basins and comes ashore in East Africa, in Iceland, and in western North America. The first system exhibits compressional features and evidences of crustal shortening, the second tensional features and evidences of crustal extension. Both are of global scale and demand a global mechanism to create them. Both are currently active, so a mechanism must be found that will create both types of features simultaneously. It is not clear that this can be accomplished either by expanding or by contracting the Earth. Another mechanism must be sought.

CONTINENTAL DRIFT

It appears likely that the first reasonable presentation of the idea of drifting continents was by Antoine Snider, who in 1858 presented maps showing the continents first in juxtaposition and then in their present positions (Figure 17.2). His purpose was to explain the similarities of the coal measures of Europe to those of North America and the correspondence of the coastlines on the two sides of the Atlantic. He did not consider a mechanism or a cause of the continental movements.

Howard Baker, a paleontologist, in the early part of this century, went beyond this and speculated about the dynamics of the system. He suggested that tidal forces caused by eccentricities in the orbits of the Earth and Venus ripped the crust off the Pacific Ocean basin and, following the idea from the British astronomer, George Darwin, formed the moon from it. The remaining protocontinent broke up, the pieces drifted apart, and the waters of the resulting oceans were captured from the breakup of the hypothetical planet now represented by the asteroids (Figure 17.3). This model was imaginative, but did not gain wide acceptance.

In 1910, F.B. Taylor, a glaciologist, related the young mountain systems surrounding the Pacific to the opening of the Atlantic. Because his emphasis was on the splendors of Tertiary mountain building, the magnitude of the proposed continental movements tended to be obscured (Figure 17.4). His mechanism was a little

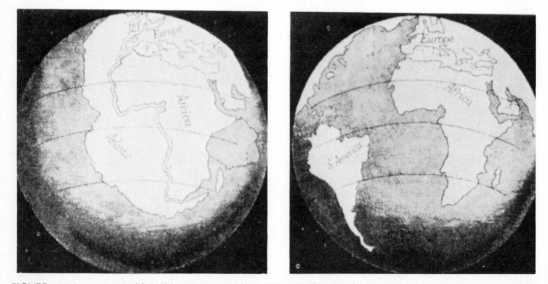

FIGURE 17.2   Maps published by Antoine Snider in 1858 to illustrate the arrangement of the continents in late Carboniferous time (about 280 m.y. ago) and at the present. (From Snider, 1858.)

FIGURE 17.3   The "displacement globe" of Howard Baker, published in 1911, illustrating the original single continent which separated into fragments in late Miocene to early Pliocene time (about 15 m.y. ago), the fragments rapidly and simultaneously moving to their present positions. (From Baker, 1911.)

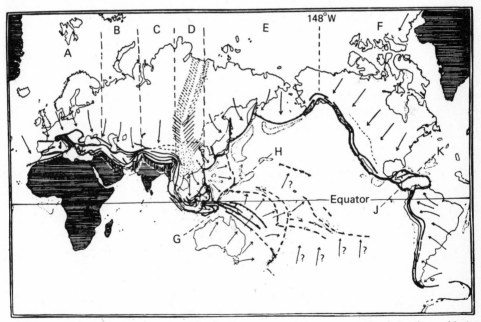

FIGURE 17.4   F.B. Taylor's 1910 map of Tertiary (60 m.y.–3 m.y.) mountain chains (heavy black lines) and crustal movements in the directions indicated by the oceans. Vertical lines mark sediment-filled foredeeps on the continents, faint dotted lines mark basins and foredeeps on the sea floor. (After Taylor, 1910.)

vague and was probably too weak. He invoked tidal action and later suggested that the capture of the moon by the Earth in Cretaceous time might have increased tidal action sufficiently to cause the continents to start sliding around.

At the turn of the century, with very little geological information available, summaries of world geology interpreted the oceans as evidence of collapse of the

Earth's crust or of submerged continents. Then Alfred Wegener, extrapolating boldly beyond the totally inadequate data on gravity at sea, concluded that the ocean floors were qualitatively different from the continents—forming, in fact, a continuing matrix of mafic (predominantly basaltic) rocks from which the silicic (more granitic) rocks of the continents protrude because they are lighter. This conclusion demanded an entirely different mechanism for the formation of the ocean basins from that required for the continents.

In January 1912, Wegener presented his ideas about continental drift at an annual meeting of the Geological Society in Frankfort am Main. Wegener was a remarkable man. He was a meteorologist by training, but early became fascinated with the idea of continental drift and was determined to prove it. He was discouraged by his old professor (and father-in-law), W. Köppen of Hamburg, who told him early in the century,

> To work at subjects which fall outside the traditionally defined bounds of a science naturally exposes one to being regarded with mistrust by some, if not all, of those concerned, and being considered an "outsider." The question of the displacement of continents involves geodesists, geophysicists, geologists, paleontologists, animal and plant geographers, paleoclimatologists and geographers, and only by consideration of all of these various branches of science, as far as is humanly possible, can the question be resolved.

Wegener made the first detailed analysis of continental drift, taking into account as much of the existing information from all branches of science as he could. He rejected the idea of permanence of ocean basins and emphasized the qualitative difference between oceans and continents. He pictured a protocontinent ("Pangaea") as breaking up at the end of Paleozoic time (Figure 17.5) with the fragments drifting slowly to their present positions and, in a later treatment, called upon thermal convection in the Earth's interior as the driving mechanism.

CONVECTION CURRENTS

The concept of convection as a driving force was further developed through the idea that radioactivity within the Earth produced the heat and that the resulting convection produced the drift of the continent. A model, developed by the geologist Arthur Holmes (Figure 17.6), of a 50 to 100-km-thick outer shell of the Earth being moved about by this convection is not too different from present ideas and was an improvement on what Wegener is usually credited with thinking—continents plowing through oceans.

Gravity measurements at sea provided the data to substantiate Wegener's concept of the qualitative difference between oceans and continents and also favored convection. The large negative gravity anomalies associated with deep-sea trenches in the East Indies (Figure 17.7) could not be explained by the water depth alone and it was concluded that they must be due to downbuckling of the Earth's outer shell, or what is termed "lithosphere" today. It should be noted that the term is ambiguous, because it is sometimes used to discriminate the solid Earth from the at-

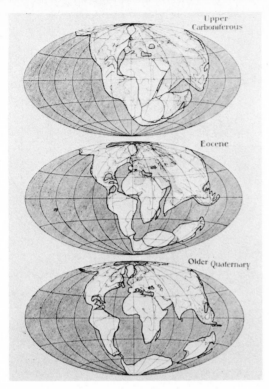

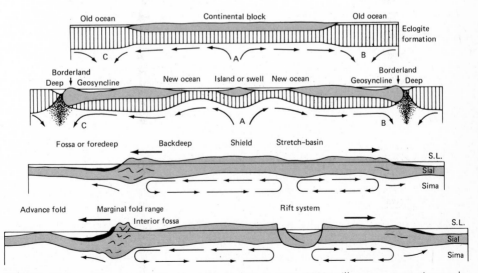

FIGURE 17.5 Drift of the continents between 280 and 3 million years ago as visualized by Alfred Wegener in 1912. (After Wegener, 1912.)

FIGURE 17.6 Diagrams published by Arthur Holmes in 1928–1929 to illustrate a convection mechanism as a driving mechanism for continental drift. Heat rising at the axes of the ocean ridges produces rifting and volcanism; heat sinking along the active continental margins produces deformation and deep earthquakes. The continents are carried along by the horizontal currents. (After Holmes, 1928.)

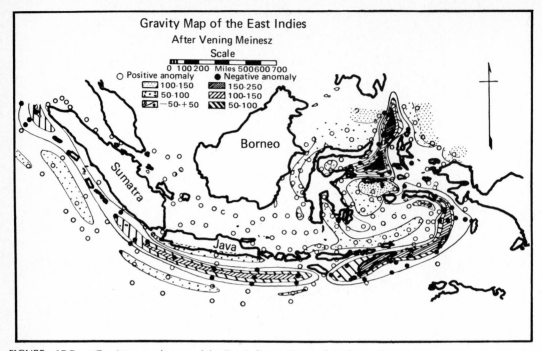

FIGURE 17.7 Gravity anomaly map of the East Indies region resulting from submarine gravity measurements by F.A. Vening Meinesz in 1923. These measurements demonstrated the presence of large negative gravity anomalies associated with the deep-sea trenches of Indonesia. The gravity anomaly values are in milligals. (After M. Ewing, 1938.)

mosphere and hydrosphere (see Chapter 3) and sometimes to designate the Earth's outer shell. The term is used in the latter sense in this chapter. A model was built to demonstrate this downbuckling, consisting of a "crust" of finite strength over a fluid substratum (Figure 17.8) subjected to horizontal pressure. With a slight initial depression to localize the deformation, it was possible to produce a downbuckle similar to that proposed. Usually it was a simple downbuckle, but sometimes it sheared as buckling proceeded and underthrusting was initiated. Since negative gravity anomalies are also found in young mountain systems, it was a logical step to superimpose Alpine mountain structures on this downbuckle or "tectogene" (Figure 17.9) and to suggest that features of this type underlay the major mountain systems as well as the island arcs and were the reason for their existence. A convection model was more attractive than compression, since it did not require such a large difference in apparent viscosity between the lithosphere and the underlying substratum.

Each of the above ideas, contraction, expansion, continental drift, and so on, had its defenders and it was difficult to choose among them on the basis of solid data rather than emotion. Arthur Holmes reflected the attitudes of the day when he commented in 1953,

I should confess that despite appearances to the contrary, I have never succeeded in freeing myself from a nagging prejudice against continental drift; in

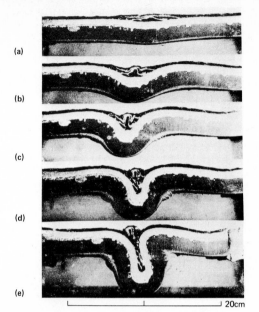

(a)

(b)

(c)

(d)

(e)

20cm

FIGURE 17.8   Model experiment of Kuenen (1936) illustrating the formation of a downbuckle or "tectogene" produced by convection inside the Earth as a result of horizontal pressure on the lithosphere. (After Kuenen, 1936.)

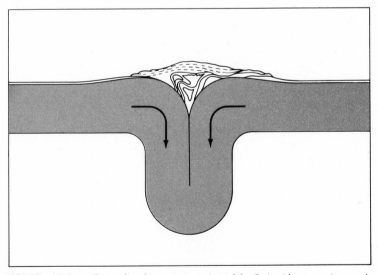

FIGURE 17.9   Generalized structure section of the Swiss Alps, superimposed on the downbuckle or tectogene and drawn with no vertical exaggeration. (After Hess, 1938.)

my geological bones, so to speak, I feel the hypothesis to be a fantastic one. But this is not science. . . . While so many contradictory voices confuse judgment, one cannot do better than commend [Carl O.] Dunbar's wise dictum that "it is unsafe to reject, a priori, either continental drift or foundering of broad land bridges."

PLATE TECTONICS MODEL

The currently favored plate tectonics model did not spring suddenly into prominence but developed as a result of earlier ideas and data. There have been many contributions, but three in particular served to make the model palatable to a majority of the earth sciences community.

LITHOSPHERE PLATES AND THE ASTHENOSPHERE

Chronologically, the first was the concept of the asthenosphere—that beneath a relatively strong outer shell (lithosphere) there exists a layer (asthenosphere) which

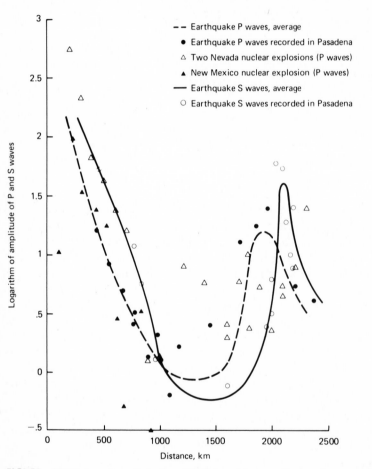

FIGURE 17.10(a)   Beno Gutenberg (1926) found that amplitudes of compressional waves decreased markedly out to a distance of about 1500 kilometers on the Earth's surface. At this distance they suddenly increase by a factor of more than ten and decrease more slowly at greater distances. Subsequently it was found that the amplitudes of shear waves behaved in the same manner. (From "The Plastic Layer of the Earth's Mantle," Don L. Anderson. Copyright © 1962 by Scientific American, Inc. All rights reserved.)

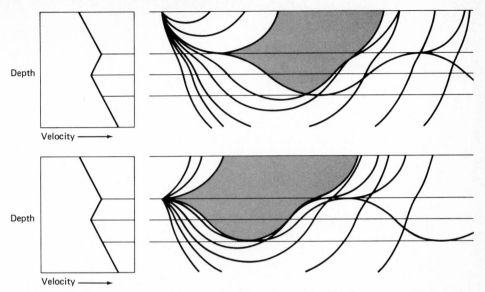

FIGURE 17.10(b)   These variations in amplitudes can be explained by the existence of a region in-side the Earth in which seismic velocities are lower than those above and below. The low velocity layer causes seismic rays to be bent, or refracted, and the energy to be greater or less at intervals on the Earth's surface. In the example shown very little energy is transmitted through the shaded region, but is concentrated on its left extreme. (From "The Plastic Layer of the Earth's Mantle," Don L. Anderson. Copyright © 1962 by Scientific American, Inc. All rights reserved.)

permits gradual movements so as to bring the Earth's surface close to hydrostatic equilibrium. This concept was first voiced in 1889, and it became apparent that something of this sort was required in order to account for the fact that gravity measurements indicated that the Earth's surface was almost everywhere near isostatic balance. It was given some substance when investigations, using body waves from earthquakes (Figure 17.10), suggested the existence of a layer at about 100 km depth in which seismic velocities were less than those in the layers immediately above and below. Studies of the differences in the energy release characteristics between shallow-focus (0-70 km) and deeper focus (70-700 km) earthquakes suggested decoupling between the lithosphere and the Earth's interior (Figure 17.11), and further studies indicated that the great earthquakes might be part of a single tectonic system. More recent studies of earthquake surface waves have revealed that the dispersion, or spreading as a result of waves of different frequencies traveling at different velocities, cannot be accounted for unless there is a layer present at depth (Figure 17.11b) in which the velocity of elastic waves is lower than in the layers above and below. Experimental petrology, coupled with field studies of rocks brought to the surface in diatremes, or diamond pipes, in southern Africa have revealed temperature anomalies and evidence of movements at depths where one would expect to encounter the lithosphere-asthenosphere boundary. Moreover, these diatremes have provided us with what may well be the first recognized samples of the asthenosphere.

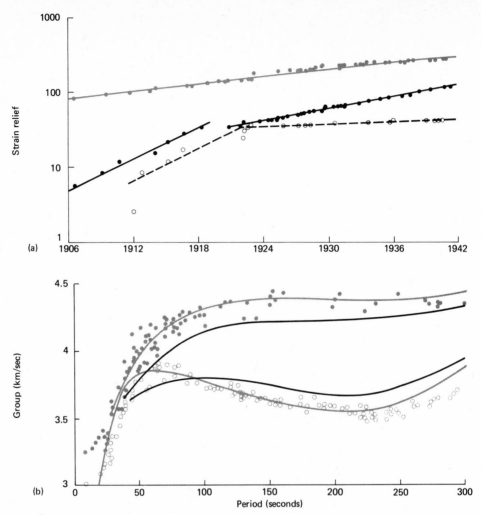

FIGURE 17.11 (a) Benioff (1954) found that there was a discontinuity in the rate of energy release by intermediate and deep-focus earthquakes (lower curves) in South America at about 1920 that was not reflected by the shallow-focus earthquakes (upper curve). This suggested that the outer shell of the Earth was decoupled from the interior. (b) In this diagram for two different kinds of surface waves, observed velocities are plotted against the period (time for a single wave to pass a receiving point). Theoretical curves for a model that includes a low-velocity layer (light lines) fit the observed data much better than a model that does not (dark lines). (From "The Plastic Layer of the Earth's Mantle," Don L. Anderson. Copyright © 1962 by Scientific American, Inc. All rights reserved.)

SEPARATION OF PLATES

The second contribution was the concept of separation of lithosphere plates along the axes of the ocean ridges. Support for this concept has come from a number of studies, principally of magnetic anomalies and of seismological phenomena.

Information from magnetic studies   During the 1950s a body of information became available on the direction of the magnetic pole preserved in rocks of various ages from the different continents (Figure 17.12). These data indicated that there had been considerable movement of the continents in relation to the magnetic pole over the course of geologic time, and, further, that there had been considerable movement of the continents in relation to each other since the end of Paleozoic time about 230 million years ago. There was much controversy about the interpretation of these data until they were supplemented by magnetic measurements in the ocean basins.

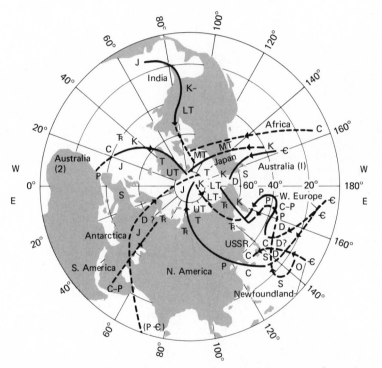

FIGURE   17.12   When paleomagnetic pole positions are measured in rocks, it is customary to map these by holding the continents in their present positions and plotting the position of the pole relative to the fixed continents with time. This produces a so-called polar wandering path. In this diagram it can be seen (a) that the pole has moved relative to any given continent through time and (b) that the polar wandering patterns for different continents are not the same. These observations are interpreted to mean (a) that the continents have moved relative to the magnetic poles, which have remained aligned approximately along the rotational axis of the Earth and (b) that the continents have moved relative to each other. Letters refer to: PЄ, Precambrian; Є, Cambrian; O, Ordovician; S, Silurian; D, Devonian; C, Carboniferous; P, Permian; TR, Triassic; J, Jurassic; K. Cretaceous; L, MT, UT, Lower, Middle, and Upper Tertiary. (After Deutsch, 1966.)

A survey made off western North America in the 1950s (Figure 17.13) revealed a pattern of linear and parallel magnetic anomalies that were continuous for long distances, but offset along fracture zones that had previously been found from echo-sounding surveys. Subsequent examination of the magnetic field in other parts of the oceans revealed that this pattern of linear anomalies was not unique to this area, but was the norm for all the ocean basins. No explanation was found for this pattern until it was coupled with another set of observations.

As early as 1906 a lava flow in France had been discovered that was magnetized in a direction opposite to that of the present magnetic field. Lavas contain magnetic minerals and as they cool in the presence of an external field, they will become magnetized in the direction of that field. Through the years additional ex-

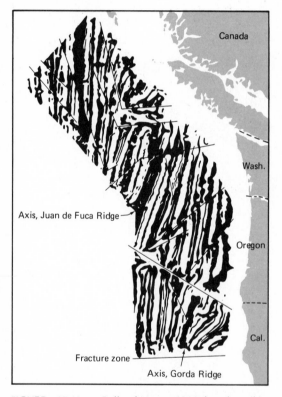

FIGURE 17.13   Raff and Mason (1961) found a striking linear pattern of magnetic anomalies in the northeastern Pacific, with discontinuities along fracture zones discovered by echo sounding surveys. The black areas represent positive magnetic anomalies (stronger than the Earth's mean magnetic field) and the white areas negative magnetic anomalies. There was no ready explanation for these anomalies when they were first discovered. (From Drake, 1970, after Raff and Mason, 1961.)

amples of this phenomenon were found, most notably in Iceland, where a number of reversals were discovered in the multilayered sequence of lava flows. It was suggested that these were due to reversals of the Earth's main magnetic field, a concept that is reasonable because the Earth's field is not a fixed property of the rocks but is generated dynamically by motions in the fluid outer core. This suggestion did not meet with instant approval, but became difficult to oppose when investigators at the U.S. Geological Survey (Figure 17.14) collected oriented samples of lavas from many parts of the world, measured the directions of magnetization and their radiometric ages, and demonstrated conclusively that the Earth's main field had reversed a number of times during the previous 3.5 million years (see Chapter 15).

Almost concurrently, other investigators were trying to find the cause of the magnetic anomalies over parts of the ocean ridge system. The conventional models for these anomalies assumed a series of magnetic blocks of varying width separated by nonmagnetic blocks. This model created problems, because it called for the presence of two entirely different suites of rocks, magnetic and nonmagnetic, to correspond with the blocks in the model, a conclusion not supported by dredging or photography, and the magnetic properties had to be rather stronger than would be expected for the rock types actually found. On the assumption that the anomalies were caused by alternating normally and reversely magnetized blocks, the ne-

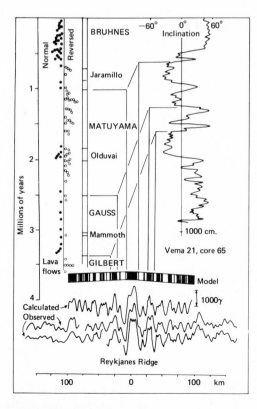

FIGURE 17.14   Measurements of the direction of magnetization and the age of lava samples from many parts of the world by a U.S. Geological Survey team indicated that the Earth's magnetic field had reversed a number of times during the last 3.5 m.y. (upper right). Similar measurements of the magnetization of deep-sea sediment cores gave the same result (upper left). Comparison with the pattern of magnetic anomalies over the ocean ridges (bottom) and calculated magnetic anomalies from models led to correlation of the time sequence of magnetic reversals, with the symmetrical pattern of magnetic anomalies on the ocean ridges and the concept of sea floor spreading. (After Drake, 1970.)

cessity for different rock types was removed and the required intensities of magnetization were halved.

The magnetic data from crossings of the ridge system had one thing in common. The anomalies on either side of the axis of the ridge were symmetrical and fell into a definite pattern of relative widths that was similar in all cases. The overall scale of the pattern might differ, but the relative widths were the same. This was not all, however; these relative widths in terms of horizontal distance from the axis of the ridge closely matched the relative times, in terms of the history of magnetic reversals that had been established. In other words, it appeared possible to relate horizontal distance from the axis of the ridge on a scale of hundreds of kilometers to time on a scale of millions of years, with the resulting inference that the age of the crustal rocks beneath the ocean floor sediments is progressively greater with distance from the ridge axis. It was subsequently found that the very weak magnetization of the sediment cores from the sea floor showed the same pattern, varying in absolute dimensions with sedimentation rate but relatively identical (Figure 17.14).

From these data and correlations it could be concluded that new crustal material was being created continuously at the ridge axis and that the older crustal materials was moving laterally away from the ridge axis at rates of a few centimeters per year. Using this information, the magnetic anomaly patterns of the sea floor could be assigned a time scale extending the chronology obtained from oriented lava flows on land.

Information from seismological studies   Although J.P. Rothe from France had been able to use existing data on the location of earthquakes in the 1950s to suggest that they were related to the axis of the mid-Atlantic ridge, it was not until the world-wide standard seismographic network (WWSSN) was installed in 1961 that detailed studies could be made of the seismicity and first motions of earthquakes associated with the ridge system.

L. Sykes of Columbia University made careful studies of the seismicity of the mid-Atlantic ridge in 1967. These earthquakes occurred at shallow depths (<60 km) and were narrowly restricted to the ridge axis or along the fracture zones that offset it. First motion studies indicated that normal faulting predominated along the ridge axis and strike-slip faulting (paralleling the Earth's surface) along the fracture zones. However, even though the fracture zones could be traced by echo sounding for some distance laterally from the ridge axis, the earthquakes were limited to the portion between the offset axes (Figure 17.15). Furthermore, studies of the first motions of the earthquakes indicated that the motion along the fracture zones was in the opposite sense from the offset of the ridge axis. This is the sense that would be expected if the two sides of the ridge were separating along an original offset. In order to differentiate these faults from ordinary strike-slip, or transverse, faults, they were termed transform faults by J. Tuzo Wilson of Canada. The seismicity and the nature of the faulting confirmed the concept of horizontal plate separation suggested by the magnetic data.

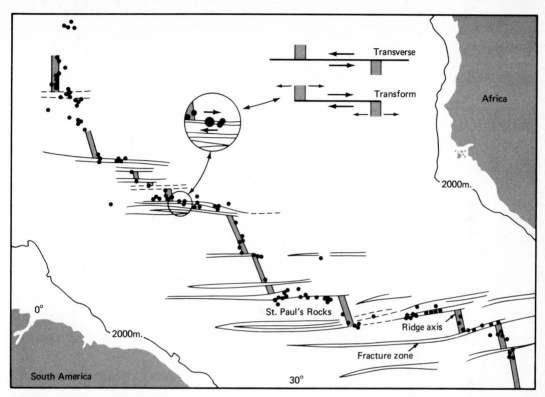

FIGURE 17.15   Earthquakes on the mid-Atlantic ridge tend to be confined to the axial trough or those portions of the cross-cutting fractures between offset segments of the axial trough. Focal mechanisms studies indicate that the direction of motion along the fractures is opposite to the direction of offset and in the direction expected, if the two sides of the ridge are separating in an east-west direction. (From Drake, 1970, modified after Sykes, 1968.)

CONVERGING PLATES

The third contribution came from studies of deep-focus earthquakes (300-700 km) in the Pacific. Such earthquakes had been identified in 1922, and in the 1930s it was pointed out that they were distributed under Japan in a zone dipping from the Nippon trench westward under China. This distribution was later found to be typical of the circum-Pacific region. Starting in 1964, a small network of seismograph stations was established in the Tonga-Fiji region of the South Pacific to study these deeper earthquakes in detail. This network allowed far more accurate determination of epicenters and foci and it was found that the earthquakes fell in a zone no more than 20 km in width that extended from near surface beneath the deep-sea trench to over 700 km beneath and behind the island arc.

An even more important result came from comparison of the character of shear waves propagated to stations in Fiji, Tonga, and Rarotonga from these deep earthquakes (Figure 17.16). For shear waves to propagate, the body must be able

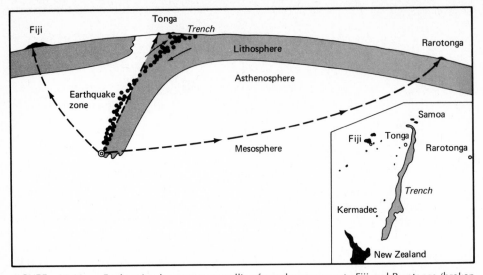

FIGURE 17.16   Earthquake shear waves travelling from deep sources to Fiji and Rarotonga (broken lines) pass through the more plastic asthenosphere and lose high frequency energy. Shear waves travelling up the seismic zone to Tonga do not show the loss of high frequency energy, suggesting that they do not pass through the asthenosphere. This has been interpreted to mean that the more rigid lithosphere is bent down and underthrust in these regions. (After Drake, 1970, data from Isaacs, Oliver, and Sykes, 1968.)

to resist the distorting forces, or have rigidity. Thus shear waves will not propagate in liquids of low viscosity, since these have no rigidity and distort readily. If the viscosity is increased, shear waves will propagate, with decreasing attenuation as the viscosity becomes greater.

Over the years seismologists have determined that in the normal case the outer 100 km or so of the Earth, the lithosphere, is rigid, and seismic waves are only modestly attenuated as they travel through it. Beneath this, in the asthenosphere, the attenuation is much greater, thus suggesting that this region is much less rigid or at least quite different in character. Farther down, the mesosphere is once again more rigid. J. Oliver and his colleagues found that seismic waves which propagated from deep earthquakes to stations at Fiji and Raratonga showed evidence of this normal distribution of properties with attenuation of the higher frequency shear waves, whereas waves that propagated upward along the seismic zone did not show any evidence of having passed through a less rigid zone. It is as if the lithosphere had broken and had been thrust under the island arc or continental margin to depths of at least 700 km. At greater depths it is difficult to determine what happens, because there are no earthquakes. Thus along the island arcs and the seismically active continental margins there is evidence of convergence of the lithosphere and extensive underthrusting in contrast to the ocean ridge system, where there is divergence and movement of material upward from the Earth's interior.

THE MODEL

In its simplest form, then, the plate tectonics model pictures the new crust (or lithosphere) actually being formed at the axes of the ocean ridges and old crust being destroyed at the island arcs and some continental margins. These are the active areas of the world today, and this activity is marked by earthquakes, volcanic activity, and relatively rapid movements of the land surface. The earthquake epicenters

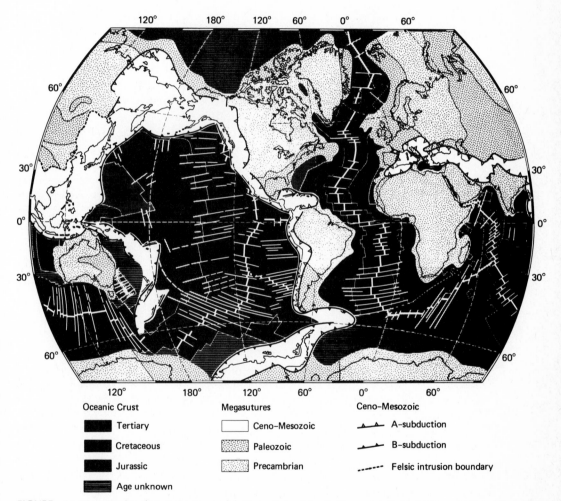

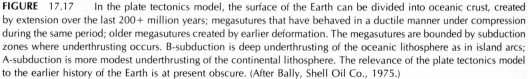

FIGURE 17.17   In the plate tectonics model, the surface of the Earth can be divided into oceanic crust, created by extension over the last 200+ million years; megasutures that have behaved in a ductile manner under compression during the same period; older megasutures created by earlier deformation. The megasutures are bounded by subduction zones where underthrusting occurs. B-subduction is deep underthrusting of the oceanic lithosphere as in island arcs; A-subduction is more modest underthrusting of the continental lithosphere. The relevance of the plate tectonics model to the earlier history of the Earth is at present obscure. (After Bally, Shell Oil Co., 1975.)

mark the edges of a small number of very large lithospheric plates moving in relation to each other (Figure 17.17). Where the plates are moving toward each other, underthrusting occurs and the compressional features associated with folded mountain systems are observed. Where they are moving apart, material comes up from depth to form new crust, and the tensional and volcanic features that we can see on the ridges or above the water surface in Iceland or in East Africa are found. Where they slide along each other, as in southern California, a shearing action occurs and the predominant feature is transverse, or strike-slip, faulting.

The number of plates and the locations of the boundaries have not remained fixed over the entire course of geologic time and although present activity can be explained in terms of a small number of plates, such explanation was not necessarily true in the distant past. It should also be noted that plate boundaries correspond to some but not all continental margins. The Pacific is ringed by seismically active belts, and it is easy to compensate for the new crust created in the ridges of the Pacific by underthrusting around the margins. The Atlantic continental margins, on the other hand, are remarkably inactive except for the Caribbean and Scotia island arcs, yet new crust appears to be generated in the mid-Atlantic ridge at approximately the same rate as in the Pacific. Since we cannot compensate for this new crust by underthrusting, we must look to another device.

Earlier we spoke of continental drift. Continental drift aroused considerable interest as a result of Wegener's work, but there were also major reservations. Most of the objections were based on the lack of a suitable mechanism and horror of the concept of continents plowing through the oceans raising mountains as bow waves and leaving ridges and rifts in their wake. Correlation of the magnetic stripes in the oceans with the chronology of magnetic reversals negated the first objection, not by providing a mechanism, but by providing strong evidence for horizontal movement on a very large scale and with time constants very close to those suggested by Wegener on the basis of data from many branches of science. Furthermore, the plate tectonics model *requires* that the continents move apart to compensate for the new crust being created in the mid-Atlantic ridge and that the rate of separation be the same as the rate of production of new crust. The second objection was removed by a change in concept from continents plowing through oceans to the concept of lithospheric plates moving in relation to each other on a viscous asthenosphere.

DEEP-SEA DRILLING: A TEST
FOR PLATE TECTONICS

Most of the support for the plate tectonics model comes from indirect geophysical measurements of one sort or another, with the principal line of evidence from the oceans, based on the notion that the magnetic stripes paralleling the ridges can be related to the chronology of magnetic reversals. This idea would be greatly streng-

thened if samples of the crustal rocks beneath the sediments could be taken at varying distances from the axes of the ridges and if the ages of the rocks corresponded
to those predicted from the magnetic data. One of the prime objectives of the deep-
sea-drilling project was to provide such data.

A great many holes have now been drilled by *Glomar Challenger* in the deep
ocean and many of these holes have penetrated through the sedimentary column
and into the underlying basaltic rocks. The basaltic rocks are usually altered and
are difficult to date radiometrically, but the sediments immediately above the basalts can be dated by examination of the microfossils contained within them. If one
makes the reasonable assumption that the basal sediments cannot be very much
younger than the basalts and further that the basalts represent oceanic crust and
were not emplaced at some time subsequent to its formation, then it is possible to
compare the rock ages with those predicted from the magnetic anomalies (Figure
17.18). The results to date indicate very good agreement between the predictions
and the dates determined from the microfossils. This is a most gratifying result and
lends great credence to the plate tectonics model and the concept of a mobile
ocean crust.

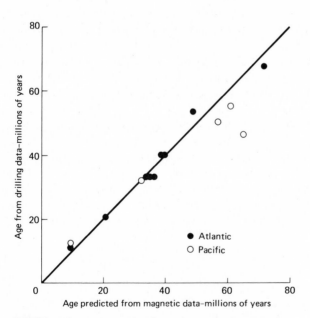

FIGURE 17.18   Age of sediment immediately above basaltic rocks versus distance to the ridge axis. If the basalt represents oceanic basement created at the ridge axis, the deep sea
drilling data confirm the concept of sea floor spreading inferred
from the magnetic anomalies. (Drilling data from Deep Sea
Drilling Reports, 1970.)

BED-ROCK GEOLOGY
OF THE OCEANS

When the idea of a deep-sea-drilling project was first put forth, one of the concepts held by many was the permanence of the ocean basins. Whereas the continents had been built up by tectonic processes through geologic time, the ocean floors were thought to have been stable and permanent. Thus holes drilled through the entire sedimentary column in the undisturbed depths of the oceans would reveal earth history back to the formation of the ocean crust. A major result of the plate tectonics model is to demonstrate the youth of the rocks and sediments that make up the sea floor. The projection of rates of movement back through time indicates that the oldest rocks and sediments on the sea floor are likely to be of the order of 200 million years, in sharp contrast to the continents where rocks are as old as 3.7 billion years.

We may use the geophysical data and the deep-sea-drilling results to construct a bedrock geological map of the ocean basins. A geological map relates time to space; that is, it depicts the distribution of rocks of different ages within the map area. If the map covers a limited area, the rock units of different ages may be sub-divided into smaller units based on changes in their character or rock type. A surficial geological map of the ocean floor would reveal mostly very young sediments and would be useful only to show lithologic differences, but a bedrock geology map is highly informative in terms of development of the oceanic areas.

On land a bedrock map would be constructed by sampling and mapping the outcrops or by examining the cuttings from boreholes. This is possible in the oceans in only a very limited way, because outcrops are limited and boreholes are few. To construct such a map (Figure 17.17), we can, however, take advantage of the systematic behavior of the magnetic measurements, their correlation with a reversal time scale, and unification of these results by the Deep Sea Drilling Project. This map is a geological map in that it relates the age of the crustal rocks and space, but it makes no mention of rock types. These should be limited in number, since, as we have seen, they are produced by the same mechanism, and where they have been sampled in the oceans and on the land, they are of similar types.

SUMMARY

1. Geological investigations on land over the years demonstrated that mountain systems were produced by compressional forces. This observation led to the idea that the earth was contracting as a result of loss of original heat.

2. The shapes of opposing shores, particularly of the Atlantic Ocean, and later the tensional nature of the ocean ridge system, led to the idea that the Earth was expanding.

3. Both the young mountain systems and the ocean ridge system are seismically and volcanologically active at the present time. Thus a model is required that will produce both types of features simultaneously. Both the mountains and the ridges are of global dimensions, so a global cause must be sought.

4. The plate tectonics model assumes that the outer shell of the Earth is made up of a small number of very large plates averaging one or two hundred kilometers in thickness. These plates move in relation to each other—converging, diverging, or in sliding contact. Seismic and volcanic activity are concentrated along the plate margins, and are presumably driven by some form of thermal convection. This model is able to explain surface geological phenomena far better than any advanced previously.

A clam, 30 cm long, picked up by the research submersible *Alvin*'s mechanical arm from a thermal area associated with the Galapagos spreading center at 2500 meters depth. (Photo by R. Ballard, Woods Hole Oceanographic Institution.)

Cycles Between Land and Sea

18

In the previous chapters we have seen that the water in the oceans is cycled through the atmosphere to the continents, where it participates in the weathering and erosion process and brings back to the ocean basins the wealth of the continents—not only mud and sand but also all the elements in solution. This state of siege and plundering of the continents by the oceans through its airborne agent, water, and the internal fifth column, life, has been going on throughout

CHAPTER

most of geological time. Surely one would expect to see the storehouse of the oceans laden with these trophies.

IS THE OCEAN GETTING SALTIER?

In 1900 John Joly argued that if we measure the total amount of sodium in the oceans and the rate of supply of sodium brought to the sea by streams around the world we could know how long the process of adding sodium to the sea had been going on. Thus we could arrive at the length of time it took for a fresh-water ocean (or at least one without sodium in it) to become roughly a 3.5 percent sodium-chloride solution. The time turned out to be less than 100 million years.

This appeared to conform to an age for the Earth determined by geophysical arguments, based on a planet cooling from a molten state, put forward by Lord Kelvin in the 1860s. It appeared then that the problem of the age of the Earth was settled—both the geophysical and geochemical arguments yielded an age for the Earth and its ocean of 100 million years or less. The only unhappy people were the paleontologists and biologists, who felt that the diversity of life and the guessed-at rates of biological speciation required a longer time.

Indeed they were right; the geophysical argument collapsed once radioactivity was discovered, because it provided an important heat source that had not been considered in the earlier cooling model. Kelvin backed off, thus leaving John Joly and his sodium argument, firmly founded on observations, in a logical limbo: Either the salty oceans were considerably younger than the Earth, or the transport of sodium was *cyclic*—i.e., sodium was being returned to the sea by processes not yet known.

Radioactivity, as we have seen, not only pronounced the death knell to Earth-cooling arguments for age determination but also provided a method for determining the ages of rocks and the Earth itself. By the systematic dating of fossil-containing rocks in the geologic column, it became clear that types of organisms in complete continuity with ocean-living types of the present time existed at least as far back as 500 million years ago. These are represented widely, implying a world-encircling salty ocean farther back than any sodium calculation would permit. Thus the recycling of sodium is required. Evidence that such recycling was a general requirement for materials brought to the sea from the continents was gathering from a variety of sources.

In the 1920s the great German chemist Fritz Haber decided that of the trace elements in seawater, gold should be extractable on an economically profitable basis. He based his decision on some early determinations of the concentration of gold in seawater, which indicated that gold might be present in seawater at levels of about one part per billion. Even at this low concentration it seemed that the extraction of gold from seawater could be an economically rewarding venture and that the supply of gold would be virtually endless. A gold concentration of one part per billion in seawater corresponds, on the average, to 4 million g of gold for every square kilometer of ocean area. Since the area of the oceans is roughly 4 million km^2, the total gold reservoir would be over 10 million million g!

After many unsuccessful attempts to extract gold from seawater, Haber decided that the reported concentrations must be too high. Subsequent determinations that he made showed that the concentration of gold in seawater is actually about 5 parts per 1000 billion, or several orders of magnitude lower than the previously accepted value. This eliminated the prospect of the economic exploitation of gold from the sea and raised the question: What is happening to all the metals brought to the sea by rivers?—essentially the same question posed by the sodium supply story.

In order to answer this, we will obviously have to track the elements in their cycles between the continents and oceans.

STREAMS AND THE MEAN RESIDENCE TIME OF ELEMENTS IN THE OCEAN

Rivers are the major carriers of dissolved material from rock weathering. As water runs downhill it is also an erosive force. Depending on the rock type, the relief of the drainage basin, and the regional climate, various quantities of detritus and dissolved material can be carried by rivers, as discussed in Chapter 13.

If we consider that only rivers supply chemicals to the sea—a situation, as we shall see, not completely true for many elements—then we can calculate how long an element remains dissolved in seawater before it is removed permanently from the ocean by deposition and burial on the sea floor. This is called the *mean residence time*. The mean residence time of an element in the ocean is the quantity in the ocean divided by the rate of supply of that element by streams, assuming that the oceanic concentration and stream flux remain sensibly constant over time. Table 18.1 gives some current estimates of stream and ocean-water composition and the calculated mean residence times of the elements in the ocean.

TABLE 18.1 Average Concentrations of Some Elements in Seawater and River Water and the Mean Residence Time of Each in the Oceans

Element	Concentration in Sea water (μg/l)	Concentration in River Water (μg/l)	Mean Residence Time in the Oceans (years)
Lithium	1.7×10^2	3	2.3×10^6
Boron	4.5×10^3	10	1.8×10^7
Sodium	1.1×10^7	6300	6.8×10^7
Magnesium	1.3×10^6	4100	1.2×10^7
Silicon	3×10^3	6500	1.8×10^4
Phosphorus	90	20	1.8×10^5
Potassium	3.9×10^5	2300	6.8×10^6
Calcium	4.1×10^5	15,000	1.0×10^6
Cobalt	0.01	0.1	4000
Nickel	0.5	0.3	7×10^4
Copper	0.25	1	10,000
Zinc	3	20	6000
Rubidium	120	1	5×10^6
Strontium	7.5	70	4×10^6
Silver	0.01	0.3	1000
Cadmium	0.1	0.1	4×10^4
Cesium	0.3	0.02	6×10^5
Barium	20	50	2×10^4
Uranium	3.3	0.3	4×10^5

What does the mean residence time tell us? First it gives us a measure of reactivity of a chemical species in the world ocean. Second, it provides us with information on how much variability to expect in the oceans. Finally, it tells us what to expect of the behavior of man-made chemicals in the oceans—in particular man-made radioactive species such as strontium-90 and plutonium, and heavy metal pollutants such as lead and mercury. However, knowledge of the mean residence time does not tell us what the mechanisms of removal are. Nor does it tell us how elements will behave that are strongly coupled to the biological cycle.

WHAT CONTROLS THE COMPOSITION OF THE SEA?

In order to understand what controls the composition of seawater, we must first recall the weathering processes, discussed in Chapter 13, by which rivers acquire their chemical burden. These processes are strongly influenced by biological reactions which form carbonic acid. This acid reacts with rock minerals to produce a solution of sodium, potassium, magnesium, and calcium bicarbonate. If sulfide minerals are oxidized, sulfuric acid is released to the ground water and stream system, and some of the bicarbonate is replaced by sulfate. The net major element composition of the average worldwide stream, corrected for supply by sea spray, is given in Table 13.2. If this solution, isolated from any detrital load, is supplied to an enclosed basin, it will lose any excess carbon dioxide "bubbled in" by respiration and will become an alkaline lake. With the action of evaporation typical of

FIGURE 18.1 A core from ancient Searles Lake in the desert of southeastern California shows evaporite deposits of halite (sodium chloride) and salts of sodium and calcium carbonate and bicarbonate (trona, nahcolite and gaylussite). The deposits were formed from alkaline lakes resulting from interior drainage and intense evaporation.

arid regions a truly alkaline lake will form. From such a lake, calcium carbonate and then sodium carbonate will eventually precipitate. This is the origin of the sodium carbonate deposits (called *trona*) and calcium carbonate deposits (called *tufa*) found in arid regions (Figure 18.1). You may recognize sodium carbonate and calcium carbonate as components of antacid pills taken for "acid indigestion."

The oceans would thus trend toward an alkaline or basic composition if a source of acid to neutralize it did not exist. Hydrogen ions are sequestered as part of the clay minerals formed in the weathering profile. Eventually this material will be washed down to the sea and actually provide some of the neutralization necessary to keep the ocean from becoming an alkaline lake. There, minerals enriched in hydrogen ions behave like "solid" acids and "neutralization" is effected. Thus the sea is made less alkaline than it might otherwise be and the cation concentrations are regulated by a set of reactions which are in some ways the reverse of the weathering reaction shown in Chapter 13.

We can summarize the reactions occurring on land and in the sea in the following way.

1. Weathering on the continents.

    Na silicate $+ H_2CO_3 =$ H silicate $+ Na^+ + HCO_3^-$

    (where H_2CO_3, Na^+ and HCO_3^- are actually dissolved and the Na silicate and H silicate are the minerals before and after the action of carbonic acid from life processes).

2. Reactions in the oceans.

    H silicate $+ Na^+ =$ Na silicate $+ H^+$

    (where the release of H^+ will combine with HCO_3^- to form H_2CO_3, which can further decompose to water and carbon dioxide and thus be available to the atmosphere and another weathering cycle via the life cycle).

Note that the continental Na silicate, which is weathered, is not the same kind of mineral as the Na silicate in the ocean, which is formed by reaction with the clay detritus supplied to the oceans. The actual sodium balance is much more complex and not yet completely worked out.

POTASSIUM

Potassium can be removed from the oceans in two ways. (1) Like sodium some will be adsorbed on clay minerals supplied to the sea by rivers. In the process it will also release mainly hydrogen ions but also calcium. (2) It will also react with submarine basalts to release primarily calcium (and possibly some sodium). This reaction is also like an ion exchange reaction, but it actually involves the formation of clay (or at least claylike) minerals, with the release of calcium and the sequestering of potassium.

MAGNESIUM

In addition to the two modes of removal identified for potassium—namely, adsorption on stream-borne clays and reaction with submarine basalts—magnesium can also be removed from the oceans as a carbonate. Although all carbonate shells of organisms are predominantly made of calcium carbonate, these can contain anywhere from 0.05 percent $MgCO_3$ (for example, foraminiferans, clams) to 15 percent $MgCO_3$ (for example, echinoderms). Because only low-magnesium calcium-carbonate tests are commonly preserved in sediments, little magnesium is permanently removed by this method.

There is, however, a secondary removal of magnesium from the sea by carbonates. Under appropriate conditions—namely, highly saline brines in lagoons strongly warmed by the sun, and in contact with aragonitic tests—a transformation occurs in which calcium carbonate (aragonite) is recrystallized to form dolomite $CaMg(CO_3)_2$. Other slower processes, not yet fully understood, also result in the formation of dolomite. In the Bahamas and several Pacific atolls dolomite is found in rocks formed throughout the geologic ages and has been observed in cores recovered by deep drilling. Clearly the process of dolomite formation in limestones, if extensive enough, could provide an important sink for magnesium, and, indeed, at some times in the past the geologic record shows that it did.

CALCIUM

Over long periods of time, the concentration of calcium in the sea reflects a balance between three processes: (1) supply from the continents by rivers; (2) release from basalt on the sea floor by reactions just discussed; and (3) deep burial of calcium-bearing skeletons deposited on the sea floor. The geologic record indicates clearly that there have been major changes in the sites of calcium carbonate deposition.

Coral reef and reef-associated deposits provide an important repository for calcium carbonate at the present time. As sea level rose at the end of the last glaciation many of these platform areas, which were high and dry during the glacial times (see Chapter 20), became inundated by water. These newly formed post-glacial shallow seas provide an environment for the accumulation of calcium carbonate where the conditions are optimal. Some of this material is swept off the platform to the surrounding deep sea, but much of it is trapped. The high rate of accumulation of calcium carbonate will continue until the platform essentially is at sea level; then further deposition there will be small.

The long-term accumulation rates on such platforms during the past few million years have been much smaller than the contemporary rates. But as we go further back in time we know that larger areas of the continents were inundated with seawater, as seen in the geological record. Large areas of relatively shallow water on submerged continental platforms were present about 100 million years ago. On these platforms vast deposits of calcium carbonate with a significant coccolith component were formed. So striking was this to the early geologists that they called it the "chalky time" or *Cretaceous* (Latin for "chalk") *Period*. This is the origin of the English White Cliffs of Dover (Figure 18.2) and the French Alpes Maritime, north of the Riviera. With such a large removal of calcium carbonate in these relatively shallow environments the supply to the deep-sea bottom must have been less if the flux of calcium through the oceans remained the same.

The opposite situation seems to have occurred between 500 and 700 million years ago, when no relict of vast calcium-carbonate deposits are found on the continents. If the flux of calcium through the oceans remained the same as today's, then the deep-sea deposits of that time must have been the main repository of calcium carbonate.

Over short periods of time, the concentration of calcium in the sea reflects a balance between two processes: (1) removal by organisms forming skeletons of

FIGURE 18.2   The famous White Cliffs of Dover are part of an extensive deposit of calcium carbonate "chalk" throughout Europe, produced primarily from coccoliths dropping to the sea floor about 100 million years ago. (Courtesy of British Travel Association.)

calcium carbonate; and (2) dissolution of these skeletons, either in the water column or on the sea floor. This view of the system allows us to answer one of the puzzling questions of marine chemistry: Why is the calcium content of the oceans so high if it is so actively removed from seawater by organisms? The answer lies in the fact that calcium carbonate dissolves in carbonic acid at sea, much as it does in limestone terraines on the continents. When a "molecule" of calcium carbonate is precipitated in the test of a marine organism, a "molecule" of carbonic acid is also formed. As these reactions occur at the ocean surface, the carbonic acid is converted ultimately to organic matter through photosynthesis; thus the surface waters of the oceans tend to remain saturated (and even, supersaturated) with respect to calcium carbonate. As the shells and organic debris sink through the oceanic water column, the dissolved oxygen is used by organisms at depth to metabolize the organic matter to produce carbonic acid: organic matter + oxygen = H_2CO_3 (carbonic acid). Since plants cannot live below the euphotic zone, this carbonic acid cannot be biologically utilized and acts instead to dissolve the calcium carbonate. Indeed the only way calcium carbonate escapes this action is by rapid accumulation on the ocean floor in waters less than 4000 m deep and in areas of high productivity, as discussed in Chapter 14. The oceans thus remain almost at saturation with respect to calcium carbonate: the more calcium carbonate is produced by organisms, the more organic matter is formed and the greater the probability that dissolution will occur at depth.

SILICON

The rate of supply of silicon to the oceans by streams is about two-thirds the flux of calcium. The dominant removal of silicon in the oceans is by organisms, just as it is for calcium. The concentration of silicon in the oceans is about one thousand times lower than calcium. Thus the silicon mean-residence time is thousands of years compared to calcium's million of years. Why is this the case?

The answer lies in the fact that the silicon dioxide tests deposited by these organisms are not influenced significantly by the production of carbonic acid in seawater that results from the decay of associated organic matter. Although the ocean is undersaturated everywhere with respect to the biogenic form of silicon dioxide, the rate of dissolution is controlled mainly by exposure time and the silicon content of the oceans and not by the production of acid (see Chapter 14). Thus the silicon concentration in the oceans is maintained at a low value by efficient removal to the sediments.

WHAT ABOUT THE TRACE METALS?

If we look at the concentrations of the trace metals in the oceans (Table 18.1) we notice one striking fact: most of them are in extremely low concentrations. This is not because more metals could not be dissolved. The expectation based on our knowledge of chemistry is that the oceans are nowhere near saturation with respect to the least soluble precipitate possible in the oceanic environment for most of the trace metals. Strong chloride or carbonate complexes with many metals are known to exist and this fact indicates that much higher concentrations of the trace metals should exist if the oceans are regularly supplied with metals by streams. Obviously removal mechanisms must exist beyond that which we expect from a model in which the oceans behave like a large beaker of salty water.

TABLE 18.2   Chemical Composition of Plankton (in units of micrograms of element per gram of dry weight of plankton)*

Element	Phytoplankton	Zooplankton
Si	58,000	—
Na	110,000	68,000
K	12,000	11,000
Mg	14,000	8,500
Ca	6,100	15,000
Sr	320	440
Ba	110	25
Al	200	23
Fe	650	96
Mn	9	4
Ti	≤ 30	—
Cr	≤ 4	—
Cu	8.5	14
Ni	4	6
Zn	54	120
Ag	0.4	0.1
Cd	2	2
Pb	8	2
Hg	0.2	0.1

* From the data of Martin and Knauer (1973).

As Table 18.2 shows, metals are concentrated from seawater by organisms. But most metals are not concentrated in organisms in relation to seawater any more than phosphorus is. In Chapter 9 we have seen that the distributions of many trace metals in oceanic profiles mimic the distributions of the nutrient elements, phosphorus and nitrogen. We know that the nutrient elements are mainly cycled in the water column and only small amounts are removed compared to the amount cycled. If the trace metals follow this path we would not expect them to be removed massively from the water column by organisms. Transport of some metals by the hard parts of organisms may, however, be effected, since these are known to accumulate in some areas of the ocean floor. Barium, for example, is associated with the highly diatom-rich sediments of the Antarctic region.

As shown in Chapter 15, separation of members of the uranium decay series nuclides occurs as a result of their chemical differences. Thorium-230, for example (see Table 15.1), produced by the radioactive decay of uranium-238 in seawater, is picked up by sedimenting particles in the oceans and deposited on the ocean floor. So thorough is this scavenging that virtually every bit of the thorium-230, as it is produced, is removed from seawater into the sediments. A similar sort of argument can be made for other natural and man-made radionuclides. What are the agents of scavenging?

Material transported from the ocean surface includes a variety of particles with large reactive surfaces. Of these, fecal pellets made of mucous-like material agglomerating shell debris and other particles provide a likely reactive surface. Reactive manganese and iron oxide particles are produced in the water column. Some of these are undoubtedly transported downward from the surface, but some are produced by the oxidation of reduced iron and manganese diffusing out of sediments, or are (perhaps) supplied by hot waters from the spreading ridge areas. These compounds are known to scavenge trace metals. The most vivid example of this must be the ferromanganese nodules in the ocean floors, where up to 1 percent nickel and copper as well as many other trace metals can be found (see Chapters 14 and 19).

Scavenging by these particles occurs even more efficiently in coastal waters, where the perpetual physical resuspension of sediments and biogenic reworking of the sediment column provides a continual flux of fecal pellets as well as oxidizable iron and manganese in solution from the sediments. Thus scavenging is virtually total in such estuarine environments.

Once the metals are deposited in the sediments, there is very little chance of their escape to the overlying water again, although some transfer within the sediment column may occur.

ROLE OF PLATE TECTONICS IN COMPLETING THE CYCLE

The sea is *not* getting saltier with time and it *is* being stripped of trace metals supplied by the continents. All the material in transit through the water column becomes a part of the deposits at the ocean bottom. This includes the detrital as well as the dissolved material. If this were the full story, we would be faced with a paradox—the whole process should stop within the next 100 million years; because then no more continental material would be available as a source. For example, at the present rate of calcium supply to the oceans, all the fossil limestone on land—the dominant source of calcium—will be used up in 100 million years. The continents themselves would all be leveled to sea level within 100 million years and would no longer be able to provide any new detrital material to the ocean if there were no new mountain building. To the ocean basin itself would be added another 400 to 1000 m of sediment (depending on the density).

All this will occur if some grand recycling mechanism does not occur. Of course the Earth boasts a much longer history than a mere 100 million years, so something is missing in our calculation. That missing ingredient is the role played by plate tectonics, discussed in the previous chapter, in recycling the ocean bottom.

The ocean deposits are reprocessed at the converging plate edges and provide the material for new mountains—thus renewing the continents—and new distributions of metals in both dispersed and concentrated forms. This latter process can even lead to the formation of ore deposits.

The cycle of continental materials supplied to the sea is then completed. What goes into the sea returns to the land again. The internal distributions within the ocean determine what will reach which continental boundary. Figure 18.3 summarizes the cycles of the elements.

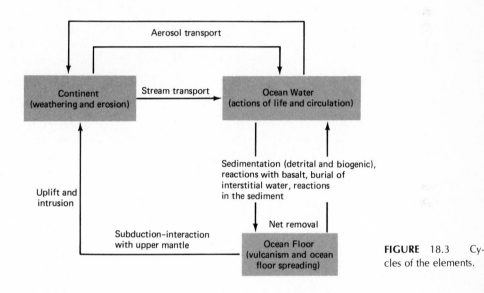

FIGURE 18.3 Cycles of the elements.

SUMMARY

1. The oceans are generally in a steady state with regard to composition.

2. A measure of the efficiency of removal of an element from the ocean is the mean residence time defined as the amount of a substance in the ocean reservoir divided by the rate of supply and also (at steady-state) rate of removal of the substance by streams.

3. Many of the elements are removed by particle association, biogenic deposition, interstitial water trapping in sediments, and reaction with volcanic rocks on the sea floor.

4. Plate tectonism returns again to the continents, through mountain building, the materials sequestered in the ocean floor.

The seaward margin of the Ross Ice Shelf, Antarctica. (Courtesy of G. Denton.)

Ocean Records of Past Climate

19

Every place on the surface of the Earth has a characteristic climate. This is a concise way of saying that when systematic observations are made over an arbitrary interval of time, every place experiences a particular range of physical conditions. Important aspects of the *atmospheric climate* at a particular locality on the ground or at sea may be described by specifying the average air temperature, precipitation, and wind velocity and by noting the annual range of these *cli-*

CHAPTER

matic indices (Figure 19.1). Similarly, every point in the sea has an *oceanic climate*, which may be specified by describing its average water temperature, salinity, and current velocity. In Chapters 4 and 6 we studied maps of currents, temperatures, and salinity describing the geographic variations in these indices of oceanic climate (Figures 4.1, 6.1). These maps, and many others like them, are prepared by averaging observations made over the past several decades— hence they describe only the *average* values of the *present-day* oceanic climate, and conceal the interesting and important fact that oceanic climate constantly undergoes change. In this chapter we will examine several examples of such changes, beginning with examples of ocean changes which have occurred over the last hundred years.

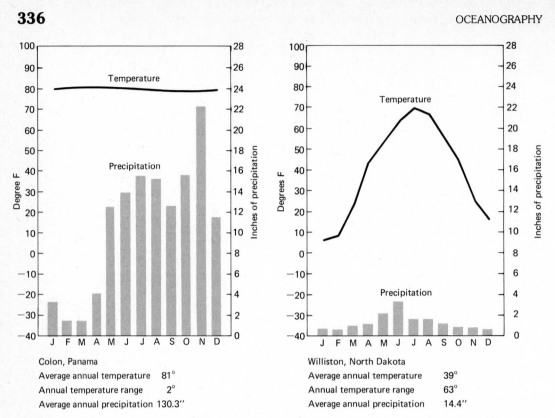

FIGURE 19.1   Climatic graphs for Colon, Panama and Williston, North Dakota. The pattern of monthly temperatures and precipitation differs noticeably in the two locations. Note that marked seasonal changes in temperature occur in Williston, which is situated in a continental interior. (From Critchfield, *General Climatology,* 2nd edition, © 1966. Reproduced by permission of Prentice-Hall, Inc., Englewood Cliffs, New Jersey.)

Although the reasons for these changes are not known, it is abundantly clear that they are intimately connected with changes in atmospheric climate and that they have an impact which is ecologically and economically significant. Later in the chapter we will examine changes of ocean climate affecting the entire ocean, and occurring on time scales longer than one century. For some of these dramatic changes, at least the outline of a scientific explanation can be given.

THE CLIMATE SYSTEM

ELEMENTS OF THE SYSTEM

The atmosphere and the ocean are the active, highly fluid parts of a global climate machine. As we have seen in Chapter 5, this machine runs on solar energy and pumps heat from low-latitude regions (which have a radiation surplus) into high-latitude regions (which have a radiation deficit). Atmospheric winds and oceanic currents are the mechanisms by which this constant, poleward energy transport takes place. If these mechanisms did not operate to balance the radiation budget at every point within the atmosphere and the ocean, the planetary system would be forced to a new equilibrium climate. Following the Stefan-Boltzman Law (Chapter 5), such a climate would be characterized by polar temperatures dramatically colder, and by low-latitude temperatures much warmer, than they are today.

Two parts of the global climate system are less active than the atmosphere and ocean, but no less important on that account. These are (1) the surface of the ice-free land, together with the animals and plants living there and (2) the bodies of permanent ice collectively known as the *cryosphere*. In polar areas, the cryosphere includes not only *ice sheets* grounded on Greenland, the artic islands of Canada, and Antarctica, but also the large masses of sea ice floating in the Arctic and Antarctic Oceans. Mountain glaciers are also considered part of the cryosphere.

INTERACTIONS

The four parts of the climate system (atmosphere, ocean, land surface, and cryosphere) are coupled together in such a way that any change in one affects the others (Figure 19.2). For example, if an increase in air temperature occurs at certain seasons in polar or mountain regions, ice there would melt and sea-level would rise. These interactions are easy to understand and predict; other consequences of the postulated initial warming are more subtle. For example, the reduction in ice cover resulting directly from a warming would decrease the reflectivity and therefore bring about an additional local warming because the local heat budget would change. This reaction, known as the albedo-snowline effect, constitutes a *positive feedback* in the climate system; that is, it would act to amplify the long-term result of any initial small change. Other parts of the climate system have negative feedback effects in which an initial change tends to be damped out by the reaction. The difficulty of evaluating the net effect of the many feedbacks in the climate system has delayed progress in understanding the processes of climatic change.

ROLE OF THE OCEAN

One of the main conclusions reached by climatologists over the past decade is that changes in the ocean play a leading role in bringing about changes in atmospheric climate. Phrased another way, we can say that interactions between the atmosphere and the ocean are major factors in controlling the year-to-year and decade-to-decade changes in the climate system. From our investigations of the heat budgets in Chapter 5 it is easy to see why this is so. In the first place, over one-half the solar radiation reaching the Earth's surface is absorbed by the sea. This absorption is primarily responsible for the existence over most of the world's oceans of a warm, mixed layer about 100 m deep, which is the atmosphere's main source of heat and moisture. Moreover, the very large amount of heat stored in the layer constitutes a thermal reservoir or buffer which must limit the rate at which long-

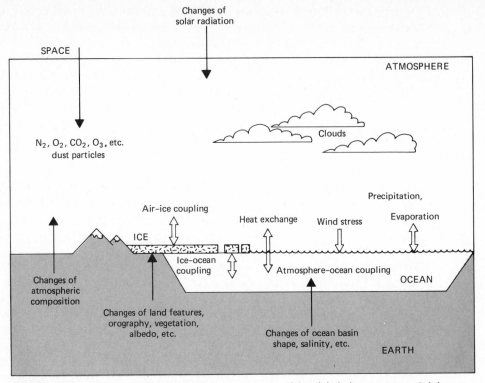

FIGURE 19.2   Schematic illustration of the components of the global climate system. Solid arrows illustrate changes that can occur in the external boundary conditions of the system. Open arrows are examples of internal processes of climatic change. (From NAS, 1975.)

term changes in the atmosphere occur. By contrast, the surface of the land stores very little heat and therefore does little to retard atmospheric change. It is for this reason that climates of regions remote from the sea exhibit wide seasonal changes in temperature, whereas areas near the sea have a more equitable climate (Figure 19.1).

ROLE OF THE CRYOSPHERE

Because snow and ice are highly reflective, the areal extent of the cryosphere exerts an important influence on how much of the sun's energy is radiated back into space. Changes in the cryosphere therefore have a significant impact on the planetary radiation budget. Such changes occur much more slowly than variations in the ocean, however, because it takes centuries or millennia to alter significantly the area of the great ice sheets. Thus the cryosphere acts as a thermal "anchor," which ultimately determines how fast a change in the climate system as a whole can occur. In dealing with short-term changes in the atmosphere and ocean we can consider that the cryosphere is constant. But as we extend our consideration to climatic changes occurring on long time scales, changes in the cryosphere are coupled with changes in the other parts of the global system.

CHANGES IN OCEANIC CLIMATE: THE TWENTIETH CENTURY

EL NIÑO

Our first example of oceanic change comes from a study of the sea-surface temperature off the coast of Peru. Monthly temperature averages (Figure 19.3) show the *normal* progression of the annual temperature cycle at this Southern Hemisphere site, with the warmest month (February) being some 3° above the coldest winter month (October). This area is influenced by coastal upwelling, hence the water temperatures here are cooler than the average ocean at this latitude (Figure 4.1). Figure 19.4 shows the month-by-month temperatures at this location, plotted as *deviations* from this average seasonal cycle. Note that some summers (December–March) are anomalously cold (for example, the summer of 1938); and that others (for example, 1925, 1965, and 1972) are anomalously warm. For reasons that are not completely understood, during such warm periods upwelling does not occur—

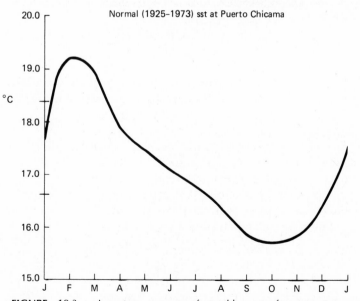

FIGURE 19.3   Long-term averages of monthly sea-surface temperatures at Puerto Chicama, Peru, an area subject to the El Niño phenomenon described in the text. Average temperatures are calculated over the period 1925–1973. (From Namias, 1976.)

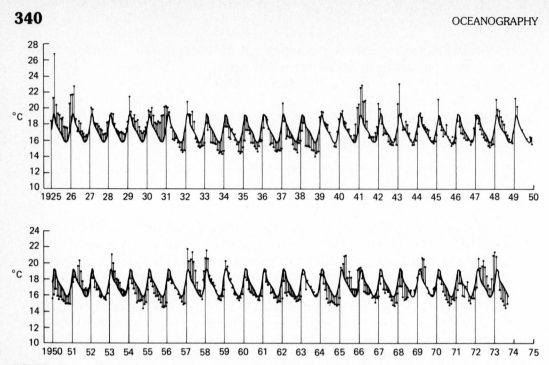

FIGURE   19.4   Graph of sea-surface temperatures illustrating the El Niño phenomenon off the coast of Peru. The smooth curve is the long-term average pattern as observed at Puerto Chicama (Figure 19.3) and at Chimbote. The long vertical lines mark January of the listed year. Points at the ends of short vertical bars indicate the average sea-surface temperature for individual months. Anomalously, warm summer (January) temperatures such as those during 1972 mark El Niño years. (From Namias, 1976.)

or does not occur with its usual intensity. Such periods of abnormally warm summer surface waters are associated with catastrophic declines in the anchovy harvest and, along the coast of Ecuador and Peru, with flood-producing rains. These economically catastrophic intervals, occuring as they do near Christmastime, have come to be called El Niño. The cause of this anomaly in ocean climate is not precisely known. What does seem certain is that in some way El Niño is linked with changes in the atmospheric pressure patterns and wind fields. One theory links El Niño to a reduced wind stress of the subtropical easterly winds, which in turn permits an accelerated equatorial countercurrent and results in diminished equatorial upwelling. These conditions are often preceded a year in advance by a weakening of the East Pacific atmospheric anticyclone—and thus offer one possible means of predicting El Niño.

THE NORTH PACIFIC

The El Niño phenomenon normally lasts only for one season. Other anomalies in ocean climate last for a number of years. One well-studied example comes from the North Pacific, where surface temperatures over large areas depart significantly from their long-term average values. The winter of 1957 (Figure 19.5) was typical of an anomaly pattern that had lasted for a decade (1948–1957). During nonsummer months this pattern was characterized by anomalously warm water in the North Central Pacific and cold water off the west coast of the United States. Starting in the spring of 1957, however, this pattern began to change; and by the spring of 1958 a different distribution of sea temperatures had replaced it. The new pattern was characterized by warm water off the western coast of the United States and

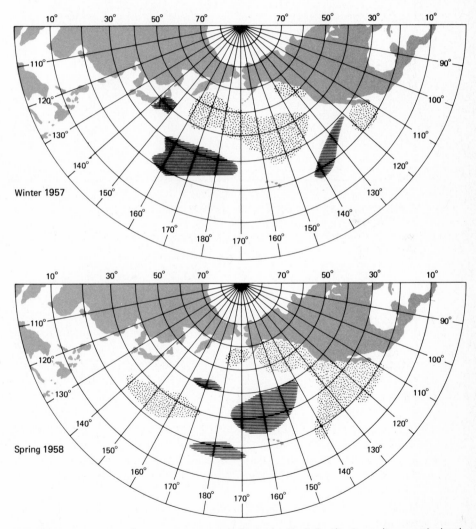

FIGURE 19.5   Sea-surface temperature anomalies in the North Pacific. Anomalies are calculated as differences from the average of the base period 1947–1966 for the winter of 1957 (*top*) and the spring of 1958 (*bottom*). Stippled areas are more than 1°F above the mean; hatched areas more than 1°F below. (From Namias, 1972.)

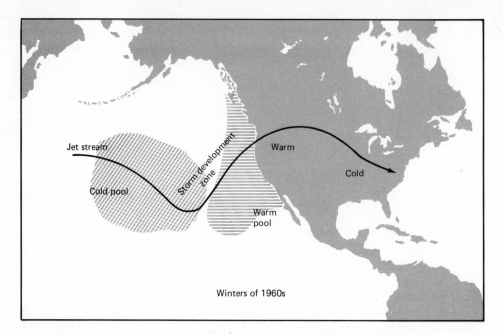

Winters of 1960s

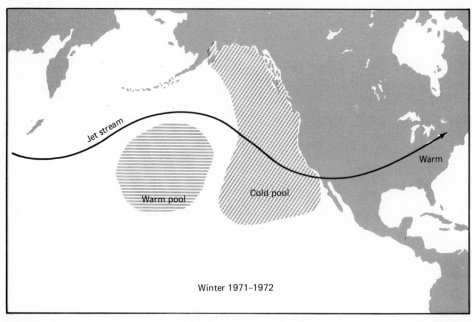

Winter 1971–1972

FIGURE 19.6   Impact of North Pacific sea-surface temperature anomalies on the winter climate of the United States. Areas of anomalously warm and cold sea-surface temperatures ("cold and warm pools") are thought to control the position of the atmospheric jet stream, which in turn influences air temperatures in eastern United States. (From IDOE Report, 1973.)

cold water in the central Pacific. The thermal effect on the atmosphere above was apparently to force a change in the position of the jet stream (Figure 19.6). As a result, weather patterns far downstream (to the east) changed simultaneously. The atmospheric data from southeastern United States (Figure 19.7) dramatically show a change of regime, with winters from 1958–1969 much colder than normal. By 1972, a climatic regime more typical of that of the 1948–1957 period had been reestablished both in the North Pacific and in southeastern United States.

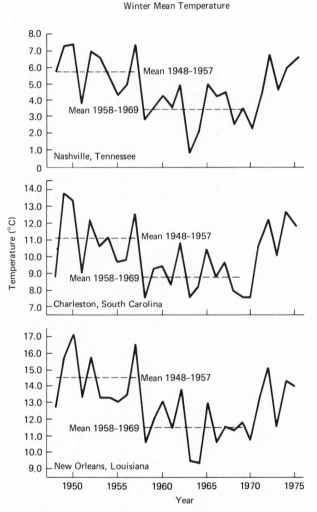

FIGURE 19.7 Winter mean air temperatures at three southeastern United States stations. (From Namias, 1975.)

THE NORTH ATLANTIC

Over wide areas of the North Atlantic Ocean annual mean sea-surface temperatures have sometimes been observed to be 1.5°C warmer (or colder) than the average. In some places these anomalies reach 3°C. Such an anomaly developed during the intervals 1902–1904 and 1913–1915. During the first interval, the temperatures in an area east of Newfoundland became warmer than usual. This pattern is known as a *warm-sea* condition. At the same time an area centered on 50°N and 30°W was colder. The opposite, *cold-sea* condition occurred between 1913–1915. Associated with these ocean temperature anomalies are changes in the annual mean barometric pressure. In the warm-sea case, the pressure of the Bermuda high is increased; the opposite is true for the cold-sea case. These conditions, generalized in Figure 19.8, are associated with changes in the boundary between the Gulf Stream and Labrador current, which, during a warm-sea condition, lies north of its usual position. As in the Pacific studies, the presence of these sea-surface anomaly patterns has been used to predict the weather downstream. When the warm-sea condition occurs, there is a significant tendency for lower barometric pressures to occur over Scotland and Scandinavia one month after the anomaly is observed. In the warm-sea case, temperature in Poland averages 2°C higher than normal for February, whereas areas near Switzerland are 3° colder following the January cold-sea condition. Although the mechanisms of these associations are not completely known, it seems clear that patterns of atmosphere circulation over Europe are significantly influenced by changing sea-temperature patterns in parts of the North Atlantic.

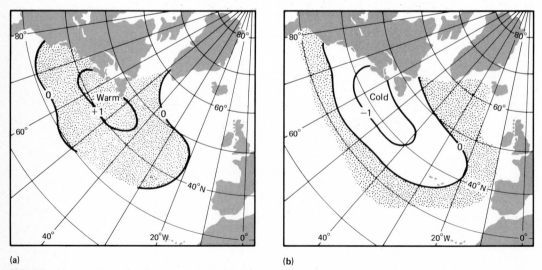

(a)   (b)

FIGURE 19.8 Generalized sea-surface temperature anomaly patterns in the North Atlantic (°C). (a) Type designated as warm-sea cases. (b) Type designated as cold-sea cases. (From Lamb, 1972.)

GEOGRAPHIC PATTERNS OF ANCIENT CLIMATE

LITTLE ICE AGE

A systematic global instrumental record is not available for periods farther back in time than about 1860. In order to reconstruct older climatic patterns, therefore, we must make do with incomplete sets of instrumental data and augment them with other kinds of observations. By compiling ship records made during the late eighteenth and early nineteenth centuries it is clear that the extent of pack ice in the North Atlantic was far more extensive than it is today (Figure 19.9). Records in the Alps and elsewhere in Europe and the United States show that winters were much colder during the same historical intervals than they are today, and that mountain glaciers generally were then well advanced beyond their present termini. From such observations a climatic period known as the Little Ice Age has been defined,

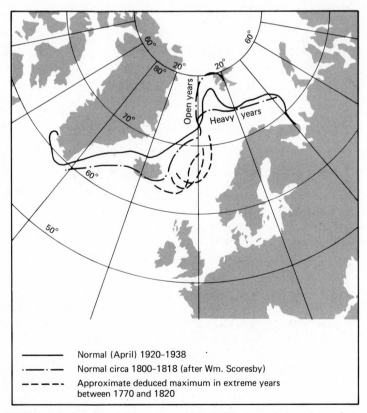

FIGURE 19.9 Maximum extent of North Atlantic spring pack ice during parts of eighteenth to twentieth centuries. (From Lamb, 1969.)

lasting roughly from 1450–1850, with temperature minima occurring in many places during the Colonial period in the United States. (General George Washington's experience at Valley Forge was no isolated occurrence, but an example of a persistent climatic regime colder than today).

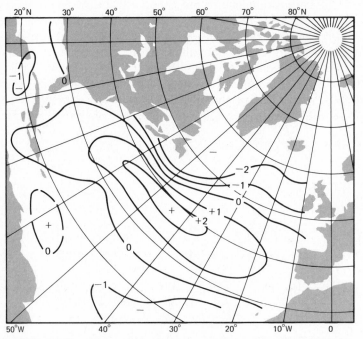

FIGURE 19.10   Sea-surface temperature anomaly pattern in the North Atlantic during the latter phase of the Little Ice Age. Temperatures plotted (°C) are obtained by subtracting the average values typical of the period 1887–1938 from average values observed between 1780 and 1840. In the area north of about 50° latitude Little Ice Age temperatures were colder than they were later in the nineteenth century. The reverse is true for most of the ocean south of that latitude. (From Lamb, 1972.)

During the latter part of the Little Ice Age, enough sea-surface temperature observations were made by British and American naval vessels to prepare a map showing the sea-surface temperature anomaly pattern (Figure 19.10) of the North Atlantic during part of the Little Ice Age. As expected, the northern part of the Atlantic was then several degrees colder than it is today. But parts of the central Atlantic during the Little Ice Age were warmer than they are today. These and other Little Ice Age data may be interpreted in terms of a change in the position of the main axis of the Gulf Stream, which during this period apparently had a more easterly course than it does today.

LAST GREAT ICE AGE

From written records we know that the conditions of the Little Ice Age lasted several centuries and were marked by advances in mountain glaciers, colder temperatures over many parts of the world's land surface, and changes in the sea-surface patterns in the North Atlantic. Going farther back in time we can no longer find direct instrumental records of climatic conditions. To reconstruct these ancient climates we must turn to natural, geological records of climate. On land, such records are obtained from plant fossils, from the position of the terminal moraines of ice sheets, and from the distribution of soil types formed under certain climatic regimes. In the ocean, we can turn to the sedimentary deposits in which microscopic fossils retain a permanent record of the geographic distribution of ancient plankton.

Studies of these marine and terrestrial records have shown that over the past two million years—a time period designated as the *Quaternary*—global climate has undergone changes far greater than anything that occurred over the past thousand years. These climatic oscillations are known as the Quaternary ice-age cycle. During any one cycle, the Earth changes from an *interglacial* state—such as the one in which we now live—to a *glacial* state, and back again. During each glacial age, the ice sheets, areas of sea ice, and mountain glaciers expanded well beyond their modern limits. These changes were particularly marked in the Northern Hemisphere, where ice sheets extended southward to cover all of Canada, parts of the United States north of New York and the Ohio River, and much of northern Europe.

Studies of the modern ocean and atmosphere have shown how the climate system balances the global radiation budget during an interglacial phase. But the geological record indicates clearly that many of the glacial phases lasted for tens of thousands of years, an interval considerably longer than the current interglacial phase, which has persisted about 10,000 years. What was the climate system like during an ice age? What geographic patterns of energy transport maintained the climate system in an ice-age equilibrium? To obtain clues for answers to these questions we turn to the ocean record of global climate 18,000 years ago, when the Earth was at the maximum of the last ice age.

To interpret the ocean climate 18,000 years ago we will first compare the distribution of ice-age planktonic fossils with the distribution of modern planktonic skeletons on the seabed. Because the distribution of modern plankton is closely related to the distribution of surface water masses, maps of the distribution of ice-age plankton provide a basis for interpreting the glacial ocean. Information shown on Figure 19.11 indicates that, during the ice age, waters of polar character covered a substantial part of the North Atlantic north of 40° and were in abrupt contact with warmer waters. Apparently, the Gulf Stream maintained a west-east course from Cape Hatteras to the shore of Spain. In the Sargasso Sea, however, the fauna of 18,000 years ago was essentially the same as it is today.

In addition to making maps of ice-age plankton distribution and deriving from them qualitative inferences about surface waters, it is also possible to estimate what

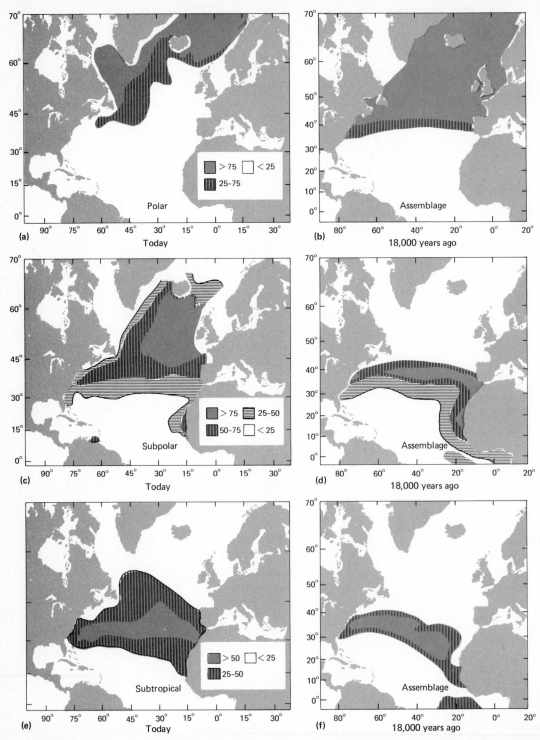

FIGURE 19.11 Maps showing the distribution of assemblages of planktonic foraminifera on the North Atlantic seabed today (a, c, e) and 18,000 years ago (b, d, f). Polar assemblage (a, b); subpolar assemblage (c, d); subtropical assemblage (e, f). (From McIntyre, Kipp et al., 1976, and Kipp, 1976.)

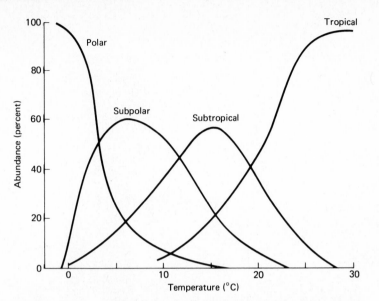

FIGURE 19.12   Idealized abundance trends of foraminiferal assemblages typical of four climatic zones. At any temperature of seawater a definite proportion of the four groups occurs. (From Imbrie and Kipp, 1971.)

the actual surface temperatures were. This can be done because the ice-age species are living today, and their responses to temperature and other physical parameters of the waters are known (Figure 19.12). A reconstruction of the surface temperature pattern of the ice-age ocean (Figure 19.13) was prepared by applying empirical equations derived from seabed distributions of modern radiolaria, planktonic foraminifera, and coccoliths. The results are perhaps more easily interpreted by examining the anomaly map (Figure 19.14). The largest changes in temperature (about 15°) were limited to areas where warm currents such as the Gulf Stream changed their flow patterns. Intermediate temperature changes (about 5°-6°C) occur in areas occupied by the eastern boundary and equatorial currents. This suggests that the rate of upwelling in these areas was greater than today, and that oceans circulated more rapidly during the ice age. In general, the subtropical gyres showed little or no change. These parts of the climate system are therefore the more stable portions, with maximum changes occurring poleward and moderate changes occurring in upwelling regions. This conclusion is supported by plotting the temperature anomaly as a function of latitude (Figure 19.15). Averaged over the global ocean, the sea-surface temperature change was only about 2.3°C. By contrast, numerical models of the atmosphere, applied to the data on ice sheets and ocean given in Figure 19.13, indicate that the average change in the atmospheric temperature (ground level temperatures over ice-free land) is about 6.5°C. This difference in the responses of the ocean and atmosphere is remarkable and tends to support the con-

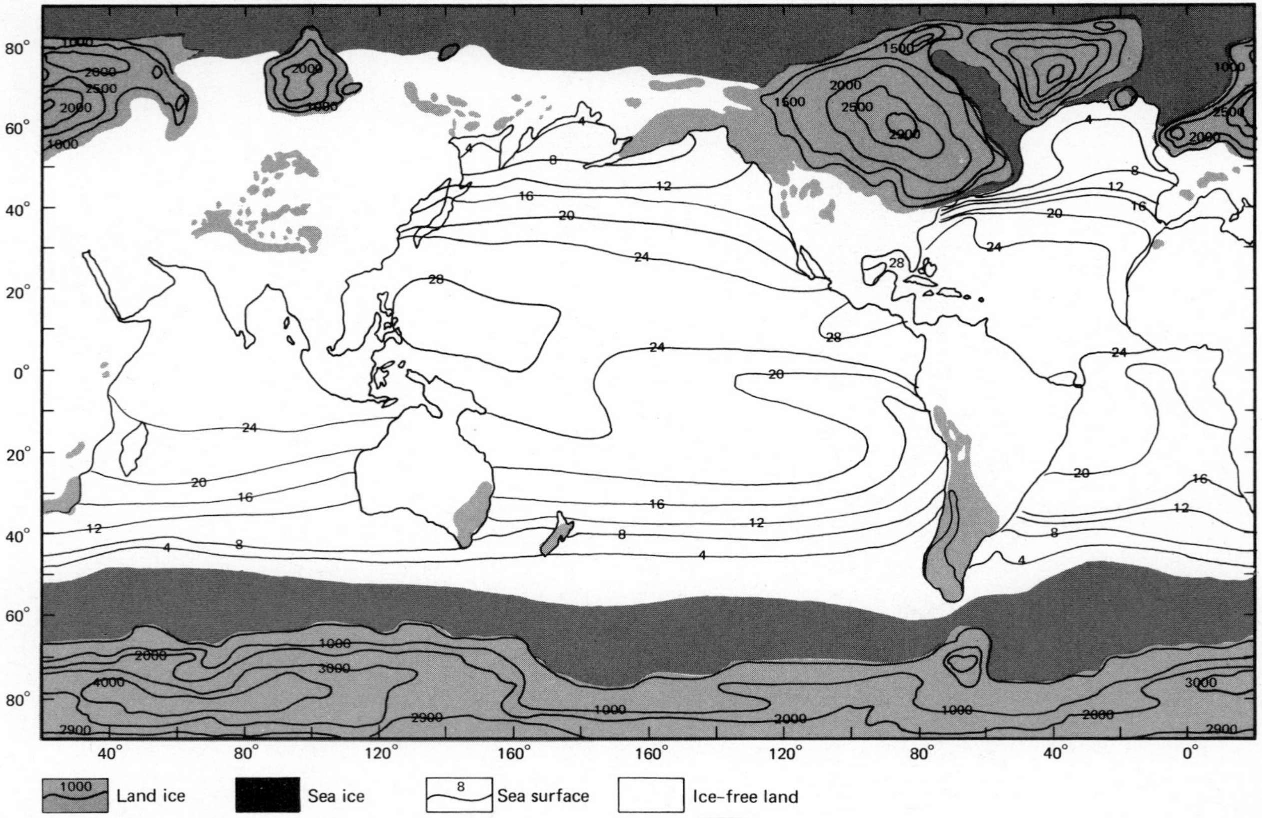

FIGURE 19.13   Reconstruction of the Earth's surface during the last ice age, 18,000 years ago. Sea-surface temperatures (°C) during an average August are estimated from the relative abundances of species of fossil coccoliths, radiolaria, and foraminifera. The extent of ice is mapped from evidence contained in geological deposits on land and in deep-sea cores. The elevation of ice sheets is calculated from physical models of ice flow. The shore line is mapped to correspond with a sea level lower than today by about 100 m. (From CLIMAP, 1976.)

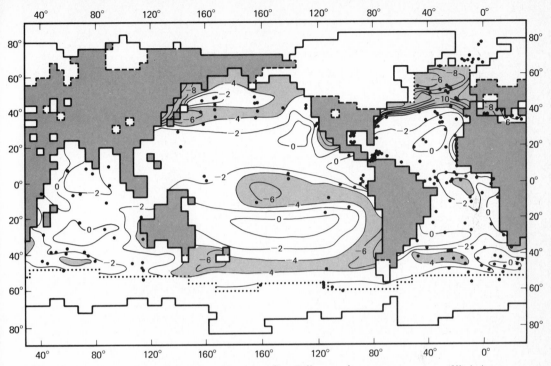

FIGURE 19.14   Ice-age sea-surface temperature anomalies. Differences between temperatures (°C) during an average August today and 18,000 years ago are contoured by light lines. Except for small areas in the subtropics, ice-age temperatures were lower. Shaded area in the ocean indicates temperature changes greater than 4°C. Outlines of continents and ice bodies conform to a grid spacing employed by numerical models of the climate system. Large dots mark the locations of cores used in reconstructing sea-surface temperatures 18,000 years ago. Dotted lines mark margins of sea ice. Dashed lines indicate ice margins on land. Heavy lines mark margins of land-based ice sheets or of the shore line in ice-free regions. (From CLIMAP, 1976.)

cept—well shown by climatic changes occurring decade by decade—that the climate system as a whole is so delicately balanced that small changes in the thermal structure of the ocean are associated with large changes in the atmosphere.

CLIMATES OF OLDER GEOLOGICAL PERIODS

Although the details of ancient ocean climates become more obscure the farther back we go in time, it is nevertheless clear that major changes in climate must have accompanied continental drift. The geometry of land and sea more than 200 million years ago had almost nothing in common with today (Figure 19.16). In the Permian Period a single supercontinent (Pangea) ran longitudinally nearly from pole to pole, surrounded by a vast and undivided world ocean. What was the climate of that world? We know little about the open-ocean climate at that time, but the continental record of climate is fairly clear. The Permain was marked by ice sheets centered near

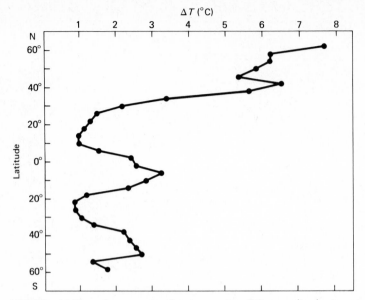

FIGURE 19.15   Ice-age sea-surface temperature (°C) anomalies for August plotted as a function of latitude. Averages are calculated by subtracting 18,000-year-old temperatures (Figure 19.13) from modern values at each oceanic grid point not covered by sea ice, summing for each latitudinal band and dividing by the number of grid points used. High northern latitudes changed the most; areas of the subtropical gyres changed very little. The areally weighted change of the world ocean was 2.3°C. (From CLIMAP, 1976.)

the South Pole. This ice sheet fluctuated, and at times covered parts of what are now Argentina, South Africa, southern India, Antarctica, and Australia. Arid regions occurred in the mid-latitudes, as shown by geological records of deserts and by deposits of salt in isolated inland seas. In low latitudes, warm-loving animals thrived in Permian seas. Clearly, pole-to-equator climatic gradients in temperature and precipitation were strong.

From the Permian onward, continental history is marked by a breakup of Pangea into separate continental blocks. Gradually, the geometry of the world ocean was transformed into the familiar pattern of our modern world. A reconstruction of the ocean as it was about 65 million years ago (Figure 19.16) catches this process about halfway. The ancestors of our modern North and South Atlantic Oceans are recognizable, although the passageway to the Pacific and the lack of connection with true polar seas probably made both the North and South Atlantic much warmer bodies of water than they are today, an inference which is supported by the geological records of these oceans. From these ocean records, as well as from the extensive records of sediments and life on land, we also know that average temperatures 65 million years ago were considerably warmer than today. Polar ice caps did not exist, and pole-to-equator thermal gradients were gentle. Oceanographically, the most striking contrast with the world of today is in the Southern Hemi-

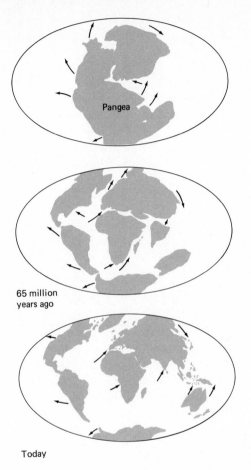

Pangea

65 million
years ago

Today

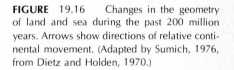

FIGURE 19.16   Changes in the geometry
of land and sea during the past 200 million
years. Arrows show directions of relative continental movement. (Adapted by Sumich, 1976,
from Dietz and Holden, 1970.)

sphere. The Drake Passage between South America and Antarctica was just beginning to form, and Australia was joined to Antarctica. Here the ocean 65 million
years ago had no east-west passage large enough to permit the formation of a west
wind drift system which dominates the climate of that hemisphere today. During
the succeeding 30–40 million years, this situation was gradually changed. About
35 million years ago a west-wind drift system associated with an Antarctic convergence developed. The formation of this circulation pattern and water structure must
be regarded as a climatically decisive event, which ushered in the ocean circulation system as we know it today and set the stage for the modern climatic regime.
Thermally isolated from the oceans of the rest of the world, the Antarctic became
the site of rapid production of cold bottom waters, with a marked impact on the
bottom waters and benthonic life of the whole world ocean. Surface waters at high
southern latitudes also cooled substantially, and on the Antarctic continent there
was a steady increase in the volume of glacier ice. Now nearly isolated from the
rest of the world, this ice mass grew until about 10 million years ago it attained

a volume close to what is has today. The Southern Hemisphere entered, and has never left, an era marked by extensive polar ice. Although the Northern Hemisphere lagged, the cooling trend was global in extent. The first ice sheets developed about 3 million years ago at high latitudes on land areas adjacent to the North Atlantic.

Why the development of Northern Hemisphere ice sheets lagged behind the Southern Hemisphere is not entirely clear. But the answer seems to lie in the contrasting geometry of land and sea. Unlike the ocean of high southern latitudes, the North Pacific and especially the North Atlantic ocean (including its extensions into the Arctic) have basin shapes that promote the advection of warm surface water into subpolar and polar regions. Perhaps for this reason also, the Northern Hemisphere ice sheets—once they did form—have since undergone a series of major fluctuations in volume and area, especially during the later part of the Quaternary period (the past 2 million years). One of the most extensive of these ice ages occurred 18,000 years ago. In the final section of this chapter, we will examine the ice-age cycles and try to understand their cause.

ICE-AGE CYCLE

Since the existence of Quaternary ice ages was discovered a century ago by Swiss and Norwegian geologists examining the records of ancient glacial deposits, scientists have speculated about the cause of these events. Two groups of theories have been advanced. One group invokes factors internal to the climate system (Figure 19.2). The other identifies various changes in the external boundary conditions as the fundamental cause of the ice-age cycle, including variations (1) in the energy output of the sun, (2) in the geometry of the Earth's orbit, which in turn influences the seasonal and latitudinal distribution of incoming solar radiation; and (3) in the content of CO_2 or volcanic dust in the atmosphere. Among these ideas only the orbital theory has been formulated so as to predict the frequency of ice-age cycles. Hence it is the only idea that can be tested geologically by determining what those frequencies are.

The major obstacle to testing the orbital theory by examining geological deposits on land is that the terrestrial record of climate is very incomplete. Each glacial advance largely removes the deposit of earlier ice sheets. But the deposits forming in the deep sea remain relatively undisturbed after they are laid down, so by using the stratigraphic techniques discussed in Chapter 15 it is possible to reconstruct a complete record of past ocean climate. One of the main tools used to build up a Quaternary climatic record is the measurement of the O^{18}/O^{16} ratio in skeletons of benthic and planktonic fossils. As discussed in Chapter 15, this ratio reflects primarily the isotope ratio in the seawater from which the skeletons are derived, a value which in turn reflects the total volume of global ice. Developed by C. Emiliani, this isotopic curve (Figure 15.10), which has proved so useful in stratigraphic correlation, is also an important climatic record.

The exact chronology of the isotope curve was uncertain for many years. In 1973, however, a Pacific core (V28-238) was studied (Figure 15.10) in which the Brunhes-Matuyama magnetic reversal was recorded, an event already dated on land by K-Ar methods as 700,000 years ago. With this chronological fix at one deep point in the core, and radiometric ages for stratographc levels 11,000 and 127,000 years old, the approximate ages of all the climatic events recorded could be estimated by interpolation. Here, the assumption is made that the sediment accumulated at a constant rate between the points for which absolute ages have been determined. Using that assumption, the ages estimated for the isotopic stage boundaries in this core can be used to date corresponding levels in all other deep-sea cores. In particular, it was possible to date the isotopic and other climatic curves obtained from two cores in the South Indian Ocean, where the sedimentation rate is high enough (about 3 cm/kyr) to permit study of the details of climate history. Two of these climatic records are shown in Figure 19.17: the isotopic curve, which reflects changes in global ice volume; and a sea-surface temperature curve derived from statistical analysis of planktonic radiolaria. The record extends back about 500,000 years, and by visual inspection can be seen to consist of a number of superposed cycles. Using suitable numerical techniques, the length of

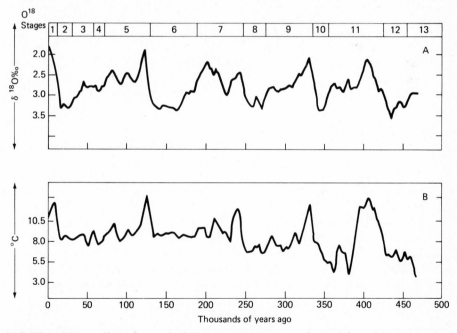

FIGURE 19.17   Climatic record of the past 500,000 years obtained from deep-sea cores in the southern Indian Ocean. Upper curve shows values of oxygen isotope ratio $\delta^{18}O$, as measured in the calcareous test of the planktonic foraminifera *Globigerina bulloides*. This curve fluctuates according to changes in the global volume of ice, becoming more positive as ice volume increases. Lower curve shows changes in summer sea-surface temperature (°C) at the core site calculated from the proportions of radiolarian species. (From Hays et al., 1976.)

four constituent cycles can be established as approximately 100,000, 42,000, 23,000, and 19,000 years (Figure 19.18). As shown in Table 19.1, these cycle lengths correspond closely to those predicted by the orbital theory. Therefore, although the mechanism by which changes in the Earth's orbital geometry influence climate is unknown, it seems quite certain that these orbital variations are an important factor in causing the succession of ice ages, at least during the past half-million years. These variations change the seasonal and latitudinal distribution of solar radiation received at the top of the Earth's atmosphere. For example, when the tilt of the Earth's axis (which is now inclined 23.5° to the vertical) decreases, more of the sun's energy is concentrated in equatorial, and less in polar, latitudes. From the

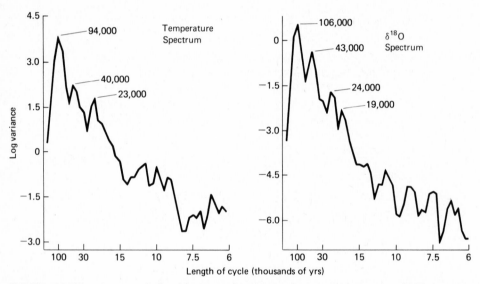

FIGURE 19.18   Spectra calculated from two climatic records (water temperature and $\delta^{18}O$) in a core from the Indian Ocean sector of the Antarctic Ocean (data shown in Figure 19.17). The total variance of each record is resolved into frequency components expressed in thousands of years per cycle. Arrows point to significant spectral peaks and indicate the average cycle length in years associated with each peak. (From Hays et. al., 1976.)

TABLE   19.1   Comparison of Astronomical Cycles with Climatic Cycles Observed in Deep-Sea Cores From the South Indian Ocean (length of cycles in years)

Astronomical Cycles		Climatic Cycles*
Effect	Length	
Eccentricity	106,000	~100,000
Tilt	41,000	42,000
Precession	23,000	23,000
Precession	19,000	19,500

* Averages of values calculated for two sets of data shown in Figure 19.18

geological record of climate, it is clear that intervals of lower tilt are associated with increases in the polar ice caps. The other orbital cycles (Table 19.1) involve changes in the eccentricity of the Earth's orbit and changes in the Earth-sun distance at particular seasons (the precession effect). Although the climatic effect of these orbital cycles is more complex than that of tilt, the correspondence between the orbital and climatic records indicate that in some way yet to be established the global climate system is sensitive to these small changes in the radiation budget.

SUMMARY

1. At any place and over any given interval of time the ocean displays a certain average value of temperature, salinity, and other physical properties. These averages constitute the ocean climate of the interval under study.

2. When records of the past several decades are studied, significant changes in ocean climate occur which are linked to changes in atmospheric circulation patterns. Of particular importance are changes in the intensity of upwelling along the coast of Peru (the El Niño phenomenon) and changes in the distribution of sea-surface temperature elsewhere in the Pacific Ocean.

3. Instrumental records indicate that during the late eighteenth and early nineteenth centuries anomalous patterns of Atlantic surface temperatures were associated with cold climates in Europe and North America (the Little Ice Age).

4. Geological records indicate that 18,000 years ago the Earth was in a great ice age involving responses of all parts of the climate system. Ice sheets covered significant parts of the Northern Hemisphere continents, sea levels were lower by about 100 m, and sea ice was more extensive than now in both Northern and Southern Hemispheres.

5. The ice-age ocean was characterized by greater upwelling along the equator and in areas of eastern boundary currents, by a more easterly course of the Gulf Stream, and by surface temperatures markedly cooler than now, except in the subtropical gyres.

6. Major changes in climate accompanied continental drift. Sixty-five million years ago the climate was considerably warmer than today, polar ice caps did not exist, and a land barrier in the Southern Hemisphere prevented the development of a circumpolar current system. Gradually, as the passage between Antarctica and Australia opened, the modern circulation pattern was established, ice accumulated on Antarctica, and global climate cooled.

7. Variation in the Earth's orbit is one factor responsible for the succession of ice ages during the last half-million years.

California shoreline south of San Francisco. (U.S. Navy Photo.)

Where the Land
Meets the Sea

20

The most common experience of man with the ocean occurs along the coasts. It is here that the power and vastness of the sea are first impressed on the observer and it is here that man most severely impacts the ocean with his activities. The coast is buffeted by the waves generated by storms at sea. These waves interact with the rocks and sediments of the coastal region to shape and erode them. The debouching of the river into the sea not only influences circulation patterns

CHAPTER

along the coast but also provides sediments that respond to this circulation as well as to the waves and tidal currents. The coastline responds to the large-scale influences of plate tectonics. The continents are also submerged or exposed to varying degrees at their edges by seawater in response to the glacial cycles. All these factors work to yield the complex features of the boundary between the continents and the oceans.

ESTUARINE CIRCULATION

Many coastal regions receiving stream flow are commonly called estuaries. Actually the distinctive qualities of estuaries are the direct consequence of the blending of seawater with fresh water. An estuary is a semienclosed body of water having a free connection with the open sea and within which the seawater is measurably diluted with fresh water deriving from land drainage. By such a definition Narragansett Bay, in Rhode Island, the Norwegian fjords, the Juan de Fuca Straits in Washington, and the mouth of the Mississippi River are all examples of estuaries. Evaporation basins such as San Diego Bay in California and the Mediterranean Sea as a whole are not estuaries, since the salinity is maintained at a level higher than the ocean, to which they connect. They are sometimes called "antiestuaries" for their water transport patterns are exactly the opposite of those for estuaries. Estuaries are generally biologically productive and are sources of a number of highly valued fisheries. They are also commonly locations of major ports and industrial complexes. Interest in understanding estuarine circulation dynamics is strongly tied to resolving these conflicting uses. Problems of pollution and other forms of human encroachment on estuarine systems are thus shared by many different interest groups.

   Estuaries have been classified in a number of different ways. One of the more useful in terms of understanding its circulation is by the amount of vertical mixing or the extent to which they are vertically stratified. They vary from highly stratified estuaries to those well mixed and with little or no vertical salinity gradient (Figure 20.1). Deep fjords, with or without a shallow sill, are commonly highly stratified

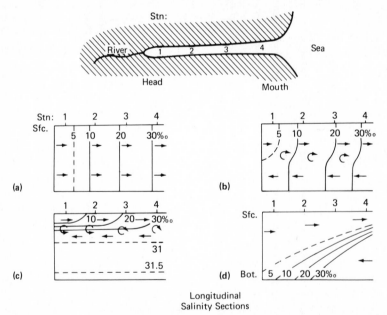

Longitudinal
Salinity Sections

FIGURE 20.1   Distribution in various types of estuaries. In a well-mixed estuary (a) the salinity is constant from top to bottom and increases seaward. At the other extreme is a fjordlike estuary (c), where there is a sharp salinity gradient with depth. A "salt-wedge" estuary typical of many rivers is shown in (d). The arrows indicate the average direction of flow in the estuary after the tidal currents have been removed.

(type C in the figure), with fresh water flowing out over a nearly uniform deep saline layer. A second type of highly stratified estuary is the so-called salt wedge (type D in the figure); the Mississippi River is an example. As the importance of tidal mixing increases, the estuary becomes less stratified. The James River in the Chesapeake Bay is perhaps the best studied example of a slightly stratified estuary (type B in the figure). All the major coastal-plain estuaries of the U.S. east coast, such as Narragansett Bay, Delaware Bay, and Chesapeake Bay, are examples of partially mixed estuaries. The vertical salinity gradient, however, can be very small, as is the case in Narragansett Bay. The extreme is the case where vertical mixing is so strong compared to the river runoff that there is no vertical salinity gradient (type A in the figure). The salinity increases from head to mouth. Shallow estuaries with strong tides can lead to well-mixed estuaries, for example, the Severn River in England.

The predominant circulation in most estuaries is the ebb and flood of the tide. Tidal currents often reach speeds of several knots and the tidal currents provide much of the energy that mixes the fresh water of the river with the salt water of the ocean. In a very general way one can predict the degree of mixing in an estuary simply by knowing the ratio of tidal flow (that is, the volume of water transported during the flood tide) to river flow (on the same time scale as the flood tide). For example, if the ratio of tidal flow to river flow is between 100 and 1000, then there is generally sufficient tidal energy to develop a partially mixed estuary. If the ratio is unity or less, as it is in the Mississippi, a salt wedge estuary is the result.

Superimposed on the to-and-fro tidal currents is a small, but very important, nontidal circulation. The tidal currents are usually an order of magnitude larger than the nontidal flow, as, for example, in Narragansett Bay, where typical tidal-current amplitudes are 20-50 cm/s and the nontidal flow is about 2-5 cm/s. However, it is the nontidal flow which transports water and other matter into and out of the estuary; whereas the tidal currents essentially move water to and fro without resulting in a net displacement.

This nontidal circulation (or net circulation) in an estuary results primarily from a combination of gravitational forces and vertical mixing. In the absence of vertical mixing there is only the gravitational force; fresh water flows out over a salt wedge, as in Figure 20.2. The complete absence of vertical mixing, such as implied in the figure, does not occur in nature. Some salt is mixed upward. Conservation of salt requires a net flow in from the ocean to equal the salt that is removed by the surface layer. With increased mixing the entering fresh water gets mixed downward. Hence, along with the vertical salinity gradient, there is also a horizontal salinity gradient, with salinity increasing toward the sea at all depths. The increased mixing also increases the volume flux in the upper and lower layers. Net velocity and salinity distribution typical of partically mixed estuaries may be seen in Figure 20.3. Note that the net speeds increase seaward. This is due to a continual entrainment of seawater along the axis of the estuary from the lower into the upper layer.

Although a conservation argument such as the above can account for the net estuarine flow, it does not explain what drives this nontidal flow. The water surface slopes down towards the sea, resulting in a pressure gradient that drives the water

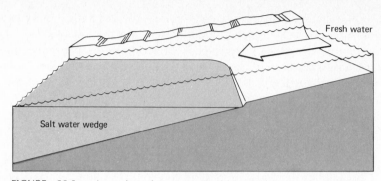

FIGURE 20.2   In a salt wedge estuary the fresh water flows over the high salinity layer, and with relatively little mixing between the two layers the boundary
between fresh and salt water is approximately at sea level.

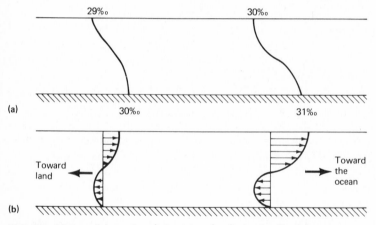

FIGURE 20.3   Schematic salinity (a) and velocity (b) distribution in a partially mixed estuary.

of the surface layer out the estuary. As shown in Figure 20.4, a reversal in pressure
gradient with depth can result from the slope of the interface between the two
layers, which is the opposite sense of the surface layer, or by the upstream decrease
in the average density of the water column. This pressure force generates a flow
up the estuary in the bottom layer.

One important effect of this nontidal circulation is that sediment is transported
up the estuary. The efficiency of such bottom transport is directly related to the
amount of nontidal circulation. Ignorance of this fact can lead to serious miscalcu-
lations regarding the problem of harbor silting. An example is provided by a coastal
engineering project aimed at ameliorating the silting up of Charleston Harbor.
Some 90,000 cubic meters of sediment had to be dredged from the channel each
year in order to keep it navigable. An attempt was made to "flush out" the channel

by increasing the river flow by diverting another river into the Charleston estuary. The increased river flow also increased the net nontidal circulation, resulted in a higher net inflow along the bottom, and increased the amount of material that had to be dredged to almost 3.5 million cubic meters per year.

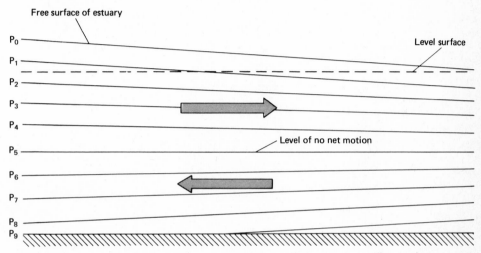

FIGURE 20.4   Schematic distribution of constant pressure surfaces in a partially mixed estuary and the net, or time averaged, motion resulting. In the upper half of the water the sloping sea surface drives the water seaward. The pressure gradient reverses in the lower half because of the changes in the density of seawater resulting from the salinity distribution.

SHAPING OF THE COASTLINE

Coastlines are not static features. Both the land and the sea are continually undergoing changes in elevation. The most obvious change is due to the tides, where coastal land is systematically exposed at low tide and submerged at high tide (see Chapter 8). Periodic storms can heighten the tidal effects.

But other changes which act on longer time scales are possible. The growth and ablation of continental glaciers directly influences sea level by storing or releasing water. Sea level at the peak of the previous glacial maximum around 18,000 years ago was about 100 m lower throughout the world, because the water from the ocean had been transferred to the continental ice sheets (Figure 20.5). Any melting of the ice sheets in Antarctica and Greenland will cause additional rises in sea level. Tidal-gauge data indicate that at least over the last 100 years sea level has been rising at a rate of about 0.2 cm a year (Figure 20.6), probably because of the melting of some ice from a portion of Antarctica.

In addition to sea-level changes resulting from ice storage and release, the continental areas burdened by ice accumulation also move up and down in re-

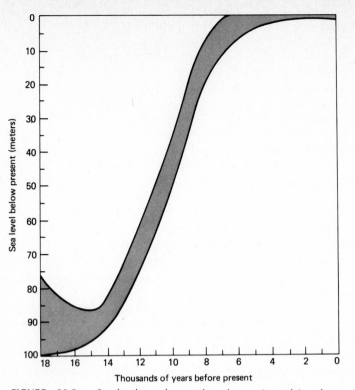

FIGURE 20.5   Sea level was lower when the continental ice sheets were greatly expanded. Beginning about 18,000 years ago there has been a long-term rise in sea level as the ice sheets shrank. This curve is based on two methods of measuring the rise in sea level, one depending on calculation of ice volumes from evidence on land of the extent of glaciation at each time, and the other on relict datable approximated shore-line deposits now preserved all along the continental shelves. (After Bloom, 1971.)

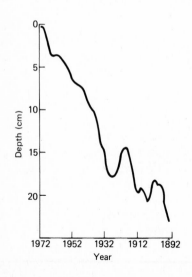

FIGURE 20.6   Tide-gauge data from the east coast of the United States shows a rise in sea level over the past 100 years of about 2 mm a year. (After Hicks, 1972.)

sponse to the addition and removal of the massive ice burden. Land that was covered by glacial ice 18,000 years ago is now recovering from the depressed elevation due to ice loading, so in some places because of this rebounding, land is rising from the sea.

In the past, sea level has changed on a worldwide scale for reasons other than ice storage and release. The worldwide changes in sea level are called *eustatic* changes. These eustatic changes in sea level can be ascribed to changes in the volume of the deep-ocean basins as the result of changes in ocean-floor spreading and oceanic-ridge formation rates. A time of increasing ocean-ridge formation would result in water lapping up on the continents, for example. The last major worldwide onlapping ended about 70 million years ago, with a few less extensive intrusions of the sea on the continents since that time (Figure 20.7).

Locally these large-scale tectonic forces involving movements of large plates in relation to each other cause elevations and warping at continental borders as well as within the continental masses. These can result in broad warps or sharp scarps along the coast.

Since both land and sea can move in relation to the other it is obvious that coastlines are not going to be static. If nothing else occurred, the glacial cycles with periodicities of the order of 100,000 years successively cause coastlines to emerge and submerge over ranges of about 100 m. This alone strongly influences the processes active along the coast and the resultant coastal patterns.

SUBMERGED COASTLINES

During the times of large continental ice-sheet presence over the last several hundred thousand years, sea level was lower. Rivers throughout the world then had a

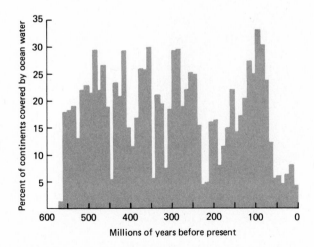

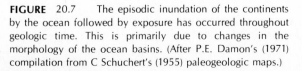

FIGURE 20.7   The episodic inundation of the continents by the ocean followed by exposure has occurred throughout geologic time. This is primarily due to changes in the morphology of the ocean basins. (After P.E. Damon's (1971) compilation from C Schuchert's (1955) paleogeologic maps.)

lower base line and therefore cut deeper into their channels. When sea level rose to higher levels as the size of ice sheets diminished, these deeply incised river channels were submerged. Chesapeake Bay is a good example of such a coastline of submergence (Figure 20.8).

In areas periodically covered by the large continental or piedmont ice sheets, such as Norway and parts of Alaska, the glaciers moved along old river channels and gouged them out. In land these are seen as high U-shaped valleys such as at Yosemite Park in California, or the Finger Lakes of New York State. Where they occur at the margin of the continent, subsequent submergence by the sea produces *fjords* (Figure 20.9).

In some areas of the world, coastal land is warping downward for tectonic reasons, thus resulting in increased submergence beyond that due to the general worldwide rise in sea level as the result of the latest major melting of the glacial ice sheets beginning about 18,000 years ago.

EMERGENT COASTLINES

Because sea level has been rising over the past 18,000 years the general trend for coastlines is to be progressively submerged. Coastlines become emergent under

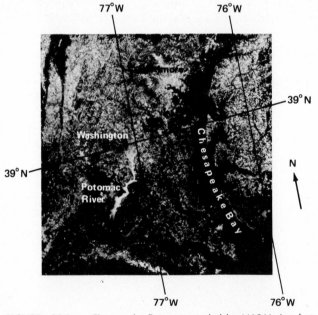

FIGURE 20.8   Chesapeake Bay as recorded by NASA's Landsat Earth Resources Satellite. The area of the photograph measures 185 km by 185 km and includes Washington, D.C. and Baltimore, Md. The Potomac River estuary shows high silt content (white appearance) mainly due to human activity. The channel-like appearance of Chesapeake Bay and the associated estuaries directly reflects the old river channels carved into the continental shelf when sea level was lower during the glacial advances. (NASA photograph.)

FIGURE 20.9 One of the classic fjords of the Norwegian coast. These features are formed by glacial gouging of old river valleys during the glacial advances of the last 2 million years. Subsequent melting of the ice and submergence by the sea results in the classic steep-walled fjords. (Photo from the Norwegian National Tourist Office.)

such a constraint only if the rate of uplift of the land transcends the rate of rise of sea level. This can occur as the result of one of two effects. (1) Land under a continental ice sheet will rebound, once the ice bearing down on it is melted. The rate of uplift and the rate of sea level rise need not be the same, since the former is controlled by the mechanical properties of the mantle and the latter is clearly controlled by the supply of water to the ocean from melting. (2) In tectonically active areas such as plate margins land is uplifted. This generally occurs episodically along fault zones, but sometimes as the consequence of a general arching in such areas. An example of this is found along the coast of California (Figure 20.10). When tectonic uplift occurs in an episodic fashion, sea cliffs usually develop which are subject to intense attack by waves until a wave-cut beach is obtained.

COASTAL FEATURES DUE TO
THE ACTION OF WAVES

The energy acquired by ocean surface from the wind is transmitted to the continental margins primarily by waves. In Chapter 8 we saw how a "sea" of waves is transformed on approaching the continental shores into the almost rhythmic beat of waves and surf on the shore.

FIGURE 20.10   The San Onofrio nuclear power plant between Los Angeles and San Diego is virtually at sea level. This picture was taken from a raised terrace which virtually encircles the reactor site. This terrace was formed at sea level and was subsequently elevated tectonically, probably all within the last million years or less. A serious problem facing the siting of nuclear power plants in California has been the concern about the possibility of tectonic instability in the near future. (Turekian photo.)

As the waves approach the land, they are refracted by the topography of the shallow water after a certain critical depth is reached, going from deep to shallow water. If a wave approaches a straight shore line with constant deepening toward the sea at all points, the wave will be bent so that it approaches parallel to the beach (see Chapter 8), virtually independent of the angle at which it originally approached the shore.

If a wave encounters offshore a rugged topography which is a continuation of the onshore topography, the wave will be so refracted that the headlands are points of concentration of wave energy, whereas the bays see the energy of the waves more attenuated (Figure 8.12). This will result in the aggressive erosion of the headlands.

Waves not only transport energy, but, at the encounter with the shore, they also physically transport water. Thus on rocky crags there is spray and general turbulence, and along beaches a breaking wave becomes swash on the beach. The waves attacking the headlands cause water to flow along shores away from the headlands toward the bays; therefore the shoreline in the bay is built out at the same time that the headland is being cut back. The end result of these two actions is a tendency toward straightening the shoreline.

Along virtually straight shore lines the direction of the long-shore currents is dictated by the direction of the waves. Sediment is also transported by these currents so that spits are developed at the terminus of the straight stretch along which the current transports sediment. Once spits develop, they can be built up to form barriers which result in the formation of lagoons between them and the mainland (Figure 20.11). Long-shore currents are not like continuous shore line rivers, however, but rather tend to break into cells so that the mass transport of water by breaking waves can have a seaward flow at the surface. These are called *rip currents* (Figure 20.12).

DEPOSITION FEATURES

When a sediment-laden stream encounters the ocean, the velocity of the stream drops sharply at its mouth and the zone of mixing of the fresh water of the stream and the salt water of the coastal ocean collaborate to cause the rapid settling out of sediment particles at the point of encounter.

What occurs next to the sediment depends on the working of many forces. If there is no continental shelf, the sediment drops to the bottom of the drowned portion of the river valley cut initially at its top during times of lowered sea stands and then moves in a cohesive way down the channel, eroding and deepening it. These sites for active turbidity currents are seen off the Congo and Magdalena Rivers (Chapter 14). Most other rivers debouch onto a shallow platform of the continental shelf.

During times of storm there is resuspension of this sediment. The amount of resuspension depends on the duration of stormy conditions, the depth of water, and the nature of the sediment pile. Resuspension of fine-grain sediment in relatively shallow waters is frequent. As the resuspended particles resettle to the bottom, the estuarine circulation discussed above tends to transport the accumulating sediment landward. Thus inlets and bays, such as Chesapeake Bay and Long Island Sound, act as sediment traps. Occasionally the trap is breached and a quantity of sediment reaches the shelf, where it is swept up and down the coast until it is either trapped again or slips off into the abyss.

DELTAS

If the rate of supply of sediments is small in relation to strong long-shore coastal and tidal currents, localized accumulations of sediments directly attributable to the stream supply are limited. On the other hand, in areas without strong tides or intense coastal long-shore currents, the stream-borne sediments build out on the shelf, thus forming what is called a *delta* (because the type-Nile delta looks like the Greek capital letter Δ in plan view, Figure 20.13). The Mediterranean Sea, the Black Sea, the Baltic Sea, the Gulf of Mexico, the Arctic Ocean, and parts of

FIGURE 20.11   Sandy Hook, New Jersey is a spit produced by sand transport by long-shore currents. New York harbor is to the north. The northward movement of the sand on the open ocean side is seen by the build up behind man-made groins (middle right of the photo) and the presence of growth bands of beach at the northern tip of the spit. (Photo courtesy of the U.S. Corps of Engineers.)

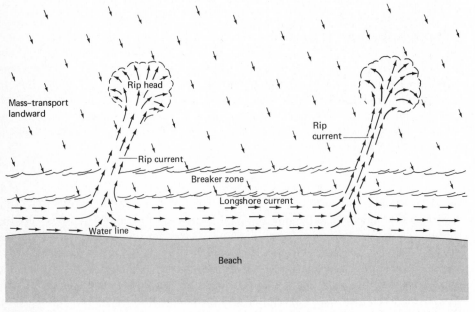

FIGURE 20.12   Diagramatic representation of rip currents showing the return of water delivered to the shore by breakers. (After U.S. Navy "Oceanography for the Navy Meterologist," 1960.)

FIGURE 20.13   The Nile Delta. The Mediterranean Sea is to the left of the picture and the Suez Canal and Red Sea to the right. The dark appearance of the delta relative to the surrounding desert is due to vegetation. (NASA photograph.)

the North Sea do not have large tidal cycles—changes in sea level are virtually imperceptible in most cases. Sediment-laden streams debouching in these bodies of water drop their burdens onto continental shelves, thus forming deltas. Thus the Mississippi (Gulf of Mexico), the Mackenzie (Arctic Ocean), the Lena (Baltic Sea) and the Rhone, Po and Nile (Mediterranean Sea), the Danube (Black Sea), and the Rhine (North Sea) Rivers each have significant deltas at their mouths.

The structure of a delta when viewed in detail is complex and its actively growing front is transitory. During times of massive discharge new channels may be cut and old ones abandoned. Most deltas have a system of distributaries, each carrying a portion of the streams' sediment and water flux (Figure 20.14). Deposition occurs at the mouths of these individual distributaries, thus forming small deltas which together make up the large delta structure.

As deltas build out on the continental shelf, the outer part must eventually reach the edge of the continental shelf or, if no clear shelf break exists, the region of rapidly increasing slope. Here the sediment pile is subject to downslope movement when it is perturbed. This release of sediments moves down canyons or as avalanches to form sedimentary layers along the continental margin, or even moves down to the abyssal plain. Commonly these are expected to behave like turbidity currents (Chapter 14).

Because sea level was lower during the last glacial age, which was at its maximum 18,000 years ago, much of the easily movable sand and clay appears to have already slumped away and the deltas at the present time are building outward without major slumps occurring.

Deltas are the depositories of fine-grain, organic rich muds derived from the erosion of upland topsoil especially during times of flood, and for this reason are prized as good farmland. The Rhine, Mekong, and Nile deltas are important agricultural areas for the countries controlling them.

FIGURE 20.14   An aerial view of the Mississippi Delta and distributary system of the Mississippi River. The river discharges through three main channels: South Pass at the center of the picture, Southwest Pass to the right, and Pass A Loutre to the left. In the distance is the Gulf of Mexico. (Photo courtesy of the U. S. Army Corps of Engineers.)

Deltaic type deposition is also closely related to the development of oil-source rocks and reservoirs, and extensive explorations of deltaic structure have therefore been made in recent years.

SALT MARSHES

In protected regions of the shore line, where strong wave action is absent, marshes develop (Figure 20.15). These marshes are important in the life cycle of many marine animals. They are formed on the mouths of many rivers protected from the open ocean by spits and in the lagoons behind barrier islands. Salt marshes appear to have their widespread distribution as a result of rising sea level since the latest glacial advance ended 18,000 years ago. During the time of rapid sea-level rise between the end of the last major glacial age and several thousand years ago, the large salt marshes did not have time to develop. But once sea-level rise slowed down considerably, salt marshes developed. The marshes extend horizontally into the lagoon on substrate that is established by the mud accumulating in the interwoven grass and roots of the marsh. They grow vertically when sea level rises and act as a record of changes occurring in coastal environments.

FIGURE 20.15 A salt marsh in Connecticut. Although the tidal cycle periodically inundates the marsh, most of the time its surface is exposed to air and fresh water. The upward growth of the marsh is controlled by the rise in sea level due to the melting of the ice caps. The rate of rise in sea level and the rate of upward growth of this salt marsh over the last 100 years are both about 2 mm per year. (Turekian photo.)

SUMMARY

1. Where fresh water encounters the sea, an estuarine circulation is set up in which salty open ocean water moves toward the shore and, diluted at the surface, returns to the ocean again. This results in a net transport of bottom sediment toward the continents.

2. Around the continental borders, tectonic, glacial and eustatic changes affect the aerial and submarine topography of the coastline as a function of time.

3. Waves acting on the coastline tend to destroy headlands and fill in bays with sediment. Long-shore transport resulting from the net movement of the ocean impinging on the shore results in the formation of spits.

4. Deltas are built up at river mouths where the supply of sediment is high and the dissipative forces related to tides and storms are relatively low.

5. Where sea level rises slowly, salt marshes develop and grow upward. They are composed of a framework of salt-tolerant grasses, but can trap sediments as well.

Offshore platform under construction in Corpus Christi

Mineral Resources of the Sea

21

It is increasingly realized that our continental mineral and energy supplies are finite; thus it is not surprising that interest has been sparked in the potential of the oceanic areas as a future source for these commodities. This chapter explores the nature of mineral resources in or beneath the sea. But before proceeding, we should look more closely into what a resource is.

CHAPTER

CONCEPT OF A RESOURCE

The term "mineral resource" is often misused, as it is equated with the existence of certain mineral commodities rather than their concentration in economic quantities. An example of this was given in Chapter 2, where the rationale for the German *Meteor* expedition was the extraction of gold from seawater. Because the quantities of gold available were far less than could be extracted economically (see Chapter 18), this cannot be considered a resource.

The U.S. Geological Survey and U.S. Bureau of Mines have attempted to clarify the term and have defined it as follows.

RESOURCE: A concentration of naturally occurring solid, liquid or gaseous materials in or on the earth's crust in such form that economic extraction as a commodity is partly or potentially feasible.

Thus the essential meaning of "mineral resource" includes the economics of extraction in the definition—mere occurrence is not enough.

Further uncertainties arise from qualifying adjectives such as "potential," "probable," or "proved." In the case of *potential* resources the degree of uncertainty, especially in the case of hydrocarbons, is very large and estimates show wide variations depending upon the assumptions made. *Probable* implies a higher degree of certainty, but here, too, the variations may be large. *Proved* resources are again an estimate, but are based on sufficient data that the existence is proved although the size may vary with different interpretations of the data.

In this chapter, we discuss the principal *resources* of seawater and the sea floor. Some of these resources have been developed in the past, others exist but have not yet proved to have sufficient economic value to create the incentives for utilization.

RESOURCES IN SEAWATER

Wherever fresh water is in short supply and people live on or near the ocean, the pure-water component of seawater itself is a potential resource. In addition, a number of chemical elements and compounds dissolved in seawater are of great potential value. The principal problem here is one of low concentration. Unless cheap energy is available, the costs of evaporation and extraction are usually prohibitive.

FRESH WATER

On ships, which do not have access to fresh water, desalinization has long been in practice and in some land areas it is quite practical. In Kuwait, for example, large quantities of natural gas are produced as a by-product of petroleum extraction. Before the advent of liquid petroleum gas tankers, there was no market for gas and much of it was burned or flared. Since there is a shortage of fresh water in Kuwait and since the gas was essentially free, it became economically viable to use the natural gas as the energy source for desalinization.

Many oceanic islands are short of water. Bermuda, for example, is almost totally dependent upon rain, which is caught and stored. When the United States air base was constructed there, it became necessary to install a desalinization plant to provide fresh water.

Other schemes have been suggested for islands with a chronic shortage of fresh water. One of the most ingenious is a scheme that scientists of the Lamont-Doherty Geological Observatory, of Columbia University, devised to respond to the chronic water shortages in St. Croix, Virgin Islands (Figure 21.1). The island lies in the trade-wind belt and each day these winds carry in the saturated air large quantities of fresh water past the island. St. Croix has a narrow shelf, and deep cold water can be found within a mile of the shore. In this plan the cold water would be pumped to a condenser in the path of the saturated air, and the fresh water that condensed on the cold pipes would be collected in a reservoir. As the cold water comes from below the photic zone, where phytoplankton deplete the nutrients, it would not be returned directly to the sea, but to a lagoon. The nutrient-rich waters would be used to stimulate the growth of plankton for mariculture (see Chapter 12).

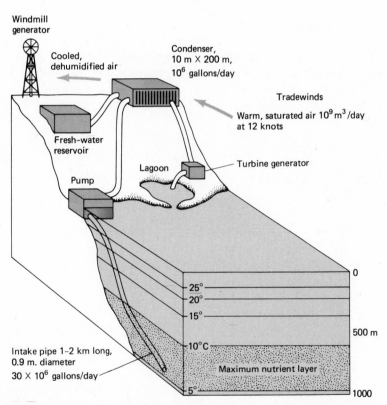

FIGURE 21.1   System proposed by Robert Gerard to extract water from the atmosphere in St. Croix, Virgin Islands. Cold water is pumped from a depth of about 1000 meters to a condenser atop a hill. Water in the saturated air condenses on the cool pipes of the condenser and is collected in a reservoir. The cooled, dehumidified air can be used for air conditioning. Part of the power required for pumping can be recovered by piping the return flow of seawater through a turbine; another part of the required power is obtained from windmills. The seawater, drawn from the deep, nutrient-rich layer can be fed into a lagoon and used for mariculture. Only the last part of this system has been tested to date. (Modified after Gerard and Worzel, 1967.)

COMMON SALT (NaCl)

First in value of the compounds extracted from seawater is ordinary salt (NaCl), which makes up 86 percent of the total dissolved salts in seawater. This is produced by solar evaporation in many parts of the world and may or may not be further refined before use.

Most of the salt used in the United States comes from underground mines, but about 5 percent has been produced by solar evaporation in the San Francisco Bay region. Salt water is run into shallow ponds and held until crystallization starts. Calcium carbonate is the first to precipitate (Figure 21.2) and most of it has crystallized by the time the brine density reaches 1.13 g/cm³. At that time the water is drawn into another pond, where calcium sulphate precipitates. Most of the calcium sulphate has crystallized at a brine density of 1.22 and the brine is drawn off again until brine density reaches about 1.26 and magnesium salts begin to precipitate. This is then drawn off and the remaining sodium chloride precipitate is commonly 99.6 percent pure and requires no further refinement for most uses.

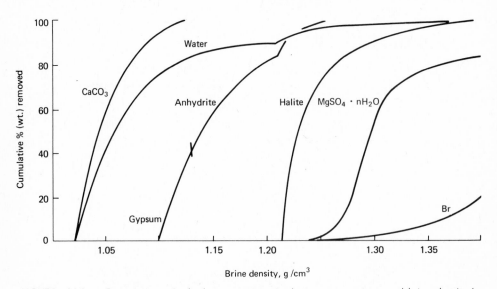

FIGURE 21.2 Precipitation of salts from seawater. As the water evaporates and brine density increases, different salts, starting with calcium carbonate, precipitate in sequence from left to right on the diagram. By drawing off the brine at different densities, the salts can be separated. (After Schmalz, 1969.)

MAGNESIUM-MAGNESIUM SALTS

Perhaps the most successful and best integrated complex for extracting resources from seawater is the Freeport, Texas, plant of the Dow Chemical Company, which was built at the beginning of World War II in response to a national need for magnesium. The location was chosen because the salinity of the Gulf of Mexico waters was respectably high, the supply of fresh water was abundant, and ready access to deep sea transportation, adequate sources of energy, and large supplies of salt, sulphur, and calcium carbonate in the form of oyster shell were available.

The plant takes in about 2 million gallons of water per minute from the Gulf. The water is first run through the power plant, serving the dual purpose of providing cooling water and warming the seawater for the processes (Figure 21.3). It is then run over calcined oyster shell, and magnesium hydroxide precipitates out. This can be utilized directly as a refractory, or converted to magnesium chloride and refined electrolytically into magnesium metal. At one time bromine was also extracted for use as an additive to gasoline, but the process has been discontinued because it became more economical to produce this from more concentrated underground brines. Besides, the demand for bromine is strongly coupled to the extent that tetraethyl lead is permitted as a gasoline additive and this is diminishing as the result of environmental concerns.

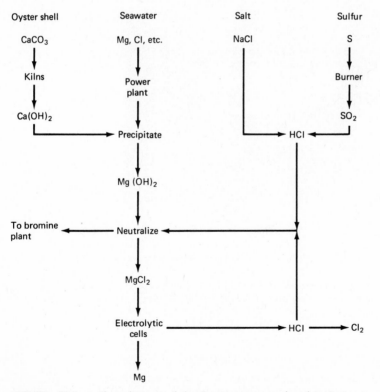

FIGURE 21.3   Flow diagram of the Dow process employed in Freeport, Texas. Raw materials are oyster shell, seawater, salt, and sulfur and the energy supply is locally derived gas. Final products include magnesium metal and salts, chlorine, and bromine.

HYDROTHERMAL BRINES

Hydrothermal brines, produced by the circulation of hot water through the rocks of the sea floor, must be classed as potential resources, since their contents have not yet been demonstrated to be economically extractable.

Such brines have been found both in the undersea and in the land portions of the world system of rifts that marks the places where lithospheric plates are separating and new crust is being formed (see Chapter 16). The best known of these places are in the Red Sea where, in the deepest portions, waters as hot as 62°C (about 144°F) are found. These waters would be unstable and would convect upward but for their high salinities, reaching 27 percent, about eight times higher than normal seawater. Their density is sufficiently high (1.18–1.23 g/cm³) that they produce reflections on echo sounder records (Figure 21.4).

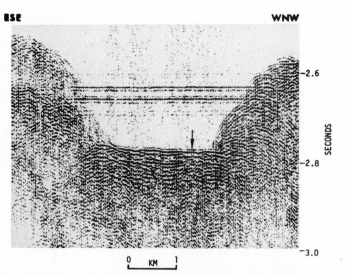

FIGURE 21.4   Seismic reflection profile across Atlantis II deep in the Red Sea showing a distinct reflection from the top of the hot brine layer. Evidence from underwater photography indicates that trash from surface vessels may contribute to the strength of the reflections as instruments have returned to the sea surface draped with newspaper shreds and assorted pieces of garbage. (After *Deep Sea Drilling Reports*, 1974, Vol. XXIII.)

These brines are of interest not only because of their high salinity, five to ten times normal, but because metals such as zinc, copper, lead, and nickel show a large relative enrichment over concentrations in normal seawater (Table 21.1). Obviously these brines are not just high-salinity normal seawater but have added constituents leached from the rock beneath the sea floor by hot solutions.

TABLE 21.1   Metals in Brines (milligrams per liter)

Element	Red Sea	Dead Sea	Ocean
Sodium	93,000	32,000	11,000
Potassium	2,000	6,000	400
Calcium	5,000	14,000	400
Magnesium	1,000	35,000	1,300
Iron	10	2	<0.004
Manganese	50	5	<0.002
Zinc	5	<20	0.003
Copper	0.3	<2	0.0003
Cobalt	0.2	—	0.00001
Lead	6	<2	0.000003
Nickel	0.3	<2	0.0005
Salinity	257‰	270‰	36‰

Metals in Red Sea Sediments

Element	Average Assay (%)	10^6 Tons in Top 10 Meters
Copper	1.3	1.06
Zinc	3.4	2.9
Silver	0.0054	0.0045
Gold	0.00005	0.000045
Lead	0.1	0.08
Iron	29	24.3
Traces of Cobalt, Nickel, Arsenic, Lead, Tin, etc.		

RESOURCES ON THE SEA FLOOR

METAL-RICH SEDIMENTS

The sediments in the deep basins of the Red Sea are similarly enriched in metals to the point that they have attracted commercial interest (Figure 21.5 and Table 21.1). At the present time porphyry copper ores averaging less than 1 percent copper are being mined in many countries. The Red Sea sediments contain about 1.3 percent copper, they contain other metals, and they can be pumped to the surface rather than blasted, carted, and crushed. If the metallurgy of extraction is not too costly, these are indeed a potential resource.

A significant feature of these sediments is their tectonic setting—in the world rift system. It is believed that the process that produces the metal enrichment is not unique to the Red Sea and evidence of hydrothermal activity has been detected in other parts of the rift system. Many of the holes drilled by the Deep Sea Drilling Project that extended through the sediments and into the underlying basaltic rock

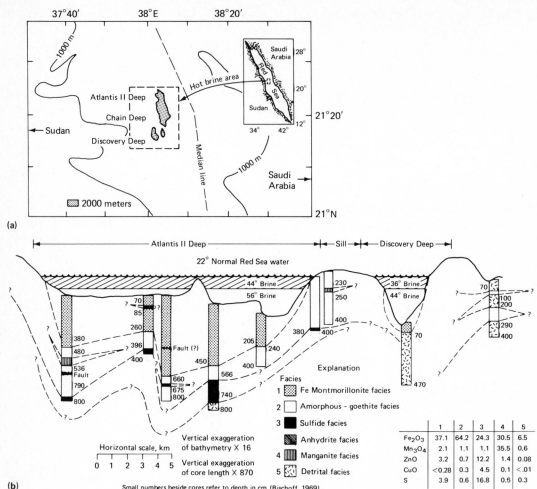

FIGURE 21.5 (a) Locations of deep brine filled-holes in the Red Sea. (b) Sediment types and metal content from cores in Atlantis II deep. (After Bischoff, 1969.)

have revealed an iron-oxide-dominated metal-rich-sediment layer immediately above the basalt. This may indicate hydrothermal reactions associated with the basalt at the time of emplacement. The precipitating iron acted as a scavenger not only for other metals derived from the basalt, but also those extracted from seawater.

FERROMANGANESE NODULES

Ferromanganese nodules must also be considered a potential resource of the sea, because they have not yet been economically utilized. They have attracted much commercial interest in many countries, and suitable extractive metallurgy has been developed, but they are not yet being mined, both for economic and for political reasons.

These nodules are brown or black concretions or incrustations made up principally of iron and manganese oxides. They carpet large areas of the sea floor, and although they vary in size they tend to be fairly uniform in any particular area (see Chapter 14). Because the nodules grow very slowly, they are found in areas where sediment accumulation rates are low. They appear to be most prevalent in the Pacific. They contain up to 36 percent manganese, but the principal interest is in copper, nickel, and colbalt contents (Table 21.2), which are present in economically interesting quantities although not in ratios that are proportional to man's need for them. They are also of interest because their chemistry and porous structure make them an excellent catalytic agent. In small pellets they could be used in place of expensive platinum catalysts to absorb the exhaust gases of automobiles.

Southeast of Hawaii (Figure 21.6) is a belt that has been estimated to contain

TABLE 21.2 Potential Metal Reserves in Manganese Nodules (assuming average diameter 3.7 cm, average abundance 18.9 kg/m², average grade [% Ni + % Cu] 22.22%, average recovery 20.25%)

Metal	Reserves in Nodules ($\times 10^6$ tons)	Reserves on Land ($\times 10^6$ tons)
Nickel	98	44.5
Copper	80	390
Cobalt	20	2.45
Manganese	2,200	5440

Adapted from Archer, 1976.

about 450 million metric tons of copper at a grade of 1 percent. This is a sizable potential resource when compared with the 1975 world production of about 7.5 million metric tons.

Many questions remain to be answered before this potential resource can be utilized. A number of mining techniques have been developed and tested, but considerable improvement in the efficiency of collecting nodules must be made if it is to be a feasible long-term operation. Extraction of the metals by chemical methods will create heavy demands for water, chemicals, and energy. A parallel problem is that of jurisdiction and protection for a capital investment, a problem that exists for all resources located outside existing recognized territorial claims (see Chapter 22).

PHOSPHORITE NODULES

Phosphate rock or phosphorite is found along many continental margins. Although some mining claims have been established, little phosphorite has yet been recovered offshore, so this must be classed as a potential resource.

Phosphorite is derived from the remains of animals preserved in the rocks or sediments. Most phosphate rocks are found along coasts where there have been rapid

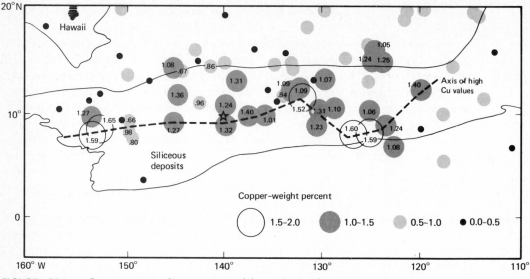

FIGURE 21.6   Copper content of manganese nodules in the Pacific Ocean southeast of Hawaii. (After Horn et al., 1973.)

changes in temperature or salinity which cause the death of large populations, and the formation of decomposing phosphate-rich matter on the sea floor. Phosphorus is widely used in fertilizers, chemicals, and detergents. It is abundant on land but unevenly distributed, so many countries must import large quantities.

CONSTRUCTION MATERIALS

Perhaps the most important solid mineral commodities currently mined from the sea bed are construction materials—sand, gravel, and shell. Over 50 million cubic yards of sand and gravel are taken off the continental shelves each year and more than 90 percent of the sand and gravel used in the Netherlands comes from the North Sea. Old shell reefs of the Gulf Coast are a major source of lime for the area and offshore shell beds are the only source of lime in Iceland, a volcanic island. Construction materials are high-bulk, low-priced commodities and the costs of shipping them long distances are prohibitive. Many of the major population centers have used up their nearby supplies on land, and real estate values in metropolitan areas often prohibit opening of new sand and gravel pits. Thus it is not surprising that communities near the coast have turned to the offshore areas, where mining and transportation costs are relatively low (Figure 21.7 pp. 388-9).

PLACER AND BEACH DEPOSITS

Heavy minerals of many types can be constructed by wave stream or current action into deposits of economic significance. This process is of some importance because of changes of sea level during the ice ages of Pleistocene time. When the ice advanced, sea level was lowered and shorelines moved seaward. As the ice melted, sea level rose and ancient beaches and river channels now were drowned beneath the waters of the continental shelves. Many minerals of value including gold, diamonds, iron, and titanium have been mined, but the only one significant economically at present is tin. For many years tin has been mined from placer deposits in channels in the Far East. These channels have been traced offshore and important amounts of the tin in Thailand and Indonesia are produced by dredging in buried offshore channels.

RESOURCES BENEATH
THE SEA FLOOR

A number of mines with onshore access extend beneath the sea. Examples are the iron mines of Belle Isle, Newfoundland, and the coal mines of Cape Breton, Canada, and of northwestern Europe. This is literally undersea mining, but it represents an incidental extension of land-based activities rather than marine mining as such.

It is difficult to recover solid materials from deep beneath the sea floor, hence it is not surprising that the principal resources extracted to date are in liquid or gaseous form.

SULFUR

One commodity extracted from beneath the sea floor is elemental sulfur. This occurs as a solid, but its low melting point makes it possible to bring it to the surface and to store it in a liquid form.

This sulfur is associated with salt domes (Figure 21.8). Salt beds have been formed in many parts of the world by evaporation of seawater and have been subsequently buried beneath thick piles of sediments. Salt is a plastic substance and flows very easily. If there are pressure irregularities in the overlying beds, the salt will move and bulge upward to form ridges or domes. If it moves to shallow depths and encounters circulating ground water, sodium chloride dissolves out and the calcium carbonate and calcium sulfate are retained to form a caprock. Anaerobic bacterial action, in the presence of hydrocarbons, reduces the sulfate to produce hydrogen sulfide. As this gas percolates upward and encounters oxidizing conditions, it forms deposits of elemental sulfur.

The sulfur is recovered by the Frasch process. Superheated water is pumped through a borehole into the caprock. This melts the sulfur, and aided by an airlift it is pumped to the surface and ashore. Here it is discharged into large bins and allowed to cool and solidify (Figure 21.9). It can then be broken up and shipped to users without further refining.

HYDROCARBONS

By far the most valuable resources recovered from beneath the sea are hydrocarbons, oil and gas. At present there are hundreds of offshore wells, and in 1975

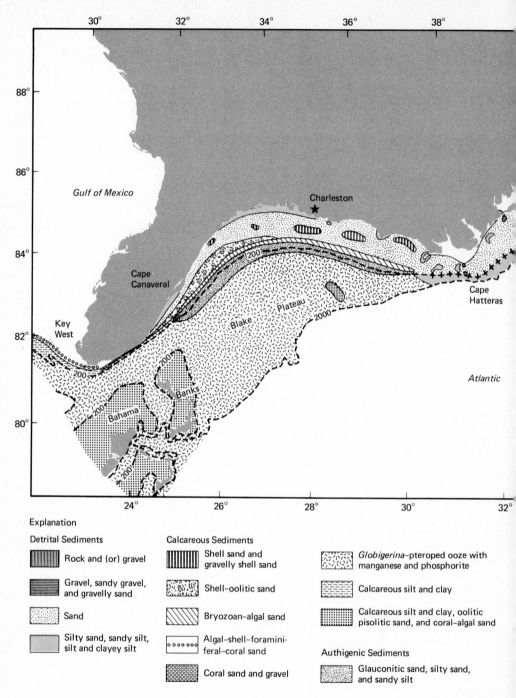

Explanation

Detrital Sediments

▦ Rock and (or) gravel

▤ Gravel, sandy gravel, and gravelly sand

▒ Sand

▦ Silty sand, sandy silt, silt and clayey silt

Calcareous Sediments

▥ Shell sand and gravelly shell sand

▒ Shell-oolitic sand

▨ Bryozoan-algal sand

○ Algal-shell-foraminiferal-coral sand

▩ Coral sand and gravel

Globigerina-pteroped ooze with manganese and phosphorite

Calcareous silt and clay

Calcareous silt and clay, oolitic pisolitic sand, and coral–algal sand

Authigenic Sediments

Glauconitic sand, silty sand, and sandy silt

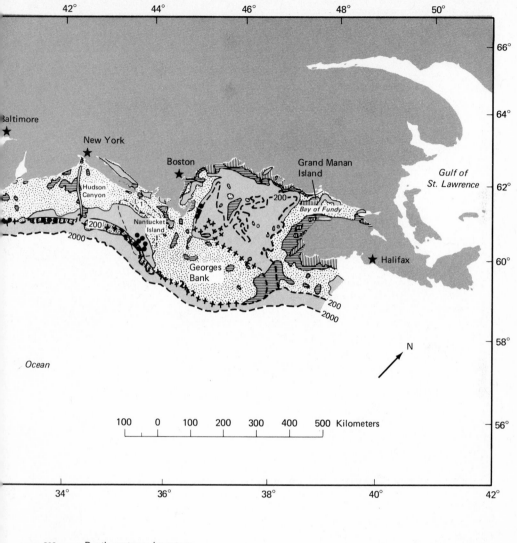

Baltimore

New York

Boston

Grand Manan
Island

Gulf of
St. Lawrence

Hudson
Canyon

Nantucket
Island

Bay of Fundy

Halifax

Georges
Bank

N

Ocean

100   0   100   200   300   400   500 Kilometers

— — 200 — — Depth contours, in meters

+ + + + +   Glauconitic sediments probably present

● ● ●   Pyrite–filled foraminiferal tests

1, boundary of zone of rounded
quartz grains; 2 limonite pellets

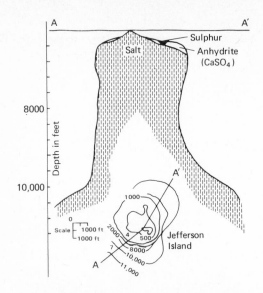

FIGURE 21.8   Cross-section of a salt dome in the Gulf Coast of Louisiana. The salt, originally in a horizontal bed, is quite mobile and moves upward as a result of sediment loading and differentials in pressure. Solution of the top of the dome removes salt and leaves behind calcium carbonate and calcium sulfate as a caprock. Native sulfur is often produced in the caprock by bacterial action.

FIGURE 21.9   Native sulfur produced from the caprock of a salt dome by the Frasch process is piped ashore and allowed to cool and harden in large bins. It is then chipped from these bins and loaded for shipment. (Drake photo.)

about 20 percent of world production came from offshore sources. Exploration holes for petroleum have been drilled in water depths exceeding 650 m (2100 feet) and production has been established in water depths of more than 125 m (about 400 feet).

A number of factors must be satisfied in order that petroleum can accumulate in economical quantities.

1. There must be a rich source of organic material of the right kind and this must be preserved until it is buried by sediments.
2. The organic matter must be subjected to the temperature and pressure conditions necessary to convert it to liquid or gaseous hydrocarbons (Figure 21.10). This is accomplished by burial beneath 1000 m or more of sediments and the increase of temperature that comes with depth.
3. The liquid or gaseous hydrocarbons must move from the fine-grained, relatively impermeable source beds to coarser grained, more permeable reservoir rocks from which they can be extracted at reasonable rates.

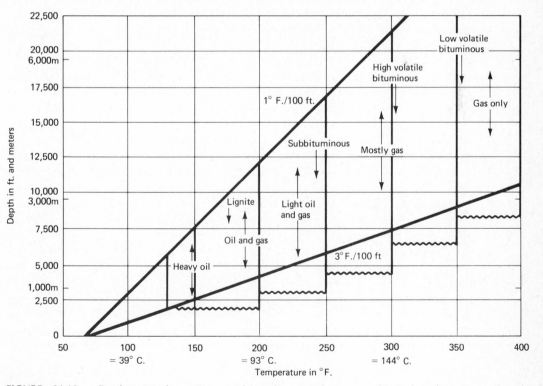

FIGURE 21.10   For the original organic material in sediments to be converted to hydrocarbons, it must be subjected to elevated temperatures and pressures. The same is true for conversion of organic debris to high-grade coal. The nature of the coal or the hydrocarbons will depend not only upon temperature and pressure conditions, but upon time as well. (After Klemme, 1972.)

4. Some kind of trap must be present to prevent the hydrocarbons from leaking out. These traps may be structural or stratigraphic (Figure 21.11), and the reservoir rock must have an impermeable cover to prevent upward migration of the fluids or gases.

5. Proper timing in the development of these various elements is essential. If, for example, the hydrocarbons move into the reservoir rock before a trap has developed, it will leak out and be lost. Furthermore, enormous quantities of hydrocarbons have been leaked into the environment because of breaching of structures by erosional, tectonic, or other natural processes, as evidenced by the offshore oil seeps in California, tar lakes and rivers in Trinidad, or the Athabaska tar sands in Canada.

The initial search for hydrocarbons naturally requires identification of a basin with a sufficient thickness of sediments that hydrocarbons can be generated. Such basins may be onshore or offshore (Figure 21.12) and even in deep water. After analysis of existing geological information, a search can be made for suitable traps using geophysical techniques, primarily the reflection seismography. Again, such traps are not limited to shallow water, because reflection records have revealed the presence of such structures as salt domes in the deepest part of the Gulf of Mexico (Figure 21.13), in the Mediterranean Sea, and elsewhere.

Having identified a trap, the question remains whether hydrocarbons are present and in sufficient quantities to cover the costs of drilling, development, production, and transportation. A discovery that would be an excellent find if on land may not be worth developing if offshore, because of the enormous costs involved. The question can be resolved only by the drill.

If preliminary drilling indicates a sufficient resource and sufficient reserves, a platform will be erected to produce the hydrocarbons (Figure 21.14). The expense of such platforms can be very high, some in the North Sea costing in excess of a billion dollars, so many wells are drilled directionally from a single platform (Figure 21.15). By this method single reservoirs, or multiple reservoirs at different depths, can be produced from one platform. Ordinarily the hydrocarbons are transported to shore by pipeline, but in some instances, where pipeline costs are prohibitive, they are stored on the platform and removed by tanker.

FIGURE 21.11 Hydrocarbons, once formed, migrate from the source beds until they are trapped in a reservoir. Traps are features that prevent further migration and may be of several types: (a) on the flanks or in the caprock of salt domes; (b) bounded by faults; (c) in folded rocks where there is closure; (d) where sedimentary structures prevent further migration.

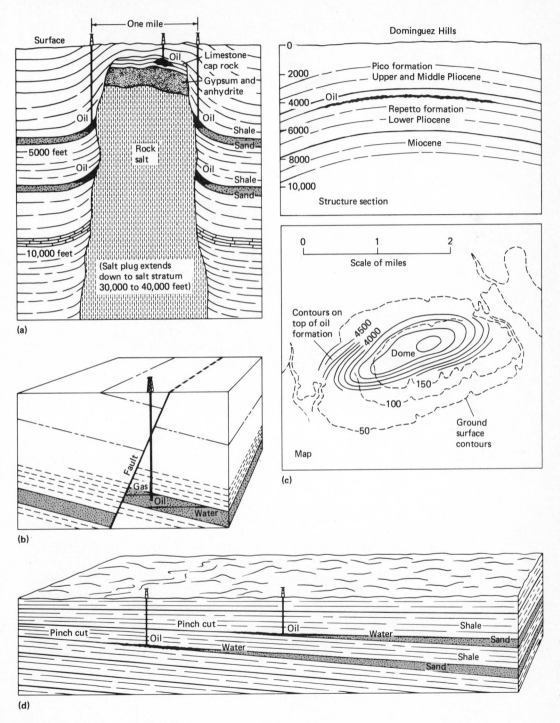

(a)

One mile

Surface

Oil

Limestone cap rock

Gypsum and anhydrite

Oil

Oil

Shale

Sand

5000 feet

Rock salt

Oil

Oil

Shale

Sand

10,000 feet

(Salt plug extends down to salt stratum 30,000 to 40,000 feet)

(b)

Fault

Gas

Oil

Water

Dominguez Hills

0

2000

Pico formation
Upper and Middle Pliocene

4000

Oil

6000

Repetto formation
Lower Pliocene

Miocene

8000

10,000

Structure section

(c)

0   1   2

Scale of miles

Contours on top of oil formation

4500

4000

Dome

150

100

50

Ground surface contours

Map

(d)

Pinch cut

Pinch cut

Oil

Oil

Water

Water

Shale

Sand

Shale

Sand

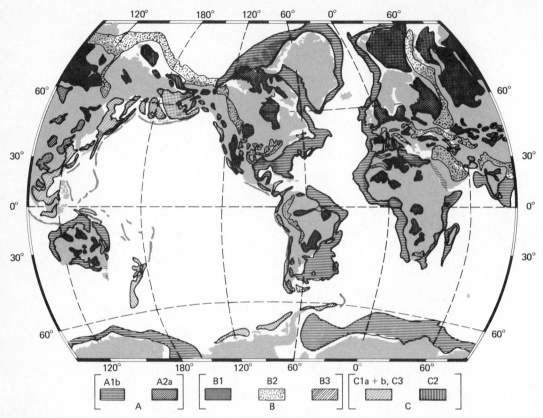

FIGURE 21.12 Sedimentary basins of the world classified by their location relative to the major structural features of Figure 17.17. The term "suture" refers to the plate boundaries. (A) basins on stable crustal rock, oceanic or continental; (B) perisutural basins, on the margins of megasutures; (C) episutural basins, perched on top of megasutures. (After Bally, Shell Oil Co., 1975.)

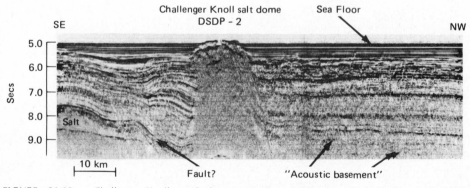

FIGURE 21.13 Challenger Knoll, a salt dome protruding slightly through the sediments in 3500 meters of water on the bottom of the Gulf of Mexico. Deep-sea drilling revealed the presence of hydrocarbons and native sulfur in the caprock of this salt dome. (From Ladd et al, 1976; courtesy of Marine Sciences Institute, University of Texas.)

FIGURE 21.14   Offshore platform off the Mississippi delta in the Gulf of Mexico. (Drake photo.)

FIGURE 21.15   The cost of offshore platforms is high, so it is customary to drill a number of directional holes from a single platform to produce a single or multiple reservoir. This platform produces from 22 individual gas wells. (Drake photo.)

SUMMARY

1. The term mineral resource refers to concentrations of naturally occurring mineral materials that can be extracted economically.

2. Estimations of the potential value of a mineral resource vary widely. In the case of hydrocarbons these variations reflect the degree of sophistication of the estimating techniques and assumptions about the nature of the sediments, conditions at depth, the geological history, and the percentage of hydrocarbons in place that can be produced.

3. The ocean contains on or beneath its floor a number of potential mineral resources, including hydrocarbons, sulfur, placer deposits, construction materials, and manganese or phosphorite deposits.

4. Some of the potential mineral resources are being utilized under present economic conditions, but others await a time when competitive sources on land are less abundant.

Recreation is one of the fastest growing uses of the shore line. (Courtesy Environmental Protection Administration.)

Man Meets the Sea

22

Although man has used the ocean since earliest time, a case can be made for the proposition that our concepts and uses of the ocean have changed more in the past 25 years than in all our previous history. In the beginning the oceans served as a source of fish and for limited coastal transportation. For some island civilizations such as those of Polynesia and Micronesia the sea was a dominant force in shaping their development. The great period of ocean exploration

CHAPTER

for the Western World began in the fifteenth century, and the rise of such nations as Spain, Portugal, the Netherlands, and England was determined in part by their ability to open trade routes and control the sea. Most major wars of recent times have been determined at sea, since he who controls the ocean controls the flow of goods. This chapter outlines some of the modern trends in man's use of the ocean and their consequences.

TRANSPORTATION

Nowhere is man's use of the sea causing a more dramatic change in social and economic patterns than in that most traditional of ocean uses, transportation and the movement of goods. Projection of ocean-borne world trade by the U.S. Department of Transportation indicates a doubling of cargo every twelve years, which means by the year 2000 the tonnage will be more than 20 times the 600 million tons carried in 1950. Air transportation plays a relatively modest role in the movement of trans-oceanic cargo. Even after excluding commodities such as oil, which are exclusively moved by ship, air transport accounted for less than one-half of 1 percent of all ocean tonnage in 1970. Air transport is about 75 times more expensive per pound than ocean transport.

Much of this increase in ocean trade is in the movement of crude oil, and because there is a major saving in costs for large bulk carriers, the number of these ships is increasing. The change has been dramatic. In 1949 the average tanker in the world fleet was 12,800 dwt (deadweight tons). ("Deadweight ton" refers to the carrying capacity of a ship; "12,800 dwt" means a ship can carry 12,800 tons of cargo.) It was 27,000 dwt in 1965 and 39,000 dwt in 1970. The largest tanker operating today is more than 300,000 dwt and the end is not in sight (Figure 22.1). The economies of scale suggest that tankers in the million deadweight-ton range may be built, and one forecast calls for the average size of the tanker fleet to be 250,000 dwt by 1985. Dry bulk carriers are also growing. Some 16 percent of these ships are now more than 80,000 dwt as compared to 2 percent in 1953. Freighter sizes are also growing, but much less rapidly. The largest freighter today is about 25,000 dwt with a draft of 35 feet. Figure 22.1 shows the trend in vessel size and capacity.

One result of this increase in trade and in size of carrier has been difficult problems in harbor development and maintenance. The location of most ports was established at least 100 years ago when vessels with drafts of more than 20 ft. (6 m) were unusual. Now all major U.S. ports are dredged to at least 30 ft. (9 m), and

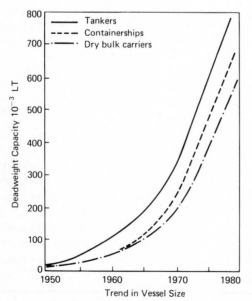

FIGURE 22.1 Historic and projected growth in size of the largest tankers bulk cargo and containerships (After Frankel and Marcus, 1972.)

many to 40 ft. (12 m). Dredging beyond 45 ft. (14 m) in New York harbor requires moving bedrock. In East Texas the dredged ship channel would have to be 28 miles (45 km) long if it were dredged to 50 ft. (15 m), 58 miles (93 km) if the channel were 70 ft. (21 m). Other areas present problems of disposition of dredge spoils or of man-made obstacles such as highway tunnels, built before anyone considered the needs for 60- to 70-ft. channels. Finally, there is the problem of finding room to maneuver these large tankers and bulk carriers. Thus some major cities and industrial areas that were built around port facilities and harbors that were adequate 100 years ago, or even 25 years ago, now find that geography is making their ports obsolete unless offshore loading facilities can be developed and material brought into port by pipeline or barge.

Even in areas where dredging is possible, there remains the problem of where to put the dredging spoils. Ship channels many feet deeper than ambient depth are in continual danger of filling. The material must be removed and a place found to dump it. Until recently much of it was dumped on land; about one-third of the half million acres of lost marshland in the United States is the result of marshes filled with dredge spoils. If the material cannot be dumped on land, it must be dumped at sea, well beyond the dredged channel and away from fishing grounds and other rich biological areas. In fact, many ecologists are unhappy to see any major offshore dumping. The texture and organic content of the spoil is probably different from the material in the area to be dumped, and the bottom dwelling organisms in the spoil are probably also different. Until anyone can predict the consequences, many ecologists would prefer to leave well enough alone unless the public interest is overriding. If Harbor A can take large ships with little or no dredging maintenance and Harbor B requires continual maintenance, they would argue that Harbor A should be developed at the expense of Harbor B even if it means industry dislocation. One consequence of the problems with maintaining ever deeper ship channels is the plan to build offshore ports. Large tankers, and perhaps bulk carriers, will anchor offshore and their cargo transferred to smaller shallower draft craft.

MILITARY POWER AND THE OCEANS

Naval battles have played decisive roles in history, from the defeat of the Persian fleet at Salamis in 480 B.C., to the British destruction of the Spanish Armada in 1588, to the battles of the Coral Sea and Midway in World War II, which helped pave the way for the blockade of Japan. As naval strategists are quick to note, it is generally not the naval battle itself which is the determining factor, but the control of the seas following the battle; this is as true today as it has been throughout the history of naval warfare.

The United States can play the role of a world power because it has sufficient naval power to make its presence felt. The military success of the 1958 decision to airlift an expeditionary force to Lebanon was ensured because of the follow-up of some 25 ships loaded with supplies. Whatever one's view of the U.S. engagement in Vietnam, the U.S. presence could not have been maintained without the nearly continuous stream of ships plowing across the ocean carrying oil, ammunition, and supplies.

Traditionally, the role of a naval power is the implicit threat of military invasion or blockade in time of peace and the insurance of the success of such operations in time of war. The mere presence of a naval force in an area (for example, the British fleet in the Indian Ocean before World War II and the U.S. Sixth Fleet in the Mediterranean in recent years) represents a critical element that must be factored into the geopolitical aspirations of all nations bordering these waters, as well as those which wish to extend their political influence to these areas. Any careful reading of political history suggests that gunboat diplomacy comes in many forms and in varying degrees of subtlety.

Once a state of warfare exists, the naval role is more easily seen. Important as the island hopping and bombings were in the Pacific in World War II in reducing Japanese ability to make war, it was steadily increasing pressure of the blockade and the destruction of Japanese shipping that brought the war machine to a halt. Some have even argued that the naval strangulation of Japanese shipping was so effective that the war would have ended before an invasion of the mainland, and thus the coup de grace of the atomic bombing of Hiroshima and Nagasaki was not necessary.

The development of the ballistic missile and nuclear submarine has added another dimension to the traditional military use of the seas. Beneath the surface of the ocean, the U.S. and the U.S.S.R. maintain fleets of nuclear submarines equipped with ballistic missiles carrying atomic warheads. These are the so-called "second strike" forces which play such an important role in U.S.-U.S.S.R. detente. The range of the missiles is several thousand miles and the combined destructiveness of their atomic armament is enormous. Their value in the present nuclear standoff is that it is impossible for either the U.S. or the U.S.S.R. to know at any given time precisely where the other's submarines are cruising. Unlike the atmosphere, the oceans are virtually opaque to radar and other forms of electromagnetic radiation (see Chapter 4). Beyond very short distances submarines below the surface can be tracked only by sound, and the present sonar devices (either echo-ranging sonars like radar, or highly sophisticated listening devices) are simply not good enough to keep track accurately, and at all times, of the other side's submarines. Thus, if either superpower should decide to launch a preemptive first strike, he might be able to destroy all the other side's land-based missiles, but he could not be assured of destroying the ballistic-missile submarines, because he would not know their location with sufficient accuracy. Thus he who launches the first strike of intercontinental ballistic missiles can be assured of having a second strike in return from the submarines beneath the surface.

SHORE-LINE DEVELOPMENT

Anyone who has had more than a passing acquaintance with land prices recently is aware that shore-line property must be very scarce, since shore-line property in the United States generally sells for appreciably more than similar land a short distance landward. However, the exact amount of shore line in the United States is not easily determined. For example, one study showed that New England has either 1395 or 6130 statute miles of shore line, depending upon whether a 3-mile or 100-foot chord was used in measuring between bays, headlands, and inlets. The same study showed that Hawaii had either 900 or 1052 miles, depending upon the use of a 3-mile or 100-foot chord. The fact that there was a large difference in New England and a small difference in Hawaii results from the highly indented glacial-formed coast of New England in contrast to the comparatively smooth volcanic-formed coast of Hawaii.

Any survey of the length of coast of the United States based on 100-foot base lines should carry a date. The number changes yearly because of cutting, dredging, filling, and building of breakwaters, jetties, and islands. Today the primary demand for shoreline on the United States is for private homes and recreation. There has been a major move of people from the interior of the country to the coasts. Megalopolitan areas from Boston to Washington and San Francisco to San Diego are developing. This, coupled with the growth of suburbia and the increased leisure of a large share of our population, has sharply increased the demand for waterfront property both for primary dwellings and for summer cottages. To meet the demand for private housing, new areas are continually opened up. In some regions the en-

FIGURE 22.2 Many estuary areas, such as the one shown here, have been cut and filled to provide housing developments. (Interior—U.S. Fish and Wildlife Service. Photo by Charles D. Evans.)

tire coastline has been transformed by dredging and filling (Figure 22.2). In San Francisco Bay 80 percent of the marsh area has been filled, mostly for housing. In principle, of course, the shore line can be increased to almost any desired length by cutting and filling, and as demands for shore line increase, pressure to increase shore frontage increases. San Diego has built two islands in the bay from the spoils of dredging the harbor.

Nearly any shore line is suitable for housing and some kinds of recreation, but there is a growing demand for beaches, both public and private, near the major metropolitan area. This demand comes at a time when our beaches are probably becoming smaller because of the entrapment of sand behind dams. In many areas, including Waikiki and Miami Beach, resort hotels have extended offshore jetties to slow the long-shore currents and to trap sand. More recently intensive efforts have been made in some areas to find sources of sand offshore and to barge or pump the sand back on the beach. Once the sand gets off the shelf and begins to pour down the submarine canyons onto the abyssal floor, it is lost. However, it seems likely that in the future, in those areas where the demand for sandy beaches is sufficiently strong and offshore sources are sufficiently close, man will be able, if he wishes, to supplement the traditional sources of sediments for beaches by feeding offshore deposits back into the system. As has occurred before, when man has intervened in a dynamic system which he understands but poorly, a few surprises are probably in store.

FISHERIES RESOURCES

Since the beginning of recorded history the oceans have been a source of food and until relatively recently the supply of fish has seemed inexhaustible. In recent years there has been an increasing number of regulations to protect the immature of the species or the spawning female. More recently compacts or regulations limit the number of fish that can be caught. Sometimes these quotas are set indirectly by limiting the fishing season or regulating the gear that can be used. In other cases quotas are set. When a given quantity of fish has been caught, the fishing season is over for another year.

The goal of present fisheries-management programs is to harvest the "maximum sustainable yield" (see Chapter 12). This means catching as many fish as possible, but not so many that the yield causes a decline in the available stock for future years. The concept of maximum sustainable yield is simple to understand in principle, but often difficult to apply in practice. Sufficient mature females must be left to supply the future stock. Immature fish should be protected and caught only when they are larger. It is probably better, however, to catch the fish as soon as possible after their maximum growth phase, because after reaching maturity they continue to add proportionately less weight for the amount of food consumed. Ecological relationships are complex and ecological balances are often delicate. On land there are abundant examples of unforeseen consequences of man's intervention; for example, by killing mountain lions to protect the deer, the deer population expands and eats all the grass, thus causing increased erosion and leaving land capable of supporting fewer deer than before the mountain lions were killed.

There is little reason to think that the ecological relationships in the ocean are any less complex. However, our knowledge of ocean ecology is generally much less advanced than terrestrial ecology. Fisheries-management programs are fraught with uncertainty. On the basis of similar data it is possible for competent scientists to reach different conclusions concerning fisheries management. Because any decision setting or changing quotas, gear regulations, or catch size can have enormous consequences to certain elements of the fishing industry, fisheries management is no place for the timid.

Several studies have been concerned with the question of the yearly production of fish in the ocean. Most arrive at a figure of between 100 and 200 million tons per year (see Chapter 12). For a period of nearly 25 years after World War II the world fish harvest increased at about 6 percent a year. Much of that recent growth was a result of opening up new areas. The great "mature" fisheries off Georges Bank and in the North Sea did not grow at a comparable rate. In these areas and in others, there is considerable evidence that certain species are already being fished to the limit, if not overfished (Figure 22.3).

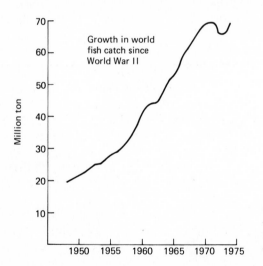

FIGURE 22.3   The rapid growth in worldwide fisheries of the post World War II era has leveled off as the total catch begins to approach the maximum sustainable yield of the oceans.

Economists argue that the present management practices for fish harvesting are inadequate to deal effectively with fisheries that are already being fished at or beyond maximum sustainable yield. As fishing competition continues, the demand for better management increases. An increasing number of countries, including the United States, have followed the lead of Peru in claiming complete jurisdiction of its offshore fisheries out to a distance of 200 nautical miles, and the 200-mile Exclusive Economic Zone has been agreed to by the participants at the Third Law of the Sea Conference. Eliminating foreign fishing fleets reduces fishing pressures at least temporarily. It has been suggested that exclusive jurisdiction also facilitates rational management of the fisheries resources, because international treaty agreements are both cumbersome and difficult to negotiate and because the coastal nation is more likely to have a sustaining interest in local fisheries management than nations with distant water fishing fleets that can move from one fishing area to another.

Although local management may work for most fish stocks (or regional agreements where the stocks lie along the boundaries of two or more nations), certain fisheries are truly international. Whales and tuna are the most conspicuous examples. These animals may spend most of their time fairly close to shore, where the food is most plentiful, but they range the world oceans. Arrangements for rational management of these species must be by international agreement.

MANAGEMENT OF MINERAL RESOURCES

The discovery of economically important mineral resources has done more to change man's ideas about the management of the ocean than any other single factor in the preceding 300 years. To date the most important of these resources has been petroleum. The first offshore oil wells were drilled by working from the shore and angling the drill head seaward. Beginning in 1948 wells were drilled in the Gulf of Mexico, and since that time wells have been sunk off the shores of many nations in all parts of the world, some 16,000 in the Gulf of Mexico alone. One recent report suggests that appreciable amounts of oil can be expected to be found off the coast of at least 70 countries. Originally all wells were drilled from fixed platforms, but as the depth of water increased it became expedient to attempt to drill from floating platforms. The *Glomar Challenger* (Figure 15.2) is an example of such a floating drill ship. Although such floating platforms are adequate for exploration, it has been necessary to date to develop all production wells from surface platforms fixed to the bottom. As oil production moves into deeper water, many expect that production wells will be completed on the sea bottom. The maximum depth of water to extract oil economically is dependent upon three major factors: technology of extraction, amount of oil in the deeper water, and development of alternate forms of energy which will compete with offshore oil. The cost of exploration and extraction increases very rapidly as the depth of water increases, costing perhaps three to five times more in 1000 m of water than in 200 m. However, a number of observers believe that once it becomes possible to complete a well on the sea bottom rather than from the surface, the cost of extraction will be nearly independent of depth; that is, it will not cost appreciably more to develop a well in 2000 m of water than in 1000 m of water. At present one can only speculate on the amount of oil to be found in deeper water, but again a number of observers are confident that we will find large petroleum deposits in the sediments draped over the continental slopes. One estimate suggests that 20–35 percent of all offshore oil will be found in water depths greater than 200 m; however, few believe that there is much oil on the deep-sea floor beyond the continental margins.

The extraction of offshore oil has grown so rapidly that any figures quoted are likely to be out of date by the time they are published, although the oil industry estimates that one-third of our oil will eventually come from offshore wells. Such a growth has not been without problems and more are in the offing. Blowouts and other accidents are apparently more difficult to bring under control offshore than on land and such accidents are a matter of increasing concern. In the Gulf of Mexico the oil wells were becoming so numerous that the Coast Guard had to step in to insure that "fairways" were left so that large ships could maneuver in and out of port. Although the fishing industry and the oil industry have made an uneasy truce, it should also be noted that to date oil development and heavy fishing have seldom coincided. If there is any place in the world where the two are not compatible and where confrontation will occur, it is on Georges Bank off New England and Canada, which is one of the most heavily fished areas in the world, as well as a potential source of large petroleum deposits.

As detailed in Chapter 21, the oceans have been a source of other mineral resources ranging from sand and gravel for building construction to manganese extracted from seawater to tin found in placer deposits off Thailand. The oceans are also a source of possible new mineral resources. There is growing interest in the ferromanganese nodules of the deep-sea floor, not so much in the manganese, but in the relatively small percentage of copper, cobalt, and nickel they contain. Although ferromanganese nodules are found at all depths, there is evidence to suggest that the percentage of copper, nickel, and cobalt increases with depth. Most if not all major deposits are found many miles from shore and in deep water. All evidence suggests that ferromanganese nodules vary greatly in concentration per unit of area and in relative abundances of different metals. Appreciable differences in metallic content have been found over relatively short distances. Thus, although ferromanganese nodules are seemingly ubiquitous over large areas of the deep-sea floor, some areas are of much greater potential economic interest than others. The company that finds a likely area for mining wants to be able to lay claim to that area for purposes of mining its mineral resources. At the moment, a number of companies and groups of companies are investing heavily in the necessary technology for both harvesting these minerals and processing them. The economic justification for these large investments will be determined in the future. Depending at least in part on the outcome of the Law of the Sea Conference, a number of international companies expect to be mining ferromanganese nodules by about 1980, probably in the Pacific.

POLLUTION

The subject of oceanic pollution probably generates more emotional debate among oceanographers today than any other single subject. As with many subjects, emotion is often a substitute for fact, since reliable data are difficult to come by. We do know that it is nearly impossible to tow a net for ten minutes along the surface in any part of the North Atlantic without finding tar or plastics, and usually both. If pollution is defined as finding measurable amounts of the by-products of our civilization, then the ocean is indeed polluted. If, however, one uses the criterion that it is not pollution until one can demonstrate a significant effect on the local ecological balance, then the case for oceanic pollution is less clear. Many argue, and with good cause, that such a stringent definition should not apply to the ocean. In the first place our ecological knowledge of the oceans is so limited that the effect could be missed. More importantly, we should not indulge ourselves with the luxury of waiting until a pollution problem exists before attempting to do something about it. One cannot talk realistically about cleaning up the oceans as one does about cleaning up a stream or an estuary.

A discussion of pollution can be logically divided into three sets of questions. (1) What kind of, and how much, material reaches the ocean and how? Does it reach the ocean by river and estuary, by sewage outfalls, by dumping of material at sea from ships, or is it transported seaward by the wind? (2) What becomes of the pollutants once they reach the ocean? How fast are they diluted to tolerable levels? What are the ways they are concentrated in the food web? In the case of organic wastes such as petroleum, DDT, and similar material, how fast are they broken down? (3) Finally, what effect does a given level of pollution have on oceanic processes? Does it inhibit growth or spawning of marine organisms? Is the pollutant so concentrated in the organisms that it is a potential health hazard to those who eat it?

Although such an outline may provide a reasonable framework for discussion, gathering authoritative information in most of these areas is difficult. We have most information on the first subject, the amount of material reaching the ocean and the route it traveled to get there, but even here our knowledge is fragmentary. A recent study of hydrocarbons in the marine environment concluded that the amount of petroleum reaching the oceans was about 6 million tons a year. For comparison the study estimated the amount that reaches the ocean from natural seepage in the ocean floor is about 10 percent of that figure. In other words, man is adding ten times more petroleum to the ocean each year than comes from natural sources. The amount that man adds comes from a number of well-known sources such as cleaning of the tanks of the tanker ships, collisions or accidents at sea, and blowouts on offshore wells. However, fully 50 percent of the oil reaches the sea *after* it has been "consumed" by man. It returns via sewer outfalls and storm drains, and it enters the ocean via the atmosphere as the smog created by automobile-exhaust emissions drifts seaward. One must always remember that we, the "consuming" public, actually consume very little in the sense that we completely alter the chemical structure of the material. Much of what we use, including petroleum, is returned to the environment in only slightly altered forms, and much of what is returned to the environment ends up in the ocean, the ultimate sink (Table 22.1).

Knowing what material is reaching the ocean is only the first step; the next question is, what becomes of it and how is it incorporated into the food web? We

TABLE 22.1 Sources of Petroleum Hydrocarbons in the Ocean

Input	Million Metric Tons per Year
Transportation	2.13
Tankers, dry docking, terminal operation, bilge, accident	
Coastal refineries, municipal industrial wastes	0.8
Offshore oil production	0.08
River and urban runoff	1.9
Atmospheric fallout	0.6
Natural seeps	0.6
Total	6.11

(From Petroleum in the Marine Environment, National Academy of Sciences, 1975.)

can often measure higher-than-normal concentrations of material in the ocean. For example, chlorinated hydrocarbons such as DDT and PCB have been measured in the surface waters of the Atlantic. The method and extent to which this material is taken up by organisms is, however, much less clear.

The third and final step in the pollution processes is even more difficult. There are any number of examples of changes in community structure in streams and estuaries which can be traced to pollution, but the number of documented cases in the ocean is generally limited to localized regions of massive oil spills or similar catastrophic events. In view of the large natural fluctuations found in biological communities, it is not easy to determine from the techniques of descriptive ecology alone what, if any, part of the observed changes is caused by pollution. Similarly, it has been difficult to document any examples of health problems arising from oceanic pollution, although there are any number of cases of such occurring in enclosed bays and estuaries, such as acute mercury poisoning in Minamata, Japan, or cases of hepatitis occurring from eating of raw shellfish from polluted areas of Raritan Bay, New Jersey.

ARE SWORDFISH SAFE TO EAT (OR WHERE DID THE MERCURY COME FROM?)

As one example of the complex nature of documenting oceanic pollution, consider the well-publicized case of mercury in swordfish which led to a ban on the interstate transport for sale of swordfish by the Food and Drug Administration. In 1970 a Canadian graduate student found that some samples of freshwater fish contained mercury levels of over one part per million. Other chemists found similar values in other areas: particularly high levels were found in swordfish. The Federal Food and Drug Administration put a ban on the sale of swordfish in the United States, and since effects of mercury poisoning are well known and documented, few people at the time seriously questioned the FDA prohibition, at least until more evidence could be gathered.

Apparently the mercury observation in swordfish took the scientific world by surprise. No one was sufficiently confident about the sources of possible mercury pollution and its pathways into the ocean to be prepared to attempt an explanation of the swordfish observations. This inability of scientists to provide a reasonable explanation of the mercury level in swordfish was perhaps at least a secondary reason for the FDA to move as quickly as it did.

The source of the mercury in the swordfish is the seawater. Swordfish are predators and live on small fish, which in turn live on smaller fish, zooplankton and phytoplankton, which in turn take up the mercury from the seawater. Somehow in this process, the swordfish concentrate the mercury, so the levels found are 10 to 100 thousand times that in seawater. How, where, and under what circumstances this concentrating mechanism occurs is not known. The level of mercury in seawater is known, and although it is a measurement that has not been made routinely until recently, the data suggest that the amount of mercury in seawater has not significantly increased in recent years. Somewhat more reassuring evidence has come from measuring the mercury content in old swordfish and tuna. Soon after the original observations, museums and laboratories were canvassed; a few measurements have been made on preserved specimens of tuna and swordfish that were caught from 25 to 90 years ago. There appears to be no appreciable difference in the mercury concentration in these samples and in modern specimens, although, because of the method of preservation, the interpretation of these results is not unambiguous.

The evidence that many find most persuasive, however, is indirect evidence: Mercury is fairly volatile; that is, it evaporates into the air. Scientists are reasonably convinced that nearly all mercury, both from natural sources such as volcanoes, as well as from man-produced sources, reaches the ocean from the atmosphere instead of flowing into the ocean from rivers. One way of determining whether the amount raining into the ocean has increased significantly in recent years is to measure the mercury content of glaciers. The topmost layer of a glacier is rain and snow that has fallen in the past few years. As you core deeper into the glacier you reach layers that were laid down in successively earlier years. Scientists have cored deep into Greenland glaciers and have samples that date back 2700 years. The data are not unequivocal, because Greenland is relatively close to the active volcanoes of Iceland, but it would appear that there has been little significant increase in the mercury fallout over this period.

These data would appear to confirm the most critical part of the scientists' arguments, which is that man's use of mercury is less by a factor of 10 than the amount of mercury that escapes each year into the environment from natural causes. One of the largest sources of man-contributed mercury is the burning of coal. Although the mercury content of coal is small, it is estimated that more than 1000 tons of mercury a year are released to the atmosphere by the burning of coal. Large as that number is, however, it is still small compared with how much mercury escapes to the atmosphere by natural causes. Ten times as much mercury is released to the atmosphere from volcanoes, hot springs, and similar phenomena as is produced by man. Thus, the consensus in much of the scientific community today is that if swordfish are unsafe to eat because of mercury content, this is not a new problem. Man's contribution of mercury to the ocean is small compared with nature's contribution. If swordfish are unsafe to eat now, they were unsafe to eat 300 or more years ago.

OCEANIC POLLUTION—IS IT REALLY A PROBLEM?

Although few observers consider ocean pollution as a major problem today, most are concerned that it may be a critical problem for tomorrow. Today a potpourri of evidence suggests that ocean pollution has some effect by whatever criteria one wishes to use. Fish have been observed with measurable amounts of DDT and PCB in their flesh. There have been fish kills in coastal waters in which pollution has been implicated. There have been marked changes in the dissolved oxygen and nutrient levels in parts of the Baltic which can be expected to have a measurable effect on the distribution of plants and animals. The stomach contents of ocean fish have been found to contain many of man's discarded artifacts. Measurable changes in the ecology of benthic communities have been observed two years after an oil spill. Probably none of these separately or in toto constitutes a serious pollution problem at present.

Whether or not you believe that ocean pollution will become a problem in the future depends in part upon your view of man's future. One school of thought claims that we cannot continue to expend our natural resources at the present rate or we will run out of these resources. It is a question not simply of expending oil and gas reserves, but of running out of important metals like copper, tin, and zinc, and even iron and aluminum. This school suggests that man will have to change his way of living if he is to survive; at least there will have to be changes in the highly developed nations such as the United States, which presently consumes about one-third the nonrenewable resources each year, although it accounts for only 5 percent of the population. If this group is correct, then we can reasonably hope that as man returns to a less resource-consuming mode of life, the chances of major ocean pollution will diminish.

Another view of the future, however, argues that by present standards of consumption, man has an almost unlimited supply of energy. Before we run out of coal, oil, and natural gas, we will develop the breeder nuclear reactor, control nuclear fusion, or find an efficient way of converting and transporting the potential thermal energy, either directly from the sun or that stored in the oceans. With an unlimited energy supply we can solve the problems of recycling limited natural materials and we can develop synthetic materials. We can continue to strive for a comfortable standard of living for all the world's population. If this is your view of the future, then it is also necessary to begin to worry about ocean pollution.

The present population of the world is about 4 billion and is growing at 2 percent a year. One projection has the world population stabilizing at 15 billion in about a century. There are many ways to estimate the effects of a nation's standard of living on pollution. None are ideal, but all conclude that pollution potential increases as the standard of living increases. Thus, the developed world with a higher standard of living contributes proportionately more to worldwide pollution. One can compare uses of fertilizers, consumption of raw materials, or solid waste disposal and arrive at different versions of the same conclusion (Figure 22.4).

A simple projection can be made by comparing energy consumption and assuming that it is a measure of pollution potential. For example, if you allow the United States a continuing growth rate of 3 percent a year per capita in energy consumption, and if you should project the rest of the world catching up to the U.S. annual energy consumption rate by the year 2000, you have a total annual energy consumption which is 15 times the projected 1980 value. The political and social problems of the world are such that the rest of the world will not catch up to the United States by the year 2000, but it is obvious that this is the goal of most of the world; and if you believe that energy is unlimited, then you might assume that it is only a matter of time. If it is not the year 2000, it will be 2084.

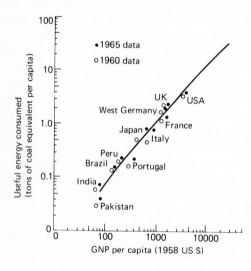

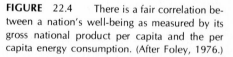

FIGURE 22.4   There is a fair correlation between a nation's well-being as measured by its gross national product per capita and the per capita energy consumption. (After Foley, 1976.)

If you will allow your imagination to project a future where the entire population of the world is consuming energy at a rate equal to or greater than that of the United States today, then perhaps you can begin to imagine the ways in which pollution may occur. Fertilizers and pesticides are two obvious examples, but consider for a moment some not so obvious examples; it has been estimated that 350,000 tons of dry-cleaning fluid and 1 million tons of gasoline evaporate each year. What is the ultimate fate of all the growth hormones fed to beef cattle? or the antibiotics made by the pharmaceutical companies? New chemical substances are being produced commercially at the rate of 400–500 a year. A population of 15 billion 100 years from now, with a standard of living equal to or better than that we now enjoy in the United States, will necessarily depend increasingly on synthetic material, intensive agricultural techniques, and complex social structures. The danger of inadvertent pollution will be much greater than it is today, and will require constant vigilance to see that we do not do something that inadvertently upsets the system.

INTERNATIONAL LAW—WHO OWNS THE OCEANS?

Formalization of the concept that the oceans and its resources belong to no one and to everyone equally is usually credited to Hugo Grotius, a seventeenth-century Dutchman. In the past, "freedom of the seas" has sometimes been interpreted as freedom of the seas for those with sufficient naval power to protect that freedom, as the United States found out in the early nineteenth century when British ships stopped U.S. merchant ships on the high seas, a practice contributing to the War of 1812. The concept that all men and nations have a certain right to use the ocean is, however, universally accepted, and these ideas were formulated in a 1958 Convention on the High Seas, which listed a number of universal rights including freedom to transport goods, lay submarine cables, and fish. The concept of universal rights was reconfirmed in one sense with the passage of a 1970 United Nations resolution which he declared "the seabed and ocean floor . . . as well as the resources . . . are the common heritage of mankind."

The Grotius concept of *res communis* was acceptable as long as the primary uses of the oceans were for communication, transporting of goods, catching a few fish from a seemingly infinite source, and scientific investigation. It was comparatively easy to agree on the few "rules of the road" for passing ships, rights of salvage, and similar maritime requirements. One of the first suggestions that Grotius would require updating came early in the twentieth century when it became apparent that the number of fur seals in the ocean was not so large; and that they could all be killed in a few years. The 1911 Fur Seal Convention between the United States, Russia, Japan, and the United Kingdom (the latter acting for Canada) was the first multinational agreement aimed at conserving an ocean resource. Since then the United States has participated in more than a dozen multinational and bilateral treaties related to ocean resource conservation.

The first major thrust for seaward extension of national jurisdiction came in 1945 from the United States, when President Truman laid claim to the nonliving resources off the U.S. continental shelf and suggested that other nations do likewise. Truman's continental shelf declaration was aimed at establishing ownership rights for the oil that many believed was to be found offshore. Three years later the United States drilled its first offshore well in the Gulf of Mexico and today some 15 percent of the world's supply of petroleum comes from the continental shelf, a figure which is expected to rise to at least 33 percent by the end of the century.

Other claims of extended jurisdiction soon followed as other nations saw different forms of wealth off their coasts. One method was to extend the breadth of the territorial sea, which traditionally has been a narrow band of water between the shoreline and the high seas. Within the territorial sea the coastal natio̶ has nearly complete jurisdiction. In the past, nations have claimed different wi̶ territorial sea, but until relatively recently nearly all claims were be̶ miles, although a nation could, and often did, effectively extend its ̶ diction by drawing its "shoreline" perimeter around offshore island̶ headlands of large open bays.

Peru looks seaward to a deep trench just a few miles offshore and has little expectation of finding large oil deposits on its narrow continental shelf, but its waters support some of the richest fisheries in the world. Perhaps inspired by the Truman declaration of 1945, Peru laid claim in 1947 to all fish off its coast to a distance of 200 miles and it did so by claiming a 200-mile territorial sea. Other nations followed, and at the start of the 1973 Conference on the Law of the Sea some eight nations claimed a territorial sea of 200 miles; such claims were aimed primarily at controlling offshore fisheries. A few nations claimed only specialized jurisdiction; for example, Iceland claimed juristiction over fisheries to 50 miles and Canada claimed 200 miles in reference to pollution regulations for those who use its polar waters.

There are few continental shelves that extend seaward of 200 miles and nearly all the fish are caught within 200 miles of shore, although some, such as tuna and whales, roam widely over the oceans in search of food. Thus, if all nations claimed jurisdiction over the resources to a distance of 200 miles, nearly all the ocean's fish and oil resources could be accounted for. However, out in the deep ocean depths, far from land, is the next of the ocean's major resources waiting to be exploited. These are the ferromanganese nodules of the deep seabed, coveted less for their manganese than for the relatively small amounts of copper, cobalt, and nickel they contain. The only way these resources could be divided between nations on the basis of extending national jurisdiction seaward is to divide the entire oceans between those nations bordering it; thus most of the North Pacific would be divided between the United States and Canada, and Mexico on the East and the Soviet Union, Japan, and China on the West. Alternately one could treat the nodules on the high seas like fish; that is, he who can harvest them can keep them. A third possibility is that the nations of the world can attempt to establish some international authority to regulate the taking of this resource and derive revenue through royalties to use for international purposes. An attempt to establish an acceptable International Seabed Authority was one of the prime goals of the 1973 Law of the Sea Conference.

Ferromanganese nodules, offshore oil, and increased fishing pressure have been the primary forces behind a continuing effort in the past 20 years to construct a "law of the sea" adequate for today. The first attempt in 1958 was primarily aimed at codifying present practices. It produced four conventions. The one on the high seas established the rights of all nations to the use of this area. A second on the territorial sea and contiguous zone spelled out the limited international rights of innocent passage through territorial seas, but significantly failed to agree on its breadth. A third on the continental shelf essentially formalized the Truman declaration by giving the resources on and under the continental shelf to the contiguous nation, but was diplomatically vague in defining the outer edge of the shelf. A fourth on fisheries was sufficiently ambiguous as to satisfy none of the major fishing nations except the United States. A second conference in 1960 failed in its attempt to resolve the breadth of the territorial sea and to produce a more acceptable convention on fisheries. The ferromanganese nodule resource was not discussed at either the 1958 or the 1960 Law of the Sea Conference.

Although the developing technology for harvesting ferromanganese nodules was the impetus for the third Law of the Sea Conference, it became apparent early in the preparation that many nations had a much more ambitious goal, namely, a complete review of all present practices and regulations. Their view prevailed and the third Law of the Sea Conference dealt with all the problems of the 1958 and 1960 conference, in addition to a number of issues that were not raised at either. This chapter is being written before the conference is completed, and there is no assurance that the conference will be any more successful than the 1960 conference in resolving the set of complex problems revolving around the antithetical concepts of the oceans as the "common heritage of mankind" and the resources of the ocean belonging to the nation off whose shores they lie. However, if agreement is forthcoming it will include (1) a territorial sea of 12 miles; (2) a provision for unimpeded passage through international straits that would otherwise be closed by a territorial sea of 12 miles; (3) a newly defined "exclusive economic zone" of 200-mile width within which the coastal nation has control of all resources; (4) some form of an International Seabed Authority with some jurisdiction over the exploitation of the mineral resources of the deep ocean floor beyond the limits of national jurisdiction; and (5) a newly defined "continental shelf" that will extend to the base of the continental slope (Figures 22.5 and 22.6). However the outer edge is defined in treaty language, it will in fact mean that the coastal nations will control the petroleum resources off their shores even if the oil is well beyond 200 miles offshore (Figure 22.7).

In many parts of the world such as the Caribbean, nations cannot extend their jurisdiction 200 miles offshore without overlapping. In these cases a "median line" or other mutually-agreed-upon means will be used to determine the boundaries between adjacent and opposite economic zones. The total area of the combined economic zones of the world is estimated to be about 36 percent of the total ocean, an area almost as large as the total land mass of the earth. Although the base of the continental slope in most parts of the world is well within the 200-mile economic zone, in a few areas of Australia, Argentina, Canada, and the United Kingdom among others, it extends well beyond that zone. Depending upon how the outer edge of the continental shelf is defined, it is expected that as much as an additional 6 percent of this new *legal* continental shelf will extend beyond the Exclusive Economic Zone.

The sixth session of the U. N. Law of the Sea Conference ended in the summer of 1977 (as this is being written). Another round of negotiations will begin in the spring of 1978. There appears to be a growing consensus on the details of the regimes that will apply to territorial seas, the continental shelves, the exclusive economic zones and the high seas. The major issue to be resolved is the regime of the deep seabed; who will control the mining of the ferromanganese nodules and how will this control be exercised? Whether the entire treaty will founder on this one issue is not clear, nor is it clear what action individual nations will take if treaty negotiations break down. Regardless of the status of the proposed treaty, however, it is equally clear that the time of Grotius is well past. As man's use of the ocean continues to increase, the legal and political regime to cope with these uses must increase in complexity and detail.

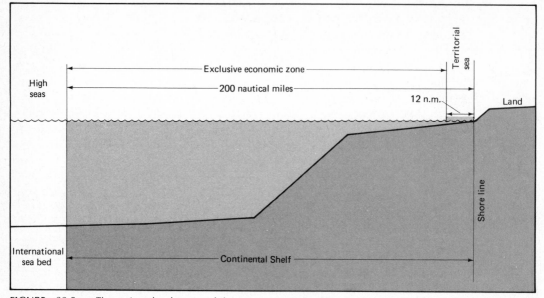

FIGURE 22.5   The various legal zones of the ocean as proposed by the Law of the Sea Conference. Note that the "legal" continental shelf does not conform with the "geological" continental shelf.

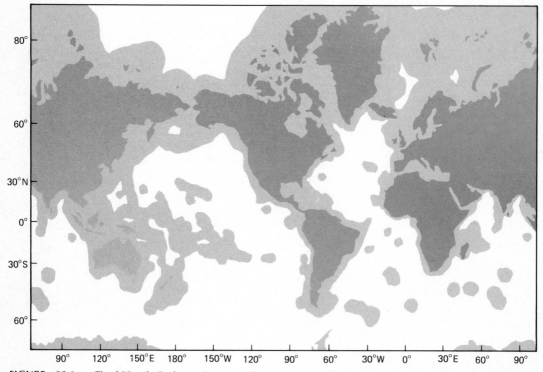

FIGURE 22.6   The 200-mile Exclusive Economic Zone as proposed by the Law of the Sea Conference. Note that every island has its own 200-mile zone, which accounts for the large area in the western Pacific Ocean.

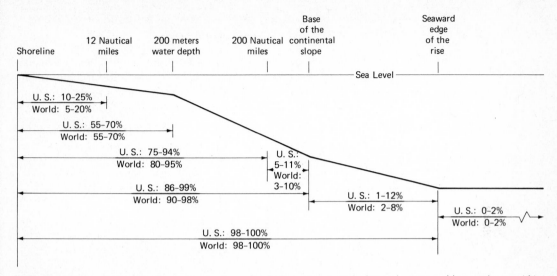

FIGURE 22.7   Estimated range of percentage distribution of potential ultimately recoverable petroleum within various boundaries offshore.

SUMMARY

1. Increased shipping and ship size require changes in our port design and the need to develop deep-water ports to handle deep-draft ships.

2. Effective control of the seas continues to be the mark of a major military power. The balance of terror in the U.S. and U.S.S.R. ballistic-missile submarines is a major factor in our present level of detente.

3. Land along the shore is among the most valuable in the United States. It is in demand for housing, recreation, and industry.

4. As our ability to catch fish increases, the need to manage fisheries based on sound ecological principles also increases.

5. The continental shelves and adjacent offshore areas contain vast amounts of oil. In the future as much as one-third of the world's oil may come from offshore wells.

6. Assuming jurisdictional problems are resolved, commercial mining of ferromanganese nodules from the deep ocean can be expected in the 1980s.

7. Plastics, oil, DDT, and other artifacts of our civilization can be found in the ocean. Except in rare circumstances it is difficult to prove that these materials have had significant impact on life in the open ocean. Whether or not ocean

pollution is a problem today, it may be a problem in the future if the world-wide use of energy and resources continues to grow.

8. As man's use of the oceans grows in scale and complexity, so does the need for resolving jurisdictional problems and developing regional and international management schemes. The Third Law of the Sea Conference, sponsored by the United Nations, was called to attempt to resolve these many issues.

Appendix I
Conversion Factors
and Constants

CONVERSION FACTORS

Fathom (fm) = 6 feet (ft) = 1.829 meters (m)
Meter (m) = 100 centimeters (cm) = 39.37 inches (in) = 3.281 ft
Kilometer (km) = 0.6214 miles (mi)
Micron (μm) = 10^{-6} m = 10^{-4} cm
Centimeters per second (cm/sec) = 0.0360 km/hr = 0.0224 mi/hr
Liter (l) = 1000 cm^3 = 10^{-3} m^3 = 1.057 quarts
Kilogram (kg) = 10^3 grams (g) = 2.205 pounds (lb)
Microgram (μg) = 10^{-6} g
Year = 31,560,000 seconds
2.303 log (base 10)x = ln (base e or natural base)x
Gram atoms (g at) = mole
Nautical mile = 6076 ft = 1.852 km = 1 minute of latitude
Knot = 1 naut. mi/hr = 1.15 stat. mi/hr = 1.85 km/hr = 51.4 cm/sec
Lux = lumens/m^2, klux = 10^3 lux

CONSTANTS

Avogadro's Number = 6.023 $\times$ 10^{23}/mole (= molecules per gram-molecular weight)
Universal Gravitational Constant = 6.670 $\times$ 10^{-8} dynes cm^2/g^2
π = 3.1416
e = 2.7182

TERRESTRIAL CONSTANTS

Mass of the Earth = 5.976 $\times$ 10^{27} g
Area of the Earth = 510.100 $\times$ 10^6 km^2
Equatorial radius of the Earth = 6378.163 km
Polar radius of the Earth = 6356.177 km
Area of the oceans = 362.033 $\times$ 10^6 km^2
Mean depth of the oceans = 3729 m

Volume of the Earth = 1.083 $\times$ 10^{12} km^3
Volume of the oceans = 1.350 $\times$ 10^9 km^3
Standard acceleration of free fall on Earth
   = 980.665 cm/sec^2
Angular velocity of Earth = 7.27 $\times$ 10^{-5}/sec

DIMENSIONAL PREFIXES

| k | kilo- | 10^3 | c | centi- | 10^{-2} | μ | micro- | 10^{-6} | p | pico- | 10^{-12} |
| d | deci- | 10^{-1} | m | milli- | 10^{-3} | n | nanno- | 10^{-9} | | | |

Appendix II
The Periodic Chart of the Elements

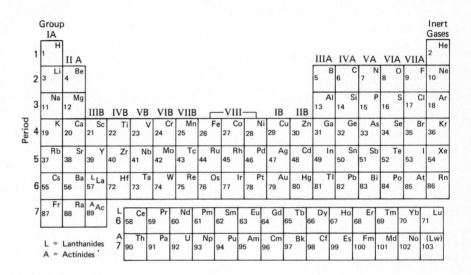

Element	Symbol		Element	Symbol
Hydrogen	H		Chlorine	Cl
Helium	He		Argon	Ar
Lithium	Li		Potassium	K
Beryllium	Be		Calcium	Ca
Boron	B		Scandium	Sc
Carbon	C		Titanium	Ti
Nitrogen	N		Vanadium	V
Oxygen	O		Chromium	Cr
Fluorine	F		Manganese	Mn
Neon	Ne		Iron	Fe
Sodium	Na		Cobalt	Co
Magnesium	Mg		Nickel	Ni
Aluminum	Al		Copper	Cu
Silicon	Si		Zinc	Zn
Phosphorus	P		Gallium	Ga
Sulfur	S		Germanium	Ge

420

Element	Symbol	Element	Symbol
Arsenic	As	Thulium	Tm
Selenium	Se	Ytterbium	Yb
Bromine	Br	Lutetium	Lu
Krypton	Kr	Hafnium	Hf
Rubidium	Rb	Tantalum	Ta
Strontium	Sr	Tungsten	W
Yttrium	Y	Rhenium	Re
Zirconium	Zr	Osmium	Os
Niobium	Nb	Iridium	Ir
Molybdenum	Mo	Platinum	Pt
Technetium	Tc	Gold	Au
Ruthenium	Ru	Mercury	Hg
Rhodium	Rh	Thallium	Tl
Palladium	Pd	Lead	Pb
Silver	Ag	Bismuth	Bi
Cadmium	Cd	Polonium	Po
Indium	In	Astatine	At
Tin	Sn	Radon	Rn
Antimony	Sb	Francium	Fr
Tellurium	Te	Radium	Ra
Iodine	I	Actinium	Ac
Xenon	Xe	Thorium	Th
Cesium	Cs	Protactinium	Pa
Barium	Ba	Uranium	U
Lanthanum	La	Neptunium	Np
Cerium	Ce	Plutonium	Pu
Praesodymium	Pr	Americium	Am
Neodymium	Nd	Curium	Cm
Promethium	Pm	Berkelium	Bk
Samarium	Sm	Californium	Cf
Europium	Eu	Einsteinium	Es
Gadolinium	Gd	Fermium	Fm
Terbium	Tb	Mendelevium	Md
Dysprosium	Dy	Nobelium	No
Holmium	Ho	Lawrencium	Lw
Erbium	Er		

Appendix III
The Geologic Time Chart

Era		Period	Millions of years before present
Phanerozoic	Cenozoic	(Quaternary)	0
			2
		(Tertiary)	
			65
	Mesozoic	Cretaceous	
			135
		Jurassic	
			200
		Triassic	
			225
	Paleozoic	Permian	
			280
		Pennsylvanian	
			325
		Mississippian	
			345
		Devonian	
			400
		Silurian	
			440
		Ordovician	
			500
		Cambrian	
			600
Cryptozoic		"Precambrian"	
			4600

References Cited

CHAPTER 2

Bowditch, N., *American Practical Navigator*. U.S. Naval Oceanographic Office, H.O. Pub. No. 9 (1966).

Maury, M. F., *The Physical Geography of the Sea, 6th edition*. New York: Harper and Brothers (1857).

Thomson, C. W., *The Voyage of the* Challenger: *The Atlantic,* Vol. I. New York: Harper and Brothers (1878).

Transactions of the American Philosophical Society, Vol. 2 (1786).

Wust, Georg, *Deutsche Atlantische Expedition "Meteor," 1925–1927*, Vol. 4. Berlin: Verlag Von Walter de Gruyter & Co. (1932).

CHAPTER 3

Bowditch N., *American Practical Navigator*. U.S. Naval Oceanographic Office, H.O. Pub. No. 9 (1966).

Heezen, B. C., M. Tharp and M. Ewing, *The Floors of the Oceans, 1: The North Atlantic*. Boulder, Colo.: Geol. Soc. Amer. (1959).

Heezen, B. C. and C. D. Hollister, *The Face of the Deep*. New York, N.Y.: Oxford Univ. Press (1971).

Sverdrup, H. V., M. W. Johnson and R. H. Fleming, *The Oceans*. Englewood Cliffs, N.J.: Prentice-Hall (1942).

CHAPTER 4

Jerlov, N. G., *Optical Oceanography*. New York: Elsevier (1968).

McLellan, H. J., *Elements of Physical Oceanography*. New York: Pergamon (1965).

Montgomery, R. B., *Deep-Sea Res., 5*, 134 (1958).

Morel, A., in *Optical Aspects of Oceanography*, N. G. Jerlov and E. S. Nielson, eds. New York: Academic (1974).

Reid, J. L., Jr., *Intermediate Waters of the Pacific Ocean*, The Johns Hopkins Oceanographic Studies. Baltimore: The Johns Hopkins Press (1965).

Worthington, L. V. and W. R. Wright, *North Atlantic Ocean Atlas*, Woods Hole Oceanographic Institution Atlas Series, Vol. II (1970).

CHAPTER 5

Bryan, K., S. Manabe and R. C. Pacanowski, *Jour. Phy. Oceanog., 5*, 30 (1975).

Budyko, M. I., *Climate and Life*. New York: Academic (1974).

423

Jerlov, N. G., *Optical Oceanography.* New York: Elsevier (1968).

Sellers, W. D., *Physical Climatology.* Chicago: University of Chicago Press (1965).

Strahler, A. N., *The Earth Sciences,* 2d ed. New York: Harper & Row (1971).

U.S. Committee for the Global Atmospheric Research Program, *Understanding Climatic Change: A Program for Action.* Washington, D.C.: National Academy of Sciences (1975).

CHAPTER 6

Duxbury, A. C., *The Earth and Its Oceans.* Reading, Mass.: Addison Wesley (1971).

Pickard, G. L., *Descriptive Physical Oceanography.* New York: Pergamon (1975).

Rossby, T., A. D. Voorhis and D. Webb, *J. Mar. Res., 33,* 355 (1975).

CHAPTER 7

Iselin, C. O'D., *Papers in Physical Oceanography, Vol. 4, No. 4* (1936).

Knauss, J. A., *Deep-Sea Res., 6,* 265 (1959).

Knauss J. A., in *The Sea,* M. N. Hill, ed. New York: Interscience (1963).

Knauss, J. A., *The Transport of the Gulf Stream,* Morning Review Lectures of the Sec. Int. Ocean. Cong., Moscow, 30 May to 9 June, 1966, UNESCO.

Richardson, P. L., A. E. Strong and J. A. Knauss, *J. Phy. Oceanogr., 3,* 297 (1973).

Warren, B. A., in *Scientific Exploration of the South Pacific,* W. Wooster, ed. National Academy of Sciences (1970).

Wooster, W. S. and J. L. Reid, Jr., in *The Sea,* Vol. 1, M. N. Hill, ed. New York: Interscience (1963).

Worthington, L. V. and W. R. Wright, *North Atlantic Ocean Atlas,* Woods Hole Oceanographic Institution Atlas Series, Vol. II (1970).

CHAPTER 8

Defant, A., *Ebbe und flut des Meers der Atmosphare und der Erdfeste.* Berlin: Springer-Verlag (1953).

Defant, A., *Physical Oceanography,* Vol. 2. Elmsford, New York: Pergamon (1961).

Gordon, R. B., *Physics of the Earth.* New York: Holt, Rinehart and Winston (1972).

Kinsman, B., *Wind Waves.* Englewood, N.J.: Prentice-Hall (1965).

Neumann, G., *An Ocean Wave Specturm and a New Method of Forecasting Wind Generated Sun,* Tech. Memo 436. Washington, D.C.: Beach Erosion Board U.S. Army Engineers (1953).

Neumann, G. and W. J. Pierson, Jr., *Principles of Physical Oceanography.* Englewood Cliffs, N.J.: Prentice-Hall (1966).

CHAPTER 9

Bé, A. W. H., *Science, 161,* 881 (August, 1968). Copyright 1968 by the American Association for the Advancement of Science.

Boyle, E. A., F. R. Sclater and J. M. Edmond, *Nature, 263,* 42 (1976).

Craig, H., *Earth and Planet. Sci. Letters, 23,* 149 (1974).

Craig, H. and R. F. Weiss, *Jour. Geophys. Res., 75,* 7641 (1970).

Fonselius, S. H., *Hydrography of the Baltic Deep Basins III,* Report 23. Lund: Fishery Board of Sweden (1969).

Redfield, A C., B H. Ketchum and F. A. Richards, in *The Sea,* Vol. 2, M. N. Hill ed., New York: Interscience (1963).

Sverdrup, H. V., M. W. Johnson and R. H. Fleming, *The Oceans.* Englewood Cliffs, N.J.: Prentice-Hall (1942).

CHAPTER 10

Heezen, B. C. and C. D. Hollister, *The Face of the Deep.* New York: Oxford (1971).

Sumich, J., *An Introduction to the Biology of Marine Life.* Dubuque, Iowa: Brown (1976).

Thorson, G., *Life in the Sea.* New York: McGraw-Hill (1971).

Weyl, P. K., *Oceanography.* New York: Wiley (1970).

CHAPTER 11

Aruga, Y., *Bot. Mag., Tokyo, 78,* 360 (1965).

Eppley, R. W. et al., *Limnol. Oceanogr., 14,* 914 (1969).

Jerlov, N. G., in *Reports of the Swedish Deep-Sea Expedition 1947–1948, 3, Physics and Chemistry No. 1,* H. Pettersson, ed. Göteborgs, Kungl. Vetenskaps-Och Vitterhets-Samhälle (1951).

Parsons, T. and M. Takahashi, *Biological Oceanographic Processes.* Elmsford, N.Y.: Pergamon (1973). Figures 11.6, 11.10 used with permission.

Purdy, E., in *Reefs in Time and Space,* L. F. Laporte, ed. Tulsa: SEPM Spec. Pub. No. 18, 10 (1974).

Riley, G., *Jour. Mar. Res., 6,* 70 (1946).

Riley, G., *Jour. Mar. Res., 6,* 111 (1947).

Russell-Hunter, W. D., *Aquatic Productivity.* New York: Macmillan (1970).

Sumich, J., *An Introduction to the Biology of Marine Life.* Dubuque, Iowa: Brown (1976).

Tait, R. and R. DeSanto, *Elements of Marine Ecology.* New York: Springer-Verlag (1972).

Thorson, G., *Life in the Sea.* New York: McGraw-Hill (1971).

United Nations Food and Agriculture Organization, *Atlas of the Living Resources of the Seas.* UNESCO (1972).

CHAPTER 12

Emery, K. O. and C. O. Iselin, *Science, 157,* 1279 (1967). Copyright 1967 by the American Association for the Advancement of Science.

Graham, M., *Sea Fisheries.* London: Edward Arnold (1956).

Holt, S., *Sci. Amer., 221,* 182 (1969)

National Science Foundation Report, *Patterns and Perspectives in Environmental Science,* 243 (1972).

Odum, E. P., *Fundamentals of Ecology.* Philadelphia: Saunders (1953).

Russell-Hunter, W. D., *Aquatic Productivity.* New York: Macmillan (1970).

Ryther, J. H., *Science, 166,* 72 (1969).

Schaefer, M. B., *Trans. Amer. Fish. Soc., 99,* 461 (1970).

Sumich, J., *An Introduction to the Biology of Marine Life.* Dubuque, Iowa: Brown (1976).

Tait, R. and R. DeSanto, *Elements of Marine Ecology.* New York: Springer-Verlag (1972).

United Nations Food and Agriculture Organization, *Yearbook of Fisheries Statistics 1970 Catches and Landings.*

United Nations Food and Agriculture Organization, *Agriculture and Fish Reports 1970.*

CHAPTER 13

Bonatti, E., *Trans. N.Y. Acad. Sci., 25,* 938 (1963).

Krinsley, D., P. E. Biscaye and K. K. Turekian, *J. Sed. Petrology, 43,* 251 (1973).

CHAPTER 14

Biscaye, P. E. and S. L. Eittreim, *Marine Geol., 23,* 155 (1977).

Biscaye, P. E., V. Kolla and K. K. Turekian, *Jour. Geophys. Res., 81,* 2595 (1976), copyrighted by American Geophysical Union.

Biscaye, P. E., *Geol. Soc. Am. Bull., 76,* 803 (1965).

Bramlette, M. N., in *Oceanography,* M. Sears, ed. Washington, D.C.: AAAS (1961).

Griffin, J. J., H. Windom and E. D. Goldberg, *Deep-Sea Res., 15,* 433 (1968).

Heezen, B. C. and C. L. Drake, in *Syntaphral Tectonics,* S. W. Carey, ed.: Univ. Tas. Dept. Geol. Pub. (1963).

Kolla, V., L. Henderson and P. E. Biscaye, *Deep-Sea Res., 23,* 949 (1976).

Peterson, M. N., *Science, 154,* 1542 (1966).

Roels, O. A., A. F. Amos, C. Garside, T. C. Malone, A. Z. Paul, and G. E. Rice, *The Environmental Impact of Deep-Sea Mining: Cruise Report on MOANA WAVE 74-2, Apr.-May 74,* NOAA Data Rep. ERL MESA-2. Boulder, Colo.: U.S. Department of Commerce, Environmental Research Labs (1975).

Turekian, K. K., in *Chemical Oceanography,* J. P. Riley and G. Skirrow, eds. New York: Academic (1965).

CHAPTER 15

Cox, A., *Science, 163,* 237 (1969).

Dietrich, G., *General Oceanography.* New York: Wiley (1963).

Hays, J. D., T. Saito, N. D. Opdyke and L. Burckle, *G.S.A. Bull., 80,* 1481 (1969).

Hays, J. D. and N. D. Opdyke, *Science, 158,* 3804 (1967).

Shackleton, N. J. and N. D. Opdyke, *Quat. Res., 3,* 39 (1973).

CHAPTER 16

Bally, A. W., *A Geodynamic Scenario for Hydrocarbon Occurrences,* Reprint of Proceedings of 9th World Petroleum Congress, Tokyo (1975).

Barazangi, M., and H. J. Dorman, *Bull. Seis. Soc. Amer., 59,* 369 (1969).

Ewing, M., and J. L. Worzel, *Geophysical Oceanographic Studies at Lamont Geological Observatory,* Selected Papers from the Governor's Conference on Oceanography, State of New York and New York State Science and Technology Foundation, 9 (1967).

First Sea Surface Gravimeter, *IGY Bulletin, 8,* 1 (1958).

Heacock, J., and Worzel, J. L., *Bull. Geol. Soc. Amer., 66,* 773 (1955).

Herrin, E., in *The Nature of the Solid Earth,* E. C. Robertson, ed. New York: McGraw-Hill (1972).

Kay, M., *Geol. Soc. Amer., 48,* 143 (1951).

Ludwig, W., J. Nafe, and C. L. Drake, in *The Sea,* Vol. 4, M. N. Hill, ed. New York: Interscience (1970).

Seely, D. R., P. R. Vail, and G. G. Walton, in *The Geology of Continental Margins,* C. A. Burk and C. L. Drake, eds. New York: Springer-Verlag (1974).

U.S. Hydrographic Office, *The Navy-Princeton Gravity Expedition to the West Indies in 1932.* U.S. Government Printing Office (1933).

CHAPTER 17

Anderson, D., *Scientific American, 207,* 52 (1962).

Baker, H., *Detroit Free Press, April 23* (1911).

Bally, A. W., *A Geodynamic Scenario for Hydrocarbon Occurrences,* Reprint of Proceedings of 9th World Petroleum Congress, Tokyo (1975).

Benioff, H., *Bull. Geol. Soc. Amer., 60,* 1837 (1949).

Deep Sea Drilling Reports, Vol. 3, 441, U.S. Government Printing Office (1970).

Deutsch, E., in *Continental Drift,* G. D. Garland, ed. Toronto: Univ. of Toronto Press (1966).

Drake, C. L., *The Geological Revolution.* Eugene, Oregon: Oregon State System of Higher Education (1970).

Ewing, M., *Proc. Amer. Phil. Soc., 79,* 47 (1938).

Gutenberg, B., *Z. Geophys, 2,* 117 (1926).

Hess, H. H., *Proc. Amer. Phil. Soc., 79,* 71 (1938).

Holmes, A., *Trans. Geol. Soc. Glasgow, 18,* 559 (1928).

Isaacs, J. E., and L. R. Oliver, *Jour. Geophys. Res., 73,* 5855 (1968).

Kuenen, P. H., *Leidsche Geol. Med., 8,* 169 (1936).

Raff, A. D., and R. D. Mason, *Bull. Geol. Soc. Amer., 72,* 1267 (1961).

Snider, A., *La Création et Ses Mystères Dévoicés.* Paris: A. Frank and E. Dentu (1858).

Sykes, L. R., in *History of the Earth's Crust,* R. A. Phinney, ed. Princeton, N.J.: Princeton Univ. Press (1968).

Taylor, F. B., *Bull. Geol. Soc. Amer., 21,* 176 (1910).

Wegener, A., *The Origin of Continents and Oceans,* English translation of original publication in German by J. G. A. Skerl. New York: Dutton (1924).

CHAPTER 18

Martin, J. H. and G. A. Knauer, *Geochim. Cosmochim. Acta, 37,* 1639 (1973).

CHAPTER 19

CLIMAP, *Science,* 191, 1131 (1976).
Critchfield, H., *General Climatology* 2nd ed. Englewood Cliffs, N.J.: Prentice-Hall (1966).
Dietz, R. and J. Holden, *Jour. Geophys. Res., 75,* 4939 (1970), copyrighted by American Geophysical Union.
Hays, J. D., J. Imbrie and N. J. Shackleton, *Science, 194,* 1121 (1976).
Imbrie, J. and N. G. Kipp, in *Late Cenozoic Glacial Ages,* K. K. Turekian, ed., New Haven, Conn.: Yale Univ. Press (1971).
Kipp, N. G., *Geol. Soc. Amer. Mem. 145,* 3 (1976).
Lamb, H. H., *Climate: Present, Past and Future,* 1. Hampshire: Associated Book Pub., Ltd. (1972).
Lamb, H. H. *World Survey of Climatology, Vol. 2.* Amsterdam: Elsevier (1969).
McIntyre, A., N. G. Kipp *et al., Geol. Soc. Amer. Mem. 145,* 43 (1976).
Namias, J., *Jour. Phys. Ocean.,* 6, 130 (1976).
Namias, J., in *The Changing Chemistry of the Oceans,* Nobel Symposium 20. Stockholm: Almqvist and Wiksell (1972).
Namias, J., in *Proceedings of the WMO/IAMAP Symposium on Long-Term Climatic Fluctuations.* World Meteorological Organization (1975).
National Academy of Science, *Understanding Climatic Change.* Washington, D.C.: (1975).
National Science Foundation, *International Decade of Ocean Exploration,* Second Report, October 1973.

CHAPTER 20

Bloom, A. L., in *Late Cenozoic Glacial Ages,* K. K. Turekian, ed. New Haven: Yale Univ. Press (1971).
Damon, P. E., in *Late Cenozoic Glacial Ages,* K. K. Turekian, ed. New Haven: Yale Univ. Press (1971).
Hicks, S. D., *NOAA Tech. Mem. NOS 12.* (1973).
Schuchert, C., *Atlas of Paleogeographic Maps of North America.* New York: Wiley (1955).

CHAPTER 21

Archer, A. A., in *CCOP/SOPAC. Tech. Bull. No. 2,* 21, G. P. Glasby and H. R. Katz, eds. (1976).
Bally, A. W., *A Geodynamic Scenario for Hydrocarbon Occurrences,* Reprint of Proceedings of 9th World Petroleum Congress, Tokyo (1975).

Bischoff, J. F., in *Hot Brines and Recent Heavy Metal Deposits in the Red Sea,* E. T. Degens and D. A. Ross, eds. New York: Springer-Verlag (1969).

Deep Sea Drilling Reports, 23, 595, U.S. Government Printing Office (1974).

Gerard, R. D., and J. L. Worzel, *Science, 157,* 1300 (1967).

Horn, D. R., M. N. Delach, and B. M. Horn, *Tech. Report Off. Int. Dec. Ocean Explor., No. 3,* National Science Foundation (1973).

Klemme, H. D., *Oil and Gas Jour., July 17* (1972).

Ladd, J. W., R. T. Buffler, J. S. Watkins, J. Lamar Worzel, and Arturo Carranza, *Phys. Earth Planet. Inter., 12,* 241 (1976).

Schmalz, R. F., *Bull. Amer. Assoc. Pet. Geol., 53,* 798 (1969).

Uchupi, E., *U.S. Geol. Sur. Prof. Pap., 475-C,* 132 (1963).

CHAPTER 22

Frankel, E. G., and H. S. Marcus, *Ocean Transportation.* Cambridge, Mass.: MIT Press (1972).

Foley, G., *The Energy Question.* Harmondsworth, Middlesex, England: Penguin (1976).

Glossary

acidic rocks   Igneous rocks that are relatively high in silica, i.e., granite, granodiorite, rhyolite.

acoustic waves   Sound waves.

adiabatic process   An energy transfer process, such as compressing water, where there is no transfer of heat energy into or out of the system.

advection, horizontal   The bulk transport of water along horizontal planes in the ocean.

advection, vertical   The bulk upwelling or downwelling of water.

anaerobic bacteria   Bacteria that thrive in the absence of free oxygen.

anti-estuary   An oceanic circulation pattern due to the formation of dense (salty cold) water such that the surface water flows into the basin and deep water flows out.

anticyclones (anticyclonic)   Current patterns in the atmosphere or ocean which move clockwise in the northern hemisphere and counterclockwise in the southern.

antinode   A point of maximum vertical displacement in a standing wave.

antipode   An imaginary point on the opposite side of the Earth found by running a line from the original position through the center of the Earth.

aquaculture (*See* mariculture.)

asthenosphere   The more plastic region beneath the lithosphere, presumably close to its melting temperature.

atmosphere   The envelope of gases surrounding a planet.

autotrophic organism   An organism which can manufacture its tissues from inorganic compounds.

basic rocks   Igneous rocks that are relatively low in silica, i.e., basalt, gabbro, diabase.

bathymetry   The measurement of depth in the oceans.

bedrock   The solid rock underlying generally unconsolidated sediment.

benthic organisms   Organisms living on or in the seabed.

biogenic debris   Dead organic tissues, more or less broken down by physical and chemical processes.

bioluminescence   Light produced by biological organisms; for example, the firefly.

biomass   The total quantity of organisms living in a defined part of the ocean at a given time.

biosphere   All life on Earth.

biostratigraphy   Stratigraphic correlation based on the study of the vertical range of fossil species.

body waves   Compressional or shear waves which propagate through an elastic medium.

brine   A highly saline solution.

Brunhes-Matuyama Reversal   Reversal of the Earth's magnetic field which occurred about 700,000 years ago.

carbohydrates   Compounds of carbon, hydrogen, and oxygen (e.g., starch, sugar, cellulose).

chemosynthesis   The manufacture of carbohydrates by organisms, which obtain the needed energy by oxidizing inorganic substrates.

431

clastic sediments Composed primarily of detritus transported mechanically to its place of deposition, i.e., sand, silt, clay.

clay mineral A class of fine-grained minerals with mica-like plating structures, generally produced by weathering or the action of hot waters on rocks.

climate The average condition of the atmosphere, ocean, and cryosphere over a defined interval of time.

climatic anomaly Deviations from the average condition.

community All the organisms living together in some defined part of the ocean.

compensation depth The depth at which, in the case of calcium carbonate, the rate of supply of shells just equals the rate of dissolution.

compressional, or *P*, waves Elastic waves characterized by volume changes and by particle motion in the direction of travel of the waves.

continental glaciers Accumulations of snow sufficient to cover large areas of a land mass and extending beyond the montane regions.

convection Mass movement, usually associated with heating, as a consequence of density differences in a gravitational field.

core of the Earth Iron-nickel metal, in part molten, making up the innermost part of the Earth.

correlation In stratigraphy, establishing the relative age of sediment sequences in different areas.

crust of the Earth The outermost layer of the Earth above the Mohorovičič seismic discontinuity.

cryosphere All of the ice and snow on the Earth.

Curie temperature The temperature above which a body can no longer be permanently magnetized.

cyclones (cyclonic) Current patterns in the atmosphere or ocean which move counterclockwise in the northern hemisphere and clockwise in the southern hemisphere.

delta A sedimentary deposit formed at the mouth of a river debouching on a continental shelf.

demersal fish Fish living on or close to the seabed.

density The mass of an object divided by the volume; under certain conditions the "weight" per unit volume.

density stratification The layering of ocean water of increasing density.

diadromous fish Fish which migrate between fresh and salt water.

diatreme A volcanic vent or pipe of explosive character.

diffusion The flux of material by turbulent or molecular processes.

dispersion The separation of waves of different wavelengths due to a dependence of sound velocity on wave length in a medium.

diurnal Daily, once in 24 hours.

echo sounding Method of depth determination in the ocean using the reflection of generated sound waves.

ecological efficiency The ratio of energy extracted *from* a trophic level in unit time to the energy supplied *to* a trophic level in unit time.

ecosystem A conceptual model stressing interactions which occur between organisms and their physical environment.

eddy A rather ill-defined term in oceanography, signifying anything from a 200 km Gulf Stream Ring to the kinds of swirls and whirls one sees in a smoke-filled room.

eddy diffusion The transport of substances or "properties" in a fluid by a mechanism simulating molecular diffusion, but at faster rates.

elastic constants   Mathematical constants that serve to describe the elastic properties of matter.

electromagnetic radiation   Energy which travels at the speed of light and encompasses wavelengths from the very short x-rays, visible light, and infrared to the very long radio waves.

El Niño   Anomalously warm surface waters off the coast of Peru during certain summers.

energy   Capacity for performing work; energy comes in many forms; for example, heat energy, kinetic energy, electromagnetic energy, chemical energy, and potential energy; can be transformed from one form of energy to another, and is conserved.

epeirogeny   Broad scale vertical movements that determine the elevation of the land surface and the depth of the sea floor.

epicenter   The point on the Earth's surface directly above the focus of an earthquake.

equilibrium tide   The tide that would exist if the tidal wave was in exact balance with the gravitational tidal forces.

estuary   A coastal region influenced by freshwater supply by streams. Circulation is generally landward at the bottom and seaward at the surface.

euphotic zone   The upper layer of the ocean in which enough light penetrates for photosynthesis.

eustatic change in sea level   Change in sea level due to increased or decreased volume of seawater in the oceans. This can either be due to ice storage on the continents or to the change in volume of the ocean basins due to large scale tectonism.

fecal pellets   Pelletized excretions of invertebrates.

ferromanganese nodule   Zoned accretions of iron and manganese oxide phases found in the oceans and some lakes.

fetch   The area over which the wind blows to generate a set of wind waves.

flux   A measure of the rate of movement of material or energy past a point; measured in units of material or energy to be transported per unit cross-sectional area per unit time.

focus   In seismology, the source of a given set of elastic waves, customarily an earthquake.

force   The cause of the acceleration (i.e., the speeding up, the slowing down, or change in direction) of material including water and air.

frequency   For periodic phenomena such as waves or vibrating strings; the number of cycles per unit time.

geochronometry   The technique of estimating the time of origin of particular sediment layers or rock bodies.

geosyncline   A surface of regional extent, usually linear, that subsides over a long period of time while accumulating sedimentary and volcanic rocks. An orthogeosyncline (true geosyncline) is a long narrow geosyncline marginal to a continental shield. Often it has two elements: eugeosyncline, in which volcanic rocks are common, and miogeosyncline, in which they are rare.

gravity anomaly   The difference between the theoretical value of gravity at sea level and the observed value.

gross primary production   The total amount of plant tissue produced by photosynthesis in a defined area and over a designated interval.

gyre   A circle described by a moving body; the large ocean gyres described by the major ocean currents are only approximately circular.

heat energy   One form of the internal energy of material; the heat energy of water proportional to the *Kelvin temperature*.

hertz (Hz)   Unit of frequency. One Hz equals one cycle per second.

heterotrophic organism   An organism which must feed on the tissue of other organisms.

hydrosphere   The realm of water including the oceans, ground water, and ice caps.

hydrothermal   Related to hot waters from the Earth's interior or the processes in which they are involved.

hypsometric curve   A plot of the percentage of elevation and depth distribution on continents and oceans.

infrared radiation   *Electromagnetic radiation* longer than the longest visible radiation (which is red) but shorter than the highest frequency radio waves.

ion-exchange reaction   The exchange of charged chemical species by certain types of solid phases such as large organic molecules and certain clay minerals.

isohalines   Lines of instant salinity.

isotherms   Lines of constant temperature.

isotope   Species of an element having, by definition, the same number of protons in the nucleus but different numbers of neutrons.

Jovian planets   The Jupiter-like low density, high mass planets beyond the orbit of Mars: Jupiter, Saturn, Uranus, Neptune.

kinetic energy   The *energy* of movement; kinetic energy of moving water is proportional to the square of the velocity.

Kelvin temperature scale   The Kelvin temperature of a body is the Celsius temperature plus 273°C.

lagoon   A small body of water connecting to the open ocean at one or a few spots. Lagoons are generally shallow.

latent heat   The heat of water vapor which is ''released'' when the vapor condenses.

lithosphere   (1) In the geochemical sense: the solid rocks of the Earth; (2) In the geophysical sense: the quasi-rigid outer layer of the Earth.

lysocline   The depth at which, in the case of calcium carbonate, a significant increase in dissolution takes place relative to shallower depths.

magnetic anomaly   The difference between the magnetic intensity of a standard Earth's magnetic field and an observed value.

magnetic polarity   The positive (north) or negative (south) character of a magnetic pole.

magnetic reversal   A change in polarity of a magnetic field, north becoming south and vice versa.

magnetic stratigraphy   Stratigraphic correlation based on the study of magnetic properties of sediments.

manganese nodule   (*See* ferromanganese nodule.)

mantle of the Earth   The rocky layer of the Earth between the Mohorovičič seismic discontinuity and the core.

mariculture   An organized effort to increase seafood production by changing the biological and physical components of the marine ecosystem.

maximum sustainable yield   The quantity of fish that can be harvested each year without continuously decreasing the supply.

mean residence time   The steady state volume or mass of a reservoir divided by the rate of supply or rate of removal of that substance.

mesosphere   The region below the asthenosphere which appears to have greater rigidity.

metamorphism   The process by which heat and pressure cause recrystallization in rocks.

mole   One mole of any chemical element or molecule is equal to the mass in grams equal to the atomic or molecular weight of the chemical species. A mole of any element, compound, or ''molecule'' contains 6.023×10^{23} atoms or molecules.

Nansen bottle A commonly used water sampler for obtaining ocean water chemical profiles. Reversing thermometers to record *in situ* temperatures are also commonly used as part of the system (see figure).

Nekton Swimming animals.

net primary production That fraction of the gross primary production which, not being used up in plant metabolism, is available for consumption by animals.

node Positions in a *standing wave* train of no vertical displacement.

normal fault A fault in which the block above the fault plane moves downward relative to the block below.

nutrient elements Elements of prime importance in the growth and maintenance of life. In seawater nitrogen and phosphorus are commonly the limiting nutrient elements.

nutrient regeneration The return of nutrients to the ocean reservoir after their incorporation into organic tissue.

ooze A descriptive term for the sedimentary accumulation of calcareous or siliceous tests.

optimum fishing rate The fishing rate which results in the maximum sustainable yield.

orogeny The deformational process through which mountains are built.

overfishing Harvesting fish beyond the maximum sustainable yield.

paleomagnetism The study of the magnetic polarization of rocks that have been preserved since their origin.

pelagic organisms Organisms living above the seabed usually used for open ocean forms.

pelagic sediment Sediment characteristic of the deep ocean environment.

photosynthesis The manufacture of carbohydrates by those organisms which obtain the needed energy from sunlight.

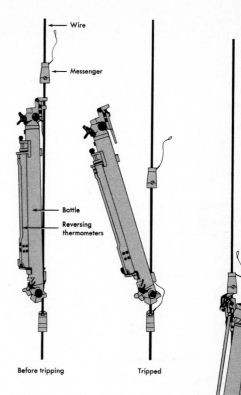

Before tripping · Tripped

After tripping

phytoplankton   Plant plankton, usually small forms floating freely in the near-surface ocean waters.

plankton   Organisms floating in the water column.

planktovores   Fish which feed on plankton.

plate tectonics   A model of the Earth's outer shell that assumes that it is made up of a small number of very large plates moving relative to each other.

polymorph   Two minerals having the same chemical composition but different basic atomic structures.

porphyry copper   Disseminated copper ore, low in grade but large in size, associated with the roots of volcanoes.

potential energy   The *energy* of displacement; the work associated with moving a fluid particle (in the absence of friction) is a measure of the change in potential energy of the particle.

precession effect   Slow changes in the Earth-Sun distance at any given season, caused by a wobbling of the Earth's axis.

pressure gradient force   A force related to the change in pressure over distance; the larger the gradient, the larger the force.

primary producers   Photosynthetic plants.

proteins   Complex compounds of carbon, hydrogen, nitrogen, oxygen (and usually sulfur) which are essential constituents of all living cells.

quaternary ice ages   Cold climates occurring during the past two million years.

radiometric age   The age of a rock determined by analysis of radioactive elements or their by-products.

red clay   A name given to the fine-grained, reddish, deep-sea, noncalcareous, nonsiliceous deposits composed primarily of clay minerals.

relative humidity   The amount of water vapor in the air compared to the amount possible if the air were saturated with water vapor; measured in percent.

resonant period   The period of natural oscillation in pendulums, springs, or water-filled basins.

salinity   A measure of the dissolved salt content of seawater commonly expressed in parts of salt per 1000 parts of seawater.

sea level   The average height of the sea averaged over a time long compared to tidal fluctuations; sea level can change over the years.

seismicity   The measure of the amount of earthquake activity in a given region.

seismic reflection profile   Profile of reflection horizons within a sediment pile obtained by using lower frequency sound waves in the manner of echo sounding.

seismograph   An instrument for recording the motions produced by an earthquake at any particular point.

seismology   The branch of science concerned with earthquakes and the elastic properties of the Earth.

semi-diurnal   Twice daily or once in 12 hours.

shallow focus earthquake   An earthquake whose center is at or close to the surface, as distinguished from a deep focus earthquake whose center is deep in the Earth's crust.

shear, or S, waves   Elastic waves characterized by distortional changes in the body through which they travel.

sill   A ridge or other below sea level geological formation which separates the water of one basin from another; as, for example, the ridge at the Straits of Gibraltar which separates the Mediterranean from the North Atlantic.

sinusoidal wave   A wave generated by plotting the value of a sine function against its angular value.

solar nebula   The primitive nebula from which the sun and planets formed.

spit   The extension of a bar or a beach as the result of long-shore currents.

stability (ocean)   As used here, a measure of the *density stratification* of the ocean waters and the resistance of the ocean waters to vertical mixing.

standing wave   A wave whose *nodes* and *antinodes* remain fixed.

stratigraphy   The geometry of layers of sediment, or the investigation of that geometry.

stratosphere   A layer of atmosphere above the *troposphere* with a constant or slightly increasing temperature.

strike slip fault   Motion of the fault is parallel to the surface of the Earth.

submarine canyon   A physiographic channel totally submerged under the sea and cut by submarine currents. (*See* turbidity current.)

surface waves   Elastic waves that propagate along the surface of the Earth. Rayleigh waves have a retrograde elliptical particle motion; love waves are horizontally polarized shear waves.

taxonomy   A biological discipline which aims to classify organisms into groups with a common ancestor.

tectogene   A large downbuckle of the crust in an orogenic belt.

tectonics   Study of the large scale structural features of the Earth and their causes.

temperature gradient   The change of temperature with depth in the Earth.

terrestrial planets   The Earth-like high density, low mass planets closest to the sun: Mercury, Venus, Earth, Mars.

test   The internal or external hard part of an invertebrate or single-celled organism.

thermal conductivity   The ability of a medium to transmit heat.

thrust fault   A fault in which the block above the fault plane moves upward relative to the block below.

tidal current   Current resulting from the advance and retreat of water from a restricted body of water as the result of tidal forces.

torque   A *force* which tends to produce a rotating motion.

trace elements   Elements existing in extremely small amounts in a system.

transform fault   Name used to describe faulting on the ocean ridges in which the direction of motion is opposite to the direction of apparent offset of the ridge axis.

transmittance   Percentage of light that penetrates a unit distance.

trophic relationship   Biological interaction having to do with the transfer of energy from one point on the food chain to another.

troposphere   That part of our atmosphere which contains nearly all of our weather; the top of the troposphere is from 10–17 km.

turbidite   The sedimentary deposit resulting from transport by a turbidity current.

turbidity current   A current of suspended sediment around water capable of causing submarine canyons and transporting continental margin deposits great distances on the abyssal plain.

ultrabasic rocks   Igneous rocks that are very low in silica, i.e., dunite, peridotite, eclogite.

ultraviolet radiation   *Electromagnetic radiation* which is shorter than the shortest visible radiation (which is violet) but longer than gamma rays and x-rays.

unconformity   A gap in the natural sequence of sedimentary layers.

underfishing   Harvesting less than the maximum sustainable yield.

upwelling   The bringing of deep cold water to the surface; almost always associated with regions of high biological productivity.

wave length   The horizontal distance between similar points on two successive waves measured perpendicular to the crest.

wave spectrum   The distribution of energy in a set of waves as a function of wave frequency.

weathering   The action of biologically produced acids in a soil to decompose rocks and produce reaction products. (*See* weathering profile.)

weathering profile   The sequence with depth of reaction products resulting from the action of life on rocks.

western boundary current   Intensified currents at the western boundaries of the ocean basins due to the Earth's rotation.

zooplankton   Animal plankton such as larvae and other non-plant organisms.

Index

439